Forests under Fire

Forests under Fire

A Century of Ecosystem Mismanagement in the Southwest

Edited by
Christopher J. Huggard and Arthur R. Gómez

The University of Arizona Press · Tucson

First printing
The University of Arizona Press

This book is printed on acid-free, archival-quality paper.
Manufactured in the United States of America
06 05 04 03 02 01 6 5 4 3 2 1

Library of Congress Cataloging-in-Publication Data

Forests under fire : a century of ecosystem mismanagement in the Southwest / edited by Christopher J. Huggard and Arthur R. Gómez.
p. cm.
Includes bibliographical references and index.
ISBN 0-8165-1775-4 (cloth : acid-free paper)
1. Forest policy—Southwest, New. 2. Forest management—Southwest, New. I. Huggard, Christopher J., 1962– II. Gómez, Arthur R.
SD565F59 2001
333.75'0979—dc21
00-010456

British Library Cataloguing-in-Publication Data
A catalogue record for this book is available from the British Library.

To Gerald D. Nash
for his pioneering scholarship
and caring mentorship
and for being a gentleman

Contents

Preface

My father was a lumberman, and his father before him. In fact, in my isolated world of southwestern Colorado it seemed to me that only one industrial endeavor beckoned to vigorous Hispanic males, and that was the lumber industry. Nurtured in the sawmill community of McPhee, Colorado, my father, Ray Gómez, trailed the lumber business out to the Pacific Northwest after World War II in hope of starting a new life among the boisterous communities like Coos Bay, Oregon. Too far removed from his beloved Southwest, he returned instead to Denver, where he engaged in every facet of the industry from sawmill production to the wholesale marketing of wood products. Like other lumber producers, he recognized the potential for economic gain, especially in the wake of unbridled postwar urbanization in the West. In the mid-1950s, Ray established his own wood-molding manufacturing plant near Durango, Colorado, a bustling Four Corners boomtown.

By 1974, the year I returned home from my overseas tour with the air force, my father had successfully managed his wood products business for nearly two decades. However, collapsed lumber markets, resulting from a nationwide recession, coupled with a gradual corporate takeover of smaller mills and factories, threatened to impinge on his own commercial endeavor. "There are only a few of us left," he lamented. "Weyerhaeuser, Georgia Pacific, and your uncle and I." This poignant though somewhat pompous remark concisely summarized the fate of all independent operators throughout the American Southwest by the late twentieth century. A failed lumber economy, worsened by federally imposed environmental standards (by-products of America's first comprehensive air and water quality legislation), reduced profits to the point that many small operators either acquiesced to well-financed and better-equipped corporations or abandoned the industry altogether.

The seed for this book sprang from the need to acknowledge the gradual disappearance of locally owned lumber mills from the southwestern landscape. It germinated, however, into a broader examination of the inherent conflict between people and their environment, in this particular case, the forest environment. The early chapters detail the forest in production and the independent owners' attempts to extract a living from the earth's bountiful harvest. The later chapters allude to efforts on the part of environmental advocates and government officials to protect an increasingly depleted resource against all threats, man-made and natural. For my part, the book is a personal tribute to the "Gómez boys"—Ray, Art, and Guillermo—and the multitude of loggers, foresters, sawmill workers, firefighters, and recreation enthusiasts who labored intently—although sometimes defiantly—to strike a more equitable balance between multiple use and conservation among the region's stately forests.

My appreciation goes to my colleague, Dr. Christopher Huggard, without whose profound conceptualization as well as organizational skills this book might never have come to fruition. His ease as an editor, moreover, has transformed random ideas and casual discussions into what we hope is a useful, readable anthology. I am ever grateful for our professional association as well as our personal friendship.

Art Gómez
Santa Fe, New Mexico

Acknowledgments

First, the editors thank the authors for their contributions to this book. Their original research offers fresh analysis of the twentieth-century American Southwest. Second, we thank the following people for their assistance: Bart Barbour, Michael Berman, Jolane Culhane, Kevin Fernlund, Al Holland, O. Orland Maxfield, Diane Sylvain, Guangqiu Xu, Liping Zhu, the staff of the U.S. Forest Service Office in Silver City, New Mexico, the Social Sciences Division of NorthWest Arkansas Community College, and the NBA. Thanks also to Dr. Christine Marín of Arizona State University's Hayden Library and to Bureau of Indian Affairs foresters David Reinhold and Morey Williams for their valuable assistance. A special note of gratitude to John and Dorothy Gómez of Cortez, Colorado, who over the years have labored to keep the memory of the McPhee community alive. Most of all, we thank our wives, Kay Pritchett and Penny Gómez, for their encouragement, loyalty, and love.

Abbreviations

AGFD	Arizona Game and Fish Department
AICA	Arizona-Idaho Conservation Act
BIA	U.S. Bureau of Indian Affairs
BLM	U.S. Bureau of Land Management
CCC	Civilian Conservation Corps
CFFP	Cooperative Forest Fire Prevention
CPA	Civilian Production Administration
DOI	U.S. Department of the Interior
D&RG	Denver & Rio Grande Railroad
EA	Environmental Assessment
EIS	Environmental Impact Statement
ESA	Endangered Species Act
FATCO	Fort Apache Timber Company
FAWLCO	Fort Apache Wholesale Lumber Company
FWS	U.S. Fish and Wildlife Service
GAO	U.S. General Accounting Office
HELP	Home Education Livelihood Program
NEPA	National Environmental Policy Act
NFMA	National Forest Management Act
NLMA	National Lumber Manufacturers Association
NOAO	National Optical Astronomy Observatories
NPS	National Park Service
OSHA	Occupational Safety and Health Administration
PARC	Predator and Rodent Control (U.S. Fish and Wildlife Service)

PRIA	Public Rangelands Improvement Act
RARE	Roadless Area Review and Evaluation
SCLDF	Sierra Club Legal Defense Fund
SEAC	Student Environmental Action Coalition
SUWA	Southern Utah Wilderness Alliance
SWFFF	Southwest Forest Fire Fighters
SWFI	Southwest Forest Industries
TVA	Tennessee Valley Authority
UA	University of Arizona
USDA	U.S. Department of Agriculture
USFS	U.S. Forest Service
VFSYU	Vallecitos Federal Sustained-Yield Unit

Introduction

Toward a Twenty-First-Century Forest Ecosystem Management Strategy

The forests of the American Southwest underwent a dramatic transformation during the twentieth century. Parklike stands of old growth ponderosa pine with expanses of grasslands beneath fifty-foot verdant canopies have been reduced to tangled, invader underbrush and puny clusters of far less majestic trees. The conversion of the vegetative landscape during the century threatened wildlife habitat and endangered species such as the elf owl, Wied's crested flycatcher, the Gila trout, and the red squirrel, among other biota. The near complete extermination of many of the Southwest's great predators such as wolves and mountain lions has also allowed for the unnatural proliferation of deer and elk, which along with domesticated cattle have overgrazed the forests and contributed to the ecological crisis.

Humans, of course, are principally responsible for this undesirable ecosystem shift. Yet, ironically, much of this ecological transformation has occurred since the beginning of the twentieth century, when foresters began implementing scientific but often short-sighted management of the national forests. The U.S. Forest Service's conservation policies, in fact, encouraged exploitation of southwestern forests and rangelands for the entire century, causing ecological problems as a result of overcutting, overgrazing, predator extermination, and fire suppression. During the second half of the century, the vast woodlands came under still greater pressures from backpackers, tourists, dirt bikers, and scientists.

Environmentalists argue that humans are destroying the environment and that there is no turning back. The multiple-use demands on forest resources, whether extractive or recreational, will continue to threaten forest ecosystems in the Southwest and elsewhere in the United States regardless of new silvicultural ecosystem management policies. Old growth forests will fall, endangered species will disappear, wildfires will burn vast areas, concessionaires will attract still more tourists, ranchers will continue to overgraze riparian zones, and all-terrain vehicular travel will carve still more ruts into eroding mountainsides. Users will continue to assault the national forests.

From this point of view, the avalanche of mistreatment will continue unabated, largely because the Forest Service's traditional decision-making process has usually favored extractor groups. Further complicating matters has been Forest Service officials' faith in their ability to manage the national forests intensively, a post–World War II policy (described as "a conspiracy of optimism" by one historian) that somehow allowed for maximum exploitation and maximum protection of forest resources.[1] The cumulative effect of these pressures culminated in a decade of frightening wildfires in the late 1980s and 1990s in the national forests and, as the 1988 Yellowstone fire demonstrated, in the national parks. The fiery devastation known as the Cerro Grande Fire that started in the Bandelier National Monument and swept through Los Alamos, New Mexico, in May 2000 illustrates the complexity of fire management on public lands. At the beginning of the twenty-first century, then, public lands management needs rethinking.

The prevailing conditions, pessimists forewarn, portend an ominous twenty-first-century future for the forestlands of the Southwest as well as the rest of the American West. Certainly, there should be concern among those interested in the maintenance of a sustainable natural resource base in the national forests. The contributors to this book realize this current concern and offer a wake-up call by showing how conflict over control of the woodland resources of the Southwest has characterized the human-forest relationship throughout the twentieth century. This traditional competition for forest resources primarily pitted ranchers and loggers against a few behind-the-scenes conservationists and preservationists until the post–World War II period. Eventually, the battle over control of the forests, which now included support for preservation of wilderness, white-water rafting, and other nonextractive uses, became too complex for one agency, the U.S. Forest Service, to unilaterally manage the national forests.

Ironically, twentieth-century federal management of the national forests, however flawed today, marked an important shift from the destructive practices of the previous three hundred years of Euro-American occupation of the North American continent. The agriculturally oriented immigrants wielded their axes as soon as they arrived in America. Wood was sustenance, used for constructing and heating homes and for cooking. Later, as the invaders moved westward, they procured timber for industrial uses such as mining and filled the rangelands with domesticated livestock. When they settled the "Great American Desert," as explorer Stephen Long called the western plains, they imported timber from the Rocky Mountain forests. Greater demands on the West's forests emerged in the late nineteenth century after logging companies exhausted the vast woodlands of the upper Midwest. Although corporate giants such as Weyerhaeuser targeted the majestic forests of the Pacific Northwest, smaller independent operators fueled the Southwest's regional timber economy. These demands meant the destruction of millions of acres of forests over the course of three hundred-plus years. In 1600 there were more than a billion wooded acres on the U.S. continent. Today there are about 730 million acres left, a figure that has remained constant, largely because of the conservation efforts of the Forest Service since 1920.[2]

In the American Southwest, first the Spanish conquest in the 1500s and then the American occupation beginning in the mid-1800s marked a clear departure in the use of the forests of the region. Rather than cut down the timber and overgraze the grasslands, native peoples lived in, on, and from the forests. The sedentary Pueblos, who cultivated maize in the Rio Grande Valley, also hunted deer and bear in forested "sky islands" in the higher elevations of the arid Southwest. The semisedentary Navajos and Apaches foraged for berries, piñon nuts, and roots while also taking game in these woodlands. As the Tewa "Song of the Sky Loom" requests from Mother Earth and Father Sky, "May [we] walk fittingly where the birds sing . . . may [we] walk fittingly where the grass is green."[3] Using the fruits of nature was a given in indigenous societies, but destroying them without proper prayer and song was heretical. As recent as the 1960s, Laguna–Santo Domingo poet Harold Bird lamented in "Death of the Woods" the Anglo mistreatment of the southwestern forests:

Jay [bird] mocks the greedy beast [logger]
Who has doomed majestic brother
His life home.
Wind tosses leaves aside as
Woodman tramps on his way,
Ax dripping oak's blood.
The forest, damp and silent,
Mourning for lost Oak.[4]

Indian peoples of the Southwest have lived a mystic communalism with the biota of the sky islands, contrary to the emergent Euro-American culture.

Even though most Euro-Americans will probably never appreciate the forests to the extent that many Indian peoples do, there is an opportunity in this century to make amends for past mistreatment of the Southwest's woodlands. The recent events concerning the Coconino National Forest north of Flagstaff, Arizona, in fact, suggest an alternative management strategy. Rather than the Forest Service single-handedly managing the forests, a coalition of policy makers known as the Grand Canyon Forests Partnership has initiated a more broadly based ecosystem management approach for a portion of the Coconino forestlands. In other words, along with the Forest Service, environmentalists and other interested parties will have greater input in the formulation of resource policies. This experiment, if successful, will point forest managers in the direction of bioregional sustainability—the management of whole ecosystems to renew forest resources—while simultaneously allowing the extraction of a portion of those resources to meet local demands. Although not as holistic (it lacks a spiritual component) as indigenous communalism with the forests, ecosystem management offers a brighter future for the southwestern forests and their users.

Destructive forest fires provided the inspiration for the development of the ecosystem management strategy in the Coconino National Forest. There, in the Hochderffer Hills, a lightning-sparked fire burned more than 16,000 acres of ponderosa pine in June and July 1996. Clearly, human mismanagement of the forests had come home to roost. Longtime logging, overgrazing, and fire suppression—elements of commodity exploitation and Forest Service management—combined to produce ripe conditions for an uncontrollable conflagration. Needless to say, hearth and home were at stake as the wall of flames rapidly scorched the earth and its pine forests on the way to crossing Flagstaff's urban interface and threatening the university-based community. Fortunately, the fire stopped short of burning down the town. But as Paul Summerfelt of the city's fire department warned: "Flagstaff has dodged the bullet many, many times. It's no longer a question of if. It's a question of when [it will burn]."[5]

This fear, which most southwesterners living adjacent to forests experience, has led to a major collaborative effort to eliminate dangerous forest conditions. Members of the Grand Canyon Forests Partnership include the Flagstaff Fire Department, Northern Arizona University, the Nature Conservancy, the U.S. Fish and Wildlife Service, the Forest Service, and independent environmentalists. The coalition has developed a plan to thin 100,000 acres of woodlands to reduce the threat of destructive fires. Using a 300-acre study plot, forest managers are cutting clusters of trees and clearing out highly flammable underbrush while also planning to reseed native grasses, limit exotic plant growth, and close some access roads. The project's goal is to thin 10,000 acres a year for the next ten years. As the acreages are thinned, the collaborators plan to reintroduce low-level fires to maintain parklike conditions that will minimize the threat of massive forest fires like the 1996 inferno and produce luscious grasses for closely managed grazing.

In the meantime, the group is searching for a local market for the timber. As Brad Ack, project director for Flagstaff, stated, "What we needed was a reverse timber sale, an antitimber sale."[6] The project will be funded by harvesting trees under sixteen inches in diameter for sale as fiber, fuel, and fence posts, unless, Ack muses, "we get a Marshall Plan for the forests." The logging community's market-driven philosophy has been excluded from the project. Because the timber industry has no input, Ack and others in the Partnership realize that "there's going to be a limit to the size of this new industry. We don't want another timber economy."[7] Whether the group will be able to create a sustainable environmental-economic plan is still uncertain.

The project members have also reevaluated the efficacy of forest preservation. When the Hochderffer fire broke out in 1996, southwestern timber companies had ceased cutting because of a sixteen-month injunction against commercial logging so biologists could evaluate the forests for Mexican spotted owl habitat. Some of the Partnership's members concluded that the general public could have identified the

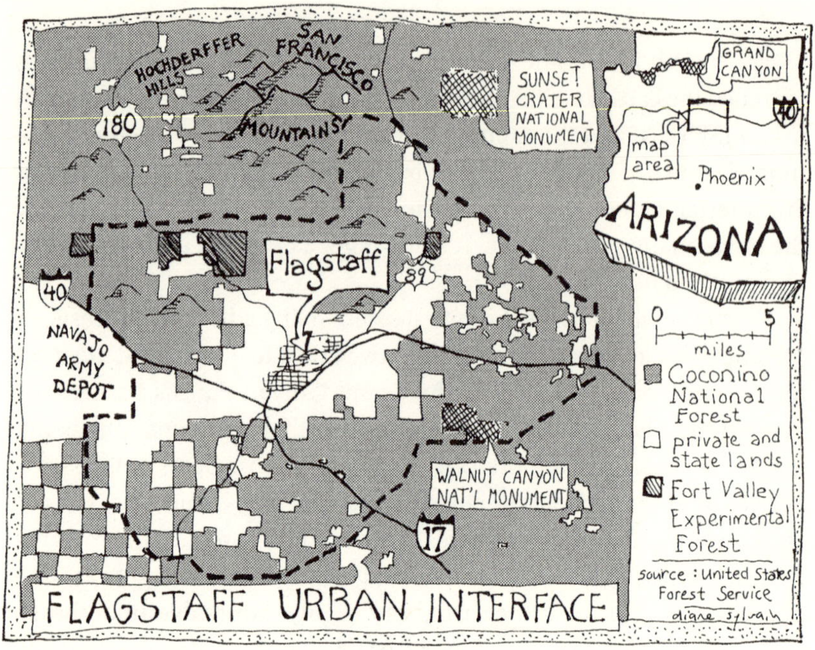

A *High Country News* map of Flagstaff, Arizona, and its forest environs, 1999. In 1996 the Hochderffer Hills wildfire burned more than 16,000 acres of timberland in the Coconino National Forest and nearly set the university community ablaze.
Courtesy Diane Sylvain and *High Country News*

cessation of logging with the fuel-load conditions that maximized the fire hazard. The environmentalists were searching for alternatives to an injunction against cutting because, as *High Country News* reporter Michelle Nijhuis wrote, another "[forest fire] disaster could turn public opinion around and bring large-scale logging roaring back to life."[8]

Instead of allowing either intensive cutting or no cutting at all, the Partnership decided on a new alternative, namely, "forest restoration" as opposed to reforestation. In addition to the above-mentioned strategies, the plan also calls for a kind of historic retention of the woodlands. When foresters cut clusters of small trees, they spare the larger ponderosas, especially the ones closest to stumps left from previous cuts.[9] These historic remnants (i.e., the stumps) have survived for decades in the dry climate of the Southwest. Reestablishing conditions of the earlier, healthier, less wildfire prone forests is made easier by the presence of old stumps, which serve as historic markers. This technique is also preferable to selective cutting because loggers do not remove or burn underbrush with the timber, as is the case with forest restoration. The result has been the clearing of hundreds of acres of forest bottoms beneath the largest trees, which now have space to grow taller. After the initial thinning is completed, the forests will better tolerate low-level natural and prescribed fires that will facilitate maintenance of more functional forests, grasslands, and wildlife habitat.

The Grand Canyon Forests Partnership is an example of the evolving collaborative nature of forest management in the Southwest and elsewhere in the United States.[10] The potential positive results excite many environmentalists and forest managers. However, some environmentalists criticize forest restoration not so much because of its goals but because of the tendency of forest managers and collaborators to believe that a strategy for managing one forest can be applied to all forests. Critics even within the Partnership warn of the dangers of a "one size fits all" management approach.

Among the detractors of the Coconino forest restoration project are Henry Carey, director of the Santa Fe, New Mexico–based Forest Trust, Doc Smith of the Northern Arizona University School of Forestry, and Martos Hoffman of the Southwest Forest Alliance. Carey warns that the Forest Service is setting a "risky precedent" by overreacting to the fire threat in the West. "This is a westwide initiative in the Forest Service. It's a program everyone's jumping on because it suits the political needs of the time; it's the new silvicultural mythology."[11] Smith argues that "this is not a prescription to manage ponderosa pine throughout its range—at all, ever, in any way . . . it isn't for the ponderosa pine everywhere."[12] Hoffman claims that the thinning model itself is flawed because ponderosas naturally grow in clusters and should be left alone. Others argue that forest restoration may work on relatively small acreages such as the lands targeted in the Coconino but not on entire national forests because of the expenses and expertise required to implement the policy. The impact of forest restoration on rangelands has not been resolved either and will remain problematic as long as the Forest Service allows too much grazing.

Despite some skepticism, the Forest Service, schools of forestry, university scientists, and environmentalists have created a viable forum for making decisions about forest management and restoration necessary to specific forests and forest conditions. This strategy is the early manifestation of regional ecosystem management, or bioregionalism. Rather than the Forest Service and logging and ranching interests exclusively implementing conservation or extractive management of the forests, which was the general trend throughout the twentieth century, coalitions of experts such as the Grand Canyon Forests Partnership are coordinating forest-specific ecosystem plans at the dawn of the twenty-first century. Forging these kinds of collaborative efforts has the potential for a more enlightened human understanding of the ecology of the Southwest's forests despite sometimes conflicting visions for future management of those forests.

This complexity points to a policy shift taking place in the Forest Service. As Thomas G. Alexander illustrates in his chapter, foresters are moving away from a functionalist approach to management that considers the multiple uses as separate but interrelated factors, to a "holistic ecosystem management" approach that recognizes the inseparable relationships among biota, whether there naturally or placed there by humans: "Rather than viewing cows [for example] as producers of red meat, . . . the Forest Service should consider them as one of the components of the ecosystem just like deer, elk, birds, plants, recreationists, and loggers." Entire ecosystems should be managed as whole organisms.

Forest management in the Southwest, however, has not always had the breadth and complexity apparent today. During most of the twentieth century, forest managers implemented conservation, or "wise use," of natural resources. With this policy, in fact, the Forest Service facilitated logging, ranching, and mining in the national forests. During the last twenty-five years of the century, however, the Forest Service made a gradual shift from conservation to ecosystem management, principally because of the influence of environmentalists and ecologists and in spite of opposition from prodevelopment politicians and sagebrushers.

This evolutionary shift in national forest management is the focus of this book. The case studies that follow examine the multiple-use management of the Southwest's forests on the public domain over the course of a hundred years. The region we target includes the states of Arizona and New Mexico as well as the southern sections of Colorado and Utah.

In the Southwest, forests vary from higher elevation stands of mixed conifer, ponderosa pine, and spruce-fir to lower elevation (below 8,000 feet) stands of piñon-juniper, Gambel and scrub oak, mountain mahogany, and various other species. Unlike the great sequoia, redwood, and spruce forests of the Olympic Peninsula, the Sierra Nevada, and the central and northern Rockies that often can thrive at lower elevations, the southwestern forests grow exclusively at higher elevations on sky islands that receive orographic rains above the arid and semiarid valleys and plateaus below about 5,000 feet. This geographical configuration has made the southwestern rangelands as significant as the timber stands themselves in the Southwest. Consequently, ranching and logging have been the principal economic endeavors in southwestern forests. And because of climate, proximity to protected antiquities such as Mesa Verde National Park and the Bandelier and Gila Cliff Dwellings National Monuments, the diversity of wildlife, and the appeal of the multiethnic makeup of the regional population, recreationists and tourists have visited the Southwest's national forests in ever increasing numbers, especially since the 1950s. Interestingly, the recreation trend reveals that the Southwest's forests were more valuable as scenery than as timber at the end of the twentieth century.

The authors focus on the human-forest relationship in about one-third of the nearly 30 million acres of national forest lands in the Southwest. Just as Congress recognized the variable uses of the national forests in passing the Multiple Use Sustained Yield Act of 1960, this anthology examines the characteristics and ecological impact of the multiple-use philosophy, with particular emphasis on logging, grazing, wildlife management, and outdoor recreation.

The first two chapters examine the private timber industry. Arthur R. Gómez and Duane A. Smith recount the community and commodity perspectives of the forest environment among different ethnic groups. Gómez, in "Industry and Indian Self-Determination: Northern Arizona's Apache Lumbering Empire, 1870–1970," examines the logging industry of Apaches and Anglos in northern Arizona. He concludes that, although several companies were successful, the ventures remained small-scale regional enterprises uninfluenced by the great timber monopolies such as Weyerhaeuser and Georgia Pacific that wielded extraordinary political and economic power in the Pacific Northwest and South.

Skillfully illustrating the impact of historic events and national legislation, Gómez details the development of the White Mountain Apaches' Fort Apache Timber Company (FATCO) and James G. McNary's Southwest Forest Industries. Gómez's depiction of the Apache-owned and -operated FATCO is especially pertinent to the current debate over minority and developing nations' rights to exploit their own natural resources in the face of environmentalist opposition. His discussion also reveals that cross-agency competition (e.g., between the U.S. Bureau of Indian Affairs [BIA] and the Forest Service) as well as cooperation have often played a significant role in resource management in the United States.

Particularly fascinating is Gómez's coverage of Apache logging in the twentieth century. During the 1910s and 1920s, he notes, Indian loggers gained first-rate lumbering and forest management training from BIA foresters. On the other hand, Gómez contrasts congressional attempts in the 1950s to "terminate" Indian sovereignty with the White Mountain Apaches' successful initiation of economic self-sufficiency and private enterprise via the Fort Apache Wholesale Lumber Company (FAWLCO). During the 1960s, President John F. Kennedy's New Trails program ignored previous calls for termination, and Indians were encouraged to promote "improved human and natural resource development on all Indian reservations." Following the demise of privately owned sawmills, the Apache lumber industry, under the umbrella of federal protection, continues to flourish to this day in northern Arizona.

Smith's study, "A Social History of McPhee: Colorado's Largest Lumber Town" demonstrates the boom-and-bust nature of the timber industry. Established for "planned obsolescence," McPhee became an Anglo-owned company town where the capitalists and their managers lived on one side of the railroad tracks and their Hispanic laborers lived on the other side. Despite segregation as well as discrimination, Hispanic culture flourished in the rural setting of the 1920s to 1940s. Logging ended in 1948, and today McPhee "has faded from both scene and memory," submerged beneath the waters of a reservoir.

These two essays reveal the overlapping of two forest management philosophies: laissez-faire exploitation and conservation.[13] The first of these concepts permeated thought and practice throughout the nineteenth century: settlers, developers, and industrialists capitalized on the timber resources, cutting trees with little regard for the land or for the future. The federal government cooperated as well by transferring millions of acres of the public domain to private ownership with no restrictions on tree cutting and grazing. And because New Mexico and Arizona were territories and not states until 1912, federal, as opposed to private, landownership determined the management principles that would dominate national forest use for the century.

By 1906, in fact, President Theodore Roosevelt and Chief Forester Gifford Pinchot had implemented the policy of conservation of natural resources under the direction of the Forest Service, an agency within the U.S. Department of Agriculture. This new philosophy, although wholeheartedly in support of development, aimed to curb waste of timber and other natural resources in the national forests, which the federal government inventoried during the first two decades of the twentieth century.[14] Renewability of natural resources (timber and range, especially) became a national imperative not only to meet domestic needs but to solidify America's position as a rising world power. In essence, the Pinchot-led movement initiated sustained-yield management (i.e., renewability of timber, grasslands, etc.) of the national forests. As environmental historian Paul Hirt has succinctly pointed out in his book *A Conspiracy of Optimism,* "the fundamental purposes of the national forests have remained largely unchanged since the turn of the century: to establish a permanent system of publicly owned forests managed by scientific experts trained to protect the forests from destruction while providing a sustained yield of renewable products and services, including wood, clean water, livestock forage, wildlife habitat, and recreation opportunities."[15]

What is not obvious from Hirt's insight is that forest managers placed greater importance on the extractive uses. During at least the first sixty years of the twentieth century, Forest Service foresters worked closest with loggers and ranchers primarily to insure that their economic needs were met. As the century advanced, other uses such as recreational development and preservation of wilderness and wildlife habitat took on greater meaning. Theoretically, since passage of the Multiple Use Sustained Yield Act of 1960, each of the uses has had equal weight and has received equal billing. Congress further defined national forest uses and how the Forest Service should manage them by passing the National Forest Management Acts of 1976 and 1990. Rather than viewing distinct uses such as logging and grazing as separate impacts on the forests, the Forest Service was now directed to manage the uses as components of a biologically diverse ecosystem, not as independent functions of the market economy, a problematic, if not improbable, goal.

In the remaining chapters of this anthology, the authors focus on the evolution of national forest management, especially since World War II, and the gradual shift from a market-driven to an ecosystem management strategy. Chapters 3 and 4 provide frameworks for understanding the two major industries of the national forests of the Southwest:

logging and grazing. Suzanne S. Forrest in "The Vallecitos Federal Sustained-Yield Unit: The (All Too) Human Dimension of Forest Management in Northern New Mexico, 1945–1998" delineates the history of logging within one particular sustained-yield unit, a post–World War II federal designation for promoting regional and multicultural exploitation of sections of the national forests. Diana Hadley examines the history of grazing in the twentieth century in "Grazing the Southwest Borderlands: The Peloncillo-Animas District of the Coronado National Forest in Arizona and New Mexico, 1906–1996."

In her study, Forrest explains why the Vallecitos Federal Sustained-Yield Unit "has been a source of almost continuous controversy" since it began in northern New Mexico in 1948. Designed to ensure local employment for villages in the area (Vallecitos, Petaca, Cañon Plaza, Servilleta Plaza, La Madera, Ojo Caliente, and Las Tablas), the project foundered in the late 1940s and 1950s because outside operators, who had capital and technological expertise, foresaw financial ruin at the hands of fickle Forest Service managers who might alter the allowable cuts at a moment's notice. Making matters worse, operators who bid on and won lumber contracts in the Vallecitos Unit ignored the intent of the federal legislation and "subjected the area residents [both Hispanic and Native American] to discriminatory and exploitative labor practices." This mistreatment inspired local Chicanos to fight back. Most visible was Reies Lopez Tijerina's 1966 "mock occupation" of a national forest campground near Vallecitos. Insisting that the federal government return the San Joaquín del Río Chama land grant to the Chicano community in accordance with the 1848 Treaty of Guadalupe Hidalgo, Tijerina and his followers generated greater ethnic awareness and inspired Forest Service officials to begin their War on Poverty training programs.

By the 1980s, when native New Mexicans were developing the viable La Madera Wood Products Cooperative to cut and sell fuelwood, *vigas,* and *latillas,* environmentalists had entered the fray. Now prepared to log their forests and sell the harvest, Hispanic operators had run into an obstacle: radical environmentalists, many who live in cities or are recent transplants from metropolitan centers in the West. These newcomers have been working to block the harvest of old growth timber at the economic expense of Hispanic logging communities in northern New Mexico. When environmentalists convinced Vallecitos loggers in the early 1990s to formulate a "culturally sensitive model of sustainable forestry," the joint project "broke down almost immediately" due to conflicts over implementation of the policy.

Telling a less controversial story, Hadley surveys the evolution of grazing policy and practices from the destructive frontier era of the late nineteenth century, to the rangeland conservation era of the first seventy-five years of the twentieth century, to ecosystem management strategies from the 1970s through the end of the century. Her focus on grazing reflects the predominance of rangelands over forestlands in the Peloncillo-Animas District of the Coronado National Forest in southeastern Arizona and southwestern New Mexico.

By 1900, the former rancher writes, "overgrazing had led to irreparable damage to the grasslands and timbered areas." Cattle companies such as the Victorio Land and Cattle Company had overstocked the Coronado National Forest, especially before systematic federal regulation began in 1905. Forest Service regulations, which established allotments and carrying capacities as well as reseeding and predator control programs, required patron ranchers to pay fees as well as conserve the rangelands. Displeased with the fees, ranchers organized livestock associations such as the South Chiricahua Cattle Raisers Association and the Peloncillo Cattle Growers Association to air grievances and influence Forest Service grassland management policies for the remainder of the century.

The coordinated efforts of the Forest Service and the cattle growers culminated in various conservation measures from the 1930s to the 1970s. The New Deal's Taylor Grazing Act of 1934 provided the bureaucratic framework for reducing wild and domestic cattle herds, determining optimal carrying capacities, and charging lessees market-value fees. In the postwar period, range managers implemented additional measures such as animal unit months, range rotation, and removal of brush and noxious weeds.

During the last quarter of the century, Hadley contends, environmentalist pressures forced rangeland managers to design "green-ranching" techniques. Congressional mandates in the National Forest Management Act of 1976 and the Rangelands Improvement Act of 1978 directed the Forest Service to pay greater attention to protection of wildlife habitat, recreation, forage, and water and soil conservation. By the 1990s, the Malpai Borderlands Group, a nonprofit organization devoted to holistic ecosystem management, had coordinated working ranchers, state and federal public lands agencies, and environmentalists in the Coronado to nurture biodiversity and rangeland sustainability.

The fight to preserve primeval forestlands in the West is not a new one. As early as 1912, preservationists lost their battle with conservationists over the Hetch Hetchy Valley in California's Yosemite. Soon thereafter, in 1916 the federal government created the National Park Service to set aside exceptional wild lands and historic and prehistoric sites. On the other hand, the Forest Service remained mainly a conservation-oriented agency; logging and grazing took top billing. During the 1920s, however, when many Americans were earning more money, had more leisure time, and were taking automobile excursions, recreational use began to take on greater importance in the national forests. Hence, there developed a movement led by forester Aldo Leopold to preserve wilderness areas, especially in the Southwest, as special designations in the national forests to serve the general public. The result, as I explain in "America's First Wilderness Area: Aldo Leopold, the Forest Service, and the Gila of New Mexico, 1924–1980," was the establishment of the Gila Wilderness Area in southwestern New Mexico.

My essay illustrates the complexity of wilderness preservation through an examination of Leopold's wilderness philosophy and the lasting impact his management ideas had on the ecology of the Gila country. Because of Leopold's fame as the author of *A Sand County Almanac* (1948) and the "land ethic" (the theoretical basis for modern environmentalism), some readers might be surprised at my conclusions. I argue that while Leopold was primarily interested in preserving ideal recreational hunting grounds in roadless areas, he also suggested that preservation of wilderness areas translated into the preservation of American democracy. In addition, some of Leopold's forest management strategies, namely, predator extermination and fire suppression, had long-term negative impacts on the ecosystems of the Gila Wilderness Area.

Wilderness advocates were able to reverse Leopold's earlier mistakes with the passage of the Wilderness Act of 1964. The post–World War II American public clearly wanted legal protection for the Southwest's wilderness areas. Likewise, local citizens and politicians of New Mexico fell in love with the Gila. As preservationists fought in the early 1950s to save Echo Park in Colorado, a grassroots contingent blocked Forest Service attempts to reduce the acreage of the Gila Wilderness Area. In the end, despite opposition from ranchers and loggers, Congress passed the wilderness law. Since that time, the Forest Service has gradu-

Radical environmentalist Susan Schock and her daughter Katy shown standing in the Black Canyon of the Gila River, Aldo Leopold Wilderness Area, Gila National Forest, New Mexico. Schock has shocked ranchers, Forest Service officers, and even other environmentalists with her candid and vociferous criticism of Forest Service mismanagement of the rangelands. Courtesy Michael Berman

ally implemented new conservation practices such as thinning deer populations and burning forest undergrowth.

During the 1980s and 1990s, however, management of the Gila Wilderness Area remained a controversial issue. Earth First!ers, the Gila Biodiversity Group, and independent environmentalists pressured the Forest Service to manage the nation's first wilderness area more holistically, because many wilderness lovers believe it represents the origins of democracy, independence, and freedom. Some wilderness advocates argue that it may already be too late to save the Gila. Radical environmentalist Susan Schock declared in 1994: "People don't know what we've lost. We [still] memorialize this shit [wilderness] as the West."[16]

Protection of America's forests has become a national obsession. The result, as John Herron shows in "'Where there's smoke': Wildfire Policy and Suppression in the American Southwest" has been a concerted effort in the twentieth century to stamp out wildfires as quickly and efficiently as possible to protect the market values as well as aesthetic values of the southwestern forests. Herron reveals that forest firefighters were too successful during the first seventy-five years of the century. Foresters were trained to save timber, a conservation policy that meant putting out fires as fast as possible through "aggressive initial attack." From the 1910s through the 1940s, resource managers and policy makers gave unequivocal support to fire suppression.

This trend continued for decades. While wartime Americans were experiencing "Meatless Mondays" and scrap rubber drives, for example, they also sacrificed to save timber: trained foresters, firefighters, and volunteers took to the skies, becoming smoke jumpers for the war cause. Parachuting into remote locations to put out fires, these brave men—Anglo, Hispanic, and Native American—risked their lives above the fiery Gila and Apache National Forests while their counterparts in the armed forces put their lives on the line in the Black and Ardennes Forests of Europe to suppress another kind of inferno.

With the unprecedented demand for lumber during the cold war construction boom, smoke jumpers remained a vital force in the firefighting arena. In the 1950s, Smokey the Bear, the most visible symbol of fire prevention in the United States, further galvanized public support for complete suppression of forest fires.

Another manifestation of this conservation philosophy was the formation in the 1960s of the Interregional Fire Suppression Crew, or Hot Shots. Unlike previous fire-fighting units composed of foresters, firefighters, and volunteers who were mobilized only when fires broke out, the Hot Shots were a permanent "rapid deployment force" devoted exclusively to fire suppression.

By the 1970s, fire suppression efforts were so successful that dangerous masses of underbrush and downed trees cluttered the forests. As a result, forest managers became alarmed at the growing number of large and hot fires breaking out in the Southwest and elsewhere. In 1972 the Forest Service introduced a regional fire plan that allowed for controlled burning, both natural and prescribed. However, this management strategy came too late, was implemented too unevenly, and has had devastating consequences. In the 1990s and early 2000 a series of massive fires burned millions of acres of national forest lands in the West and claimed more than fifty lives and hundreds of homes. Researchers have concluded that "fire exclusion . . . is a clear recipe for disaster." The Park Service's attempt in May 2000 to initiate a prescribed burn in the Bandelier National Monument reveals the complexity of fire management and the potential dangers inherent in the southwestern firescape; more than 250 homes in Los Alamos, New Mexico, burned in the state's largest recorded fire in history. Forest Service and Park Service adjustments to this new burn philosophy will undoubtedly be a great challenge for twenty-first-century fire managers.

Thomas G. Alexander, in "Struggle in an Endangered Empire: The Search for Total Ecosystem Management in the Forests of Southern Utah, 1976–1999," reveals the dangers inherent to conservation of the national forests from a functionalist point of view. For Alexander, "public demand to preserve both amenity [recreation] and economic [logging and grazing] values through fire suppression" has been the principal cause of Forest Service mismanagement and has led to destructive beetle infestations and out-of-control wildfires in the Dixie National Forest and elsewhere in Utah.

Further complicating forest management in the Intermountain Region of the Forest Service, Alexander demonstrates, are the overwhelming numbers of recreationists. The increasing pressures from national forest visitors have forced the Forest Service to hire and place more personnel in the field, but they have also introduced a new controversy over use. Many campers and hikers who perceive wilderness in aesthetic terms condemn the activities of off-highway travelers and snowmobilers who view wilderness as a source of entertainment and thrills. As a result, traditional battle lines between protectionists and extractors have been reinforced, with one user group urging protection of pristine forests and the other advocating the development of the forests as a natural Disneyland.

The key to solving these problems, Alexander argues, is "holistic ecosystem management." Rather than allowing former loggers and ranchers to make primary policy decisions, as had been the case in many national forest offices during the twentieth century, hydrologists, wildlife managers, ecologists, soil scientists, and other trained ecosystem managers also ought to be participating in important decisions concerning the public lands. Although the Forest Service has been gradually implementing ecosystem management for nearly two decades, reallocation of funds within the agency would accelerate the shift from the functionalist to the ecosystem management strategy. Rather than dedicating 65 percent of the Forest Service's budget to the federal, regional, and supervisory offices, a greater percentage of federal funds should be funneled into the field, where scientists can put it to good use to improve the health of national forests.

The impending shift from functionalist to ecosystem management has fueled great controversy for the Forest Service. One discordant characteristic of this change concerns endangered species. In "Biopolitics: A Case Study of Political Influence on Forest Management Decisions, Coronado National Forest, Arizona, 1980s–1990s," Paul W. Hirt reveals the incredible complexity of national forest management. University of Arizona scientists, the Forest Service, the U.S. Fish and Wildlife Service, Apaches, politicians, and environmentalists tangled over the construction of astronomical observatories on Mount Graham in the Coronado National Forest in the very heart of the endangered red squirrel's habitat. Raging since the late 1980s, this litigious political and environmental battle—known as "squirrels versus 'scopes"—has not yet been resolved. It illustrates that the Forest Service still supports development in the face of the Fish and Wildlife Service's mandate to protect endangered species, and it discloses the extreme measures that university administrators and professors can take to protect their observatories.

Several powerful individuals and organizations have been drawn into this increasingly contentious struggle. Senator John McCain (R-Ariz.), a U.S. presidential hopeful in 2000; Representative Morris Udall (D-Ariz.), former chair of the House Interior Committee; former University of Arizona President Jack Schaeffer, an investor in one of the Mount Graham telescopes; the Apache Survival Coalition, which has spiritual claims to the mountain; the Sierra Club Legal Defense Fund; and Earth First!ers have all played roles in the litigation process. Each interest group has supported or opposed injunctions and congressional bills and riders related to the construction of the telescopes. In this fracas, university scientists denied that the red squirrel habitat was endangered, while radical environmentalists destroyed expensive equipment after two hundred old growth spruces and firs were cut to make room for the first observatory. As a participant in the conflict, Hirt concludes that "political considerations often exercise a controlling influence over biological decisions."

Hal K. Rothman concludes the volume with his epilogue, "Seeing the Forest Not for the Trees: The Future of Southwestern Forests in Retrospect," a commentary on the shift from extractive to recreational use of the national forests of the Southwest. Arguing that postindustrial economics plays into this transformation in use, Rothman contends that forest managers better get used to recreationists and tourists. This growing and affluent constituency is becoming aware of its forest needs and already far outnumbers the traditional localized logging and grazing interests. When recreational users realize their political power, Forest Service managers are going to have to act; as Rothman states, "Forests in the southwestern United States have become more valuable as scenery than as timber." For locals used to industrial production of forest resources, it's a very bitter pill to swallow when hikers, snow-boarders, skiers, and other recreationists are given equal consideration in evolving management strategies. Some sagebrushers have even been willing to perpetrate acts of violence against forest managers and the forest environment itself. The postindustrial world of the twenty-first century does not promise to be any more peaceable than the industrial world of the twentieth century.

Despite the current controversies surrounding the forests of the Southwest, forest managers are headed in the right direction. Entire ecosystems, with their plants and animals, are becoming increasingly important as the primary foci of the multitude of experts involved in forest management. No doubt there will be setbacks—summers disrupted by wildfires, endangered species that lose more of their habitat, and riparian zones eroded because of logging, grazing, and recreational activities. Yet, when compared to the market-oriented practices of foresters during the early part of the twentieth century and even into the post–World War II period, policy makers of the last quarter of the century have considered the impact of treating one component of a forest on its other components, whether trees, grasses, squirrels, or owls.

This ecosystem management trend suggests a positive future (although a politically contentious one) for the forests and their inhabitants in the Southwest. Whether humans who visit national forests in ever-increasing numbers can use their intelligence to ensure that the rest of Earth's inhabitants—whether floral or faunal—survive remains the greatest question. For nature's sake, and our own, we can only hope for greater human understanding of individual forest biospheres.

Ecosystem management offers an alternative solution to more than a century of gross mistreatment of the national forests. Rather than managing the national forests in the twenty-first century principally to extract their main components, forest managers will be working to nurture the biotic components to foster biodiversity and species interdependence. For this holistic management strategy to succeed, the Forest Service will have to continue to engage in forest-management partnerships, scientific inquiry, and skillful diplomacy. Our national forest heritage is at stake, and it's time we realized it.

Christopher J. Huggard
Fayetteville, Arkansas

NOTES

1. Paul W. Hirt, *A Conspiracy of Optimism: Management of the National Forests since World War Two* (Lincoln: University of Nebraska Press, 1994).

2. Douglas W. MacCleery, *American Forests: A History of Resiliency and Recovery* (Durham, N.C.: Forest History Society, 1994), 5.

3. See Bertha P. Dutton, *American Indians of the Southwest* (Albuquerque: University of New Mexico Press, 1983), 27.

4. Quoted from ibid., 50.

5. Michelle Nijhuis, "Flagstaff Searches for Its Forests' Future," *High Country News*, March l, 1999, 8.

6. Ibid., 9.

7. Ibid., ll.

8. Ibid., 9.

9. It should be noted that forest restoration is not the same as the selective cutting that many logging companies have implemented in the national forests. During selective cutting, loggers usually cut the largest trees for harvest, and they do not clear out the underbrush and dead trees, fuel for the wildfires. Hence, selective cutting, although greatly superior to clear-cutting, is a very different animal from forest restoration.

10. For more information on collaborative public lands management in the West, see *High Country News*, October 12 and November 28, 1998, and February 1, March 1, August 30, and November 22, 1999.

11. Nijhuis, "Flagstaff Searches," 10.

12. Ibid., ll.

13. See Donald Worster, *Dust Bowl: The Southern Plains in the 1930s* (Oxford: Oxford University Press, 1979).

14. See R. McGreggor Cawley, *Federal Land, Western Anger: The Sagebrush Rebellion and Environmental Politics* (Lawrence: University Press of Kansas, 1993), 15–33.

15. Hirt, *A Conspiracy of Optimism*, xix.

16. Quoted from Tony Davis, "A Struggle for the Last Grass," *High Country News*, May 2, 1994, 14.

Forests under Fire

1

Industry and Indian Self-Determination

Northern Arizona's Apache Lumbering Empire, 1870–1970

ARTHUR R. GÓMEZ

The Four Corners subregion of the American Southwest is a topographic phenomenon where the states of Arizona, Utah, New Mexico, and Colorado join borders at a common point. A unique feature of the greater Colorado Plateau, the Four Corners is a visual labyrinth of undulating land surfaces punctuated by the Colorado River and its principal southern tributaries: the San Juan, Little Colorado, and Gila Rivers. The Colorado Plateau region contains an estimated 108 million acres of land, comprising portions of western Colorado, northwestern New Mexico, northern Arizona, and southeastern Utah.

Although home to nearly one million ethnically diverse residents, only five of the Colorado Plateau's numerous municipalities contain populations in excess of 25,000 people. More than 65 percent of the region's residents are Anglo American, while

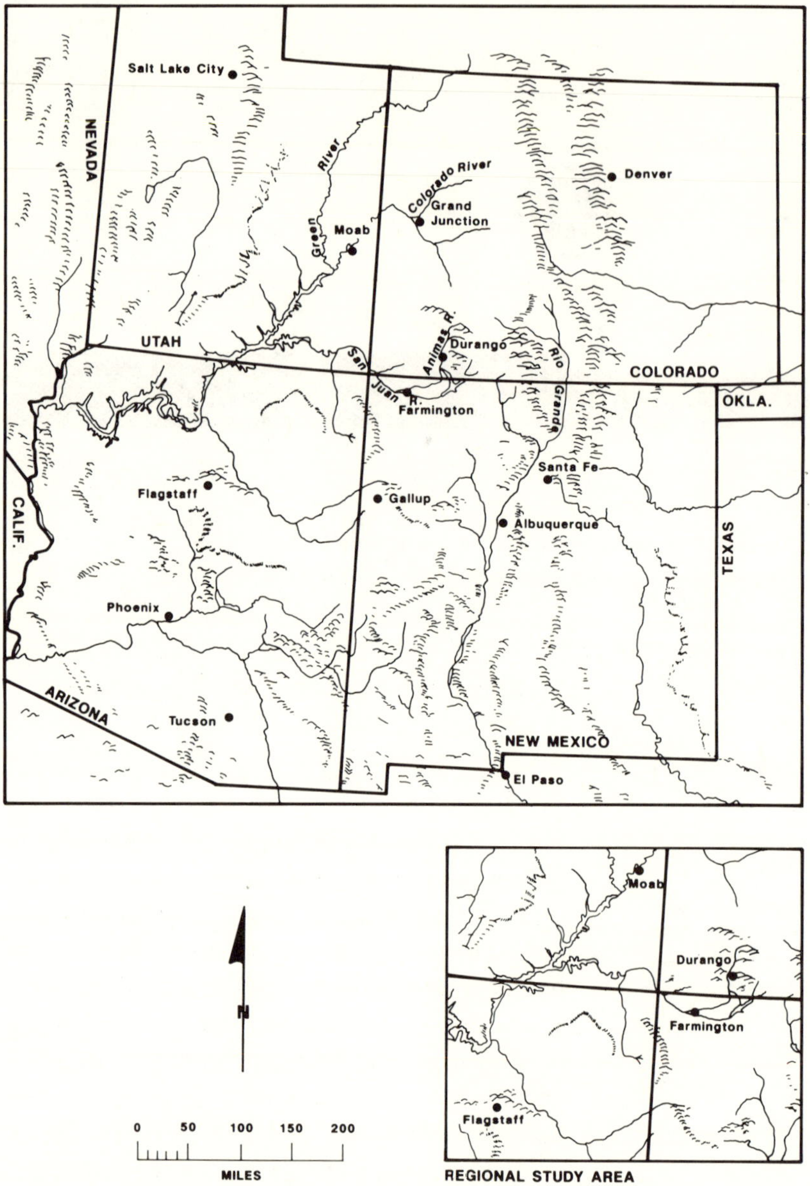

The Four Corners region.

Map by Ernesto Martínez, National Park Service, Santa Fe, New Mexico

24 percent American Indian, 10 percent Hispanic, and 1 percent African American inhabitants make up the remainder of its population. Federal ownership and stewardship dominate the plateau region, with 85 percent of the total landmass in the public domain. The Colorado Plateau encloses twenty-seven national parks, seventeen national forests, and thirty-six Indian reservations, as well as 32 million acres managed by the Bureau of Land Management (BLM).

Silhouetted along the southernmost edge of the Colorado Plateau, rising in prominent relief above its otherwise austere surroundings, is the Mogollon Rim. Nearly 30 million years in the making, this salient uplift dominates the landscape. Christened in honor of Capt. Gen. Juan de Ignacio Flores de Mogollón of New Mexico in the early 1700s, its lush valleys and stately woodlands lured Hispanic and Anglo American pioneers to its embrace throughout the eighteenth and nineteenth centuries. First Lt. Edward Fitzgerald Beale of the Army Corps of Topographical Engineers recorded one visual description of the area, calling it "the most beautiful region I ever remember to have seen in any part of the world." Virtually uninterrupted forests of "gigantic pines," accentuated by well-watered valleys, left a favorable impression on the young army officer.[1]

The seemingly boundless timber resources that so impressed Lieutenant Beale during his reconnaissance were destined to fall under the federal stewardship of the U.S. Forest Service (USFS) and the U.S. Bureau of Indian Affairs (BIA) as the twentieth century dawned. Continuous federal presence in northern Arizona gave rise to an industry that for more than a century forged a distinct regional character within the Colorado Plateau. With the city of Flagstaff as the epicenter, outlying lumbering communities—encircled by six national forests and two Indian reservations, which, combined, totaled more than 20.6 million acres of prime, merchantable timber—made the manufacture of wood products their sole commercial endeavor. Utilizing the ethnic vitality of its local population, northern Arizona's timber industry, not unlike mining ventures in the southern portion of the state, appeared to be eclectic from its inception. Accordingly, the multiethnic Colorado Plateau region quickly established preeminence in the state's wood products industry.

The Mogollon Rim not only formed the core of northern Arizona's vast timber country, it also defined the ancestral home of the plateau's earliest occupants. The Apaches, or southern Athabaskans (a.d. 1000–

1500), occupied lands in northern and central Arizona once inhabited by prehistoric communities. These early agricultural people—the Salado, Hohokam, and Mogollon—apparently abandoned the Salt and Gila Valleys coincident with the Apache appearance.

Shortly after their arrival, the southern Athabaskans split into several smaller groups, each developing in relative isolation from the others. This social fragmentation produced seven major distinct Apache groups or bands: the Jicarilla and Mescalero Apaches, who reside in the present state of New Mexico; the Lipan Apaches and Kiowas of eastern Texas; the Chiricahua Apaches, who were later removed to western Oklahoma; the Navajos of northern Arizona and northwestern New Mexico; and the western Apaches. At present, four bands—the White Mountain, San Carlos, Cibecue, and Aravaipa—occupy territory in northern and central Arizona between the Little Colorado and the Gila Rivers.

While it appears that the western Apaches may not have exploited the commercial value of the forest, they clearly held their timbered lands in the highest regard as a perennial source of food and shelter. The Apaches, noted renowned ethnologist Morris Opler, often attached a spiritual importance to the trees themselves. For example, Apache legend holds that the aspen is the mother of the forest. When all other vegetation is destroyed by fire or other natural causes, the aspen is the first tree to grow back in clusters as if to protect the younger coniferous saplings that eventually reforest the denuded area.[2]

The relationship of the western Apaches to their forested environment was anchored in pragmatic necessity rather than economic opportunism. However, Anglo Americans who arrived at the Colorado Plateau by the late 1800s promptly appraised the monetary potential of northern Arizona's virgin timber belt. Accordingly, the first recorded commercial exploitation of the Mogollon Rim began with the U.S. Army and the establishment of Camp Apache in 1870 (renamed the Fort Apache Indian Reservation).

Fifteen years after the establishment of the San Carlos Division in 1874, a southern extension of the Fort Apache Agency that housed an estimated five thousand western Apaches, Lt. Col. Simon Synder urged construction of an agency sawmill, noting: "There is not a single building at this point; troops and stores are under canvas during the entire year and suffering from the intense heat." Within months, Capt. John Bullis, Synder's replacement, had ordered the agency's first steam-powered sawmill built, locating it fifty miles northeast near the crest of the Nantanes Ridge. Bullis used armed guards to supervise and train Apache prisoners in the operation of the new sawmill. Thus, virtually with the dawn of the timber industry in northern Arizona, Apache labor—albeit under coercion—produced the first rough-cut ponderosa pine used to construct the army's military post, buildings for the new San Carlos Agency, the Indian school, and ancillary boys' and girls' dormitories. In addition to supplying the army's needs, the "agency sawmill" provided lumber products to Globe, Miami, and Peridot, nearby mining communities.[3]

Coincident with the army's arrival in northern Arizona, Mormon pioneers initiated a steady migration from southern Utah into the Little Colorado River Valley. Charged with expanding the church's influence beyond Utah's Wasatch Valley, the Latter-day Saints traversed the Colorado River at Lee's Ferry (near present Lake Powell) to establish a chain of communities from present Pipe Springs to Springerville, Arizona. Mormon settlers wasted no time in exploiting the timber resources readily available among the pleasant surroundings of the White Mountains. Mormon lumbermen logged timber and milled lumber in countless small communities—Snowflake, Heber, Woodruff, Pinedale, Lakeside, and Show Low, among others—located below the Little Colorado. They transported their fabricated wood products to St. George, Utah, for use in the construction of the temple that to the present day remains that city's most prominent landmark.[4]

The somewhat contentious transition from military supervision to civilian stewardship of Apache timberlands took effect in the early twentieth century. In one final exercise of presidential authority in 1909, Theodore Roosevelt directed the recently established U.S. Forest Service—designated the Bureau of Forestry in 1905—to manage all Indian-owned timber resources. Thus by executive order, reservation timberlands were considered to be public domain. Initially, this cooperative arrangement seemed acceptable to the U.S. Department of the Interior (DOI), under whose authority the BIA operated. The Indian agency not only expressed its willingness to embrace the precepts of planned forest management, but it even sponsored Ranger Theodore Swift of nearby Tonto National Forest to undertake a preliminary survey of timber resources located on the San Carlos Reservation. Indian agent Luther Kelly voiced his personal endorsement of the federal partnership in his annual report: "The agreement to have the Forest

Service take charge of forests on Indian reservations is such a wise provision." Kelly noted further, "It seems to me there are so many localities [on the reservation] not adapted to farming that could be utilized for tree planting."

During the William Howard Taft administration, however, differences in management objectives between the two federal agencies polarized their leadership. Essentially a conservation agency, the Forest Service cultivated timber exclusively for commercial sale to manufacturing concerns located outside of the federal reserve. "The philosophy of the Forest Service was of a piece of the Progressive Era," observed one historian. "It emphasized concepts of the greater good as prescribed by the scientific and responsible few." In contrast, the BIA, whose principal responsibility was the economic well-being of the reservation and its residents, early on invited sawmill activity and other related manufacturing endeavors onto Indian lands.

These managerial disparities were further amplified when Secretary of the Interior Richard A. Ballinger squared off against his counterpart in the Department of Agriculture, Gifford Pinchot. Ballinger, a western-educated devotee of natural and mineral resource exploitation, was the diametric opposite of Pinchot, an ardent conservationist and Roosevelt appointee, particularly in his views toward timber resources management. Committed to restoring the preeminence of the DOI in the supervision of public lands, Ballinger challenged the legality of Roosevelt's 1909 mandate and won. The issue of jurisdiction over Indian timberlands was just one of numerous heated confrontations between the political rivals, the sum of which resulted in President Taft's dismissal of Pinchot from office.

On June 25, 1910, Congress authorized the BIA to regulate timber sales on Indian lands with the stipulation that proceeds from such sales benefit the tribes. One month later, Taft revoked Roosevelt's executive order and reinstated the supervision of Indian-owned timber into the hands of the BIA's Branch of Forestry. By the authority specified in these two momentous decisions, the BIA prepared to place Arizona's Apache-owned forest resources into full production.[5]

Technological change and Anglo insistence on a planned but systematic removal of natural and mineral resources on all Indian lands throughout the American West propelled the residents—Indian and non-Indian—of the Mogollon Rim toward a symbiotic economic coalition. By the mid-1930s, Washington had authorized tribal leaders to join federal administrators in devising a more bureaucratic approach to developing the commercial and recreational uses of the forests of northern Arizona. During the summer of 1910, the BIA appointed forester J. P. Kinney to supervise all Apache timber resources in northern Arizona. Kinney in turn commissioned a detailed commercial timber survey for both Fort Apache and San Carlos.

The task of assessing the commercial potential of the vast blanket of virgin ponderosa pine in the White Mountains and along the Nantanes Ridge rightfully fell to Kinney and his able associate, George A. Gutches. Kinney, a law school and forestry graduate from Cornell, and Gutches, who had derived his cumulative experience in the southeastern woodlands, were two of only a handful of professionally trained foresters employed by the BIA at that time. Within months after his inaugural visit to northern Arizona, Gutches detailed his assessments in two separate reports to the BIA.

In his June 18, 1910, report, entitled "Care and Protection of Timber on Fort Apache Indian Reservation," Gutches noted that the reservation divided naturally into five timber species. First, there was Engelmann spruce, which covered the upper slopes of Mount Baldy and Mount Ord at elevations exceeding 9,200 feet. "The area covered by merchantable timber is very broken and mountainous," Gutches said. "Logging will be difficult and it is very doubtful if this timber will have a sale value, at any price, for some years." Second, Gutches observed mixed types of timber on the lower slopes of Mount Baldy and Mount Ord between 8,600 and 9,200 feet. The forester estimated that about 35,000 acres of Engelmann spruce, yellow pine, and Douglas fir were available, but he cautioned, "Logging would be very expensive over this area on account of the roughness of the country and the scattered condition of the merchantable timber. The timber has no sale value at the present time."

Gutches's economic evaluation of the third type of timber—yellow pine—was far more optimistic. He observed that the species covered 75 percent of the reservation (675,000 acres) between altitudes of 7,000 and 8,600 feet. "The yellow pine is the principal lumber tree on the reservation, and is the only tree that could be considered as having a sale value at the present time," he advised. Impressed with the average size and diameter of these trees, the forester estimated that most of the lumber produced from them would result in 30 to 50 feet of "clear lengths."

While Gutches concluded that the timber on Fort Apache Reservation had commercial potential, he warned that marketing the product would present certain problems. "There are large flat mesas north of the North Fork of the White River, east of Cooley's Ranch, and east of Bonito Creek," the inspector noted. "The rest of the area containing timber of this type is rough and broken and will be difficult for logging." Gutches recommended constructing a railroad spur on the reservation to handle the logging. Finally, Gutches tendered one last, prophetic comment: "It is recommended that the Indian Office make every effort possible to have that part of the Fort Apache Indian Reservation included in the Sitgreaves and Apache National Forests restored to its former legal [BIA] status."[6]

Gutches reported similarly encouraging prospects for San Carlos timber while noting that it was not as abundant as that of the White Mountains. In his report dated July 30, 1912, Gutches calculated the timber resources on San Carlos to be an estimated 221 million board feet of merchantable ponderosa pine. "The local demand at Globe, Clifton, Morenci, and neighboring mining towns is sufficient to consume this timber if a sale could be made." However, in a cautionary note the district forester judged that only 30 percent of the estimated 70,000 acres of forest on San Carlos appeared to be merchantable (trees at least 85 feet high and 24 inches in diameter).

Thus, efficient management practices were imperative for the San Carlos Apaches to meet current demands for lumber products and still take into account the sustainability of future forests. Gutches therefore imposed cutting restrictions on all prospective San Carlos logging operations. He recommended that no trees under 18 inches be harvested unless injured or dying. Gutches provided a maxim for area loggers: "If 33 ⅓ percent of the log is sound, it should be considered merchantable." It appeared that from the outset production of wood products on San Carlos vis-à-vis Fort Apache would be greatly impeded by the limitations of the resource itself. This inherent problem forced San Carlos tribal leaders in later years to reevaluate their commitment to forest production in deference to the more dominant White Mountain Apache timber industry.[7]

Presumably, George Gutches had only vague knowledge of the impending legislation in Washington, D.C., that enabled the commercial sale of timber on the Fort Apache Reservation. Nevertheless, both the congressional act of June 25, 1910, that sanctioned the sale of standing timber on Indian lands and President Taft's repeal of the Roosevelt Proclamation of 1909 hastened implementation of the astute forester's recommendations. In September 1911 Commissioner of Indian Affairs Cato Sells and Acting Secretary of Agriculture B. T. Galloway announced the region's first joint timber sale. The proposed sale offered ponderosa pine from approximately 85,000 acres on the Fort Apache Reservation as well as timber from the adjacent Sitgreaves National Forest. A recurring need to attract skilled sawmill operators to the reservation to assure success in both the production and sale of manufactured wood products preoccupied federal advisors. To the dismay of the BIA and the White Mountain tribe, their initial attempt to sell 330 million board feet of timber received no bids.

That BIA managers expressed concern over the need to attract lumbering operations in many respects mirrored the direction of the entire western lumber industry at the advent of World War I. America's entry into the war greatly accelerated forest management. Fearing that war demands would cause a national shortage of lumber products, both the Forest Service and the BIA reexamined the resources that could be made accessible to private manufacturers. The vast virgin stands of northern Arizona pine represented a virtually untapped reserve of timber. While early BIA efforts to market Indian-owned timber in the White Mountains were unsuccessful, the decades following World War I ushered in an era of intensified timber production. Mormon, Hispanic, African American, and eventually Apache lumbermen who for decades had made the timber-rich environment of the Mogollon Rim their home hoped to take advantage of these new economic opportunities.

America's entry into the war in 1917 stimulated a nationwide demand for forest products. Wartime requests for barracks construction materials on virtually all military training facilities and the demand for lumber to build ships and planes precipitated a welcomed boom throughout the industry. Prior to the Great War, most logging operations had been confined to the Great Lakes Region and selected areas within the Southeast and Pacific Northwest. Although some activities took place on federally owned Indian reservations, there was only a handful of consequential timber operations active in the Southwest before this time. The war, however, stimulated the demand for Indian timber from western reservations.

Ready markets for Arizona pine coincided with a flurry of legislation designed to attract non-Indian business interests to both San Carlos and Fort Apache. In 1918 Congress authorized BIA superintendents administering Indian reservations to assume responsibility for forest protection as well as timber sales. The same legislation mandated the BIA to educate the Apaches to conserve as well as develop their forests. Meanwhile, Carl Hayden, Arizona's newly elected congressman, endeavored to pass regulations to give non-Indian operators access to tribal lands for grazing, mining, and logging.

During these years, BIA agents on San Carlos assessed the commercial value of the forested acreage of the Nantanes Ridge more in terms of its grazing than its logging potential. Throughout the 1920s both Apache and non-Indian cattlemen generated impressive incomes in the cattle industry. In 1927 an estimated nine thousand cattle ranged across the San Carlos Reservation.

The proven monetary success of the San Carlos cattle industry directly affected the future of the agency's forestry program. Range areas often abutted timber stands; therefore, BIA administrators customarily ordered prescribed burns to clear away underbrush for new grasses. Moreover, grazing stock inflicted irreparable damage to tree seedlings. In light of the potential for stock raising on San Carlos, BIA officials advised tribal leaders to utilize their forested lands for grazing rather than for timbering purposes. When a twelve-member tribal council was created in 1934, the San Carlos Reservation phased out all non-Indian leases to grazing lands.[8]

In contrast to San Carlos, the Fort Apache Agency, left holding its September 1911 offer without a bidder, seemed more determined than ever to market White Mountain timber as the reservation's principal economic commodity. After two failed attempts to execute a timber sale prior to World War I, Superintendent William McPherson advised Commissioner Sells to advertise the timber for a third time. Not surprisingly, Sells expressed reluctance to do so. Nevertheless, Forest Service and BIA administrators in Washington feared that wartime demands had the potential to cause a national shortage of wood products. Therefore, Sells recanted and opened White Mountain Apache timber reserves to private enterprise.

In August 1917 the two federal agencies again jointly advertised the sale of more than 635 million board feet of virgin ponderosa pine. The notice specified 235 million board feet available in the Sitgreaves National Forest, while the remaining 400 million board feet would come from the Fort Apache Reservation. The addition of 70 million board feet of tribal-owned timber to the original 1911 offer, BIA forestry experts judged, would offset the anticipated cost to build a railroad spur from

Holbrook into the White Mountains to gain access to the timber unit. Besides the requirement to construct a railroad, the sale prospectus also mandated the installation of a sawmill at Cooley, Arizona. The BIA managers expected the purchaser to locate a manufacturing plant "within the body of timber to be logged." This way, producers could haul finished lumber from the reservation to the Santa Fe Railroad terminus in Holbrook for redistribution to Pacific Coast and midwestern wholesalers. The contract proposed a twenty-year period of logging beginning in January 1920. This allowance of a two-year grace period provided ample time to construct the proposed railroad.[9]

For all of the sale's attractiveness, the government received only one bid, that of the recently organized Apache Lumber Company of nearby Flagstaff. The principal organizers of the company were Thomas E. Pollock, an able financial executive, and Amasa B. McGaffey of Albuquerque, a well-seasoned sawmill operator. On September 5, 1917, Tom Pollock, his brother Robert, and McGaffey incorporated the Apache Lumber Company, which they capitalized at $1 million. The company also funded the Apache Railway Company, more enticing to BIA foresters, for an additional $1 million.

In submitting their bid, Pollock and McGaffey agreed to build a narrow-gage spur into the White Mountains. They also promised to relocate their sawmill from Flagstaff onto the reservation. The BIA agreed not to demand lease payments from the firm for the mill site provided that Pollock and McCaffrey employ Apache laborers on a regular basis. On October 31, 1917, Secretary of the Interior Franklin K. Lane approved the contract. The Indian Service had at last consummated the Unit #1 timber sale, the first of many future sales on Apache-held lands.[10]

The two signatories concluded the Unit #1 timber sale in February 1918, but cutting operations did not begin for two more years. The BIA required the delay to allow the Apache Lumber Company ample time to construct the railroad that all believed was vital to a successful contract. After the agency's failure to negotiate its first sale, BIA Forestry Chief Kinney had journeyed to Fort Apache. His investigative report noted that the proposed sale area held sufficient timber to offer an additional 100 million board feet for commercial purposes. "To justify the very heavy expenditure involved in the construction of this main [railroad] line," Kinney advised, "it is necessary that the purchaser be assured a large block of timber." The forester proposed that the amount of timber offered in the Unit #1 sale be substantially increased.

Undaunted by the task at hand, Pollock and McGaffey executed plans to complete the spur from Holbrook to Cooley, seventy-three miles to the southeast. The Santa Fe Railroad not only endorsed the plan but loaned the Apache Railway Company $300,000 in cash toward construction. In compensation, Santa Fe officials negotiated a twenty-year contract to purchase 400,000 railroad ties from the proposed sawmill. Though labor shortages, inclement weather, and disputes over rights-of-way hampered progress, the company's Mexican and Navajo track laborers reached the Apache reservation in February 1919. With deft skill and accuracy, Don Cooley, son of the celebrated frontier scout for whom the mill site was named, drove in the last ceremonial spike to inaugurate the opening of the Apache Railway Company.[11]

Amasa McGaffey designed the sawmill. First, he cleared away sixty acres, trenched out the log pond, and set up a portable sawmill for the lumber required to construct the commissary, a schoolhouse, a small hotel, and fifty-six single-room dwellings to house the sawmill workers and their families. Hailed as the most up-to-date facility of its day, the Cooley Mill used electricity, not steam, to power the entire operation. A small generating plant provided sufficient energy for three motor-driven band saws, each with an aggregate capacity of 175,000 board feet per day. McGaffey wisely installed a planing mill to surface the rough-cut pine boards before shipment to national markets. The company predicted the mill would produce an estimated 40 million board feet per annum employing only a single shift.

In time, the township of Cooley boasted a modern hospital, a bakery, and even a city park. The combined investiture of the Apache Lumber Company in the sawmill and its ancillary company town amounted to $3.4 million. Added to the $1.7 million to build the Apache Railway, Thomas Pollock and associates spent more than $5 million before cutting a single commercial log of Fort Apache timber. When the mill began operations in the fall of 1919, company success rested entirely on its loggers and sawmill operators to exploit timber quickly and efficiently.[12]

By all accounts, the Pollock-McGaffey undertaking was doomed to fail from the outset. By the time the logging and sawmill operations reached full capacity, World War I had ended. America's "return to normalcy" resulted in a steady decline in the industry nationwide; simultaneous with a worldwide recession in 1921, lumber sales plummeted. Flagstaff mills suffered a 50 percent reduction in sales. Overproduction and ever-constricting markets left lumberyards with a surplus of finished wood products and no outlet for distribution. Ironically, the Cooley sawmill's state-of-the-art technology worked to the company's detriment. During it first full year of operation, the new facility added 50 percent more lumber to northern Arizona's pre-World War I cut (80 million board feet). Bewildered, McGaffey and Pollock found that their lumberyard had a 20-million-board-feet surplus of unsold lumber. To make matters worse, the BIA announced that the Apache Lumber Company was in default of its timber contract because it failed to harvest the requisite 30 million board feet of tribal-owned timber per annum. Financially ruined, the Cooley sawmill ceased operations, and the heavily indebted Apache Railway Company listed its properties for sale. Pollock retired to his ranch near Flagstaff, no doubt devastated by the projected net loss of $8 million. Despite the best intentions of the federal government, huge parcels of mature ponderosa pine on the Apache reservation stood unattended.

In December 1923 the Forest Service filed an appraisal of the defunct Apache Lumber Company. The document estimated that the Cooley Mill had processed about 100 million board feet of lumber from Unit #1 before its deactivation in the winter of 1921. Foresters determined that the company had harvested 50 of the 400 million board feet of timber offered under contract. While the federal agency attributed closure of the Apache Lumber Company partly to "inefficient management," it concluded that the sawmill failure was attributable to "a major slump in the lumber market beginning in 1920."[13] Pollock and McGaffey, meanwhile, saw no recourse but to advertise the sawmill and the Apache Railway Company for immediate sale. Just weeks after Northern Arizona Securities publicized the sale, a Louisiana-based firm made inquiries.

Indiana-born financier James G. McNary, a former French teacher, newspaper publisher, and banker, along with William M. Cady, a scrapping woodsman from Alexandria, Louisiana, expressed interest in the Cooley Mill. The unlikely business partners had shared lumbering experiences in the dense pinewoods of the Southeast in the early 1900s, where the dictum "clear cut and clear out" dominated the industry. With timber stands in the Southeast nearly exhausted and virgin stands of ponderosa pine left abandoned in northern Arizona, the lumbering magnates turned their covetous eyes toward the Mogollon Rim.

In November 1923 McNary and Cady informed the BIA of their

James G. McNary, president of Southwest Lumber Mills, ca. 1940. This portrait was taken at the height of the former Albuquerque banker's personal and financial success in the northern Arizona lumber business. Courtesy Martha Chilcote Collection, Pinetop, Arizona

intent to purchase the Cooley Mill and the Apache Railway. In his letter to Commissioner Charles Burke, McNary requested that the Unit #1 contract be modified to enable his company to harvest Apache-owned timber in sufficient quantities to establish a double-shift policy at the mill, install a box factory, and expand planing operations. These modifications, McNary insisted, would double production capacity to 100 million board feet per annum over the next ten years. Additional timber allocations would, McNary insisted, enable the W. M. Cady Lumber Corporation to recoup at least a portion of its initial investment. In their response, BIA foresters offered an added incentive. They promised Cady Lumber access to the even larger Maverick Mountain Unit (600 million board feet) if the owners agreed to lengthen the Apache Railway line.[14]

With a modified contract in hand and BIA assurances of sufficient timber to process lumber well into the next decade, McNary and Cady bought out the Apache Lumber Company for a fraction of Pollock and McGaffey's original investment. The 1.5-million-dollar purchase included the Cooley property and the Apache Railway as well as the Flagstaff Lumber Company, which in the long term proved to be a practical acquisition. For the time being, however, Cady and McNary shut down the Flagstaff facility in order to concentrate fully on the revitalization of the Cooley sawmill, renamed McNary in early 1924. The Cady Lumber Corporation quickly spent $3.5 million to install an all-electric plant with three band saws, the first of its kind in the region. To this it added a box factory, a new planing mill, and lumber storage units. To assure maximum productivity, Cady Lumber imported nearly eight hundred skilled lumbermen from Louisiana—most of them African American—to operate the new facility. Though alien to the northern Arizona industry, the recent immigrants were, in Cady's opinion, "the best sawmill labor in the world."[15] In July 1926 the McNary sawmill reported an annual capacity of 100 million board feet.

The company also saw great potential for profit among the town's numerous amenities. The company owned one of the largest mercantile outlets in Arizona, transacting a reported $60,000 a year in sales. Cady Lumber also owned all employee housing, sufficient to accommodate the community's two thousand–plus residents. The town of McNary had its own public school, an electric power station, and an efficient water and sewer system. The Apache Railway provided both freight and passenger service to and from Holbrook, with wagon service to the agency at Fort Apache. Unlike their neighbors, McNary residents enjoyed regular deliveries of coal, heating oil, and livestock.

Besides the reservation operation, the company revitalized the obsolete facility in Flagstaff. Cady overhauled the mill, upping its daily capacity to 240,000 board feet, most of the inventory being converted into "box shook" (shorter pieces of waste material salvaged to make shipping crates for agricultural produce). It soon became obvious to

Tree felling with a crosscut saw, ca. 1935. The Cady Lumber Corporation expressed particular satisfaction with African American loggers like the man shown above. James McNary and William Cady imported hundreds of black laborers from Louisiana to northern Arizona because of their exceptional skill in both logging and sawmill operations. This lumberjack has just sawed a large ponderosa pine. Courtesy Martha Chilcote Collection, Pinetop, Arizona government foresters that the amount of timber remaining in the Unit #1 cut could not sustain both the McNary and the Flagstaff sawmills. The initial success of the Cady Lumber Corporation was, however, short-lived as the near total collapse of the nation's timber industry beginning in 1929 hastened the demise of yet another lumbering endeavor on the Fort Apache Indian Reservation.

In a letter dated March 28, 1928, James McNary detailed the financial woes of his company to the BIA with the hope that the Indian agency might reduce stumpage prices for reservation timber. According to McNary, Cady Lumber lost $169,063 from sawmill operations in addition to a deficit of $14,567 sustained from the Apache Railway Company. While McNary fixed some of the blame squarely on the BIA for offering "inferior grade timber" from Unit #1, the financier attributed the recent failure to "an overall depressed market for processed lumber."[16] While not totally unsympathetic to the company's plight, the BIA did not reduce stumpage rates for the remaining Unit #1 harvest. Meanwhile, company sales dwindled from $200,000 a month during late 1929 to an anemic $30,000 by early 1930. The beleaguered Cady Lumber Corporation struggled four more years before passing into receivership on December 7, 1934.

The presidential elections of 1932 had a decided impact on the future of forest management on Arizona's Apache reservations. Franklin D. Roosevelt's pledge of a "new deal" for the American people included the nation's American Indian constituency. FDR's selection of Indian rights advocate Harold Ickes as secretary of the interior and Ickes's subsequent appointment of John Collier as commissioner of Indian affairs initiated a policy of joint stewardship of Indian-owned natural resources. According to one scholar, "As the chief architect of the Indian New Deal, Collier actively sought to modify and diversify the government's role in administering a more benevolent and responsive Indian policy."

Collier espoused political autonomy among the nation's Indian tribes as the best framework for economic success on the reservation. Collier's philosophy, embodied in the Indian Reorganization Act of 1934, guaranteed more direct tribal participation in decisions affecting reservation economic development. Implicit in Collier's new dictum was the production and sale of Indian timber. Accordingly, newly formed tribal councils at San Carlos and Fort Apache assumed a new proactive posture with respect to all future timber activity.

In response to Washington's call for more effective forest management, BIA superintendents reevaluated the prospects for commercial harvesting. A new cadre of conservation-minded BIA foresters replaced their more tradition-bound, profit-driven mentors. These new managers promptly formalized plans for future timber production on reservation lands based on the innovative "sustained-yield" concept. Sustained yield implies that stands of trees are a renewable resource and therefore capable of continuous cutting every twenty-five years. In this manner, forest managers ensured maximum continuous production of forest products with minimal impairment to the resource. The objective was to achieve a balance between net growth and net harvest.

With requisite forest management plans in place, the two Apache agencies, Fort Apache and San Carlos, solicited local lumbermen to harvest and process Indian timber. Foremost among the respondents were the descendants of Mormon immigrants to the Little Colorado River Valley in the late 1800s. The names Webb, Petersen, and Porter were the most evident among the numerous logging and sawmill families who earned their livelihood among the woodlands of northern Arizona.

A pair of veteran loggers from Lakeside and Pinedale, Arizona, were the first of the independent operators to arrive on the scene. Henry "J. H." Webb, born in Chihuahua, Mexico, to Mormon parents, immigrated to the White Mountain region, where he honed his skills as a logger and sawmill operator. He was the first of the "independents" to seize the opportunity to log timber on the Apache reservation. In the spring of 1935, Webb entered into an agreement with Southwest Sash and Door, a wholesale lumber outlet in Phoenix, to cut lumber for sale throughout the metropolitan area. Webb wrote to Senator Carl Hayden asking the powerful politician to intercede on his behalf. He requested that a small block of Fort Apache timber, suitable to meet his commitments to the Phoenix wholesalers, be made available for bid. Citing eighteen years of logging experience, most recently on the Apache reservation, Webb touted himself as a reliable contractor.

Webb's seemingly innocuous request met with unexpected resistance from Superintendent William Donner. Apparently, the Indian agent hoped to build a second agency sawmill near Carrizo Creek, near the recently completed State Highway 60, to supply the construction needs of the Phoenix metroplex. "By selling small blocks of timber to small portable operators," Donner argued, "we are curtailing the few available resources for Indian labor." Webb countered that there were

A typical portable circular-type sawmill operation, 1956. A. G. "Golden" Petersen's sawmill, located about two miles north of Whiteriver, Arizona, typified the hundreds of portable, two- to three-man operations that dominated the northern Arizona landscape. By the mid-1980s virtually all of these small, privately owned sawmills had ceased operations and disappeared. Courtesy Martha Chilcote Collection, Pinetop, Arizona

ample timber reserves on the Fort Apache Reservation to satisfy the needs of both the independent operators and prospective Indian-owned businesses. "There are eight portable sawmills in operation on [the adjacent] Sitgreaves National Forest," Webb rebutted.[17]

It appears that Webb's persuasiveness, coupled with congressional endorsements from Hayden and others, convinced the BIA to announce the sale of a narrow strip of timber known as the North Boundary Unit. The June 30, 1936, advertisement said the unit held an estimated 2.5 million board feet of timber on about 980 acres. The fingerlike unit contained stands of easily accessed ponderosa pine. On August 17 the BIA awarded Henry Webb and his partner, A. G. "Golden" Petersen, owner of a portable sawmill capable of producing 10,000 board feet per day, a two-year contract.

By all accounts, the Petersen-Webb coalition was a winner. Local lumbermen recognized Petersen, who owned several sawmills in northern Arizona, as perhaps the most knowledgeable portable sawmill operator in the business. His highly modernized mills proved capable of cutting logs equal in width to those processed at McNary's remodeled band saw mill. Webb and Petersen anticipated no problems in meeting the terms of their government contract. Even Superintendent Donner acquiesced, saying, "Both Petersen and Webb have excellent reputations in this district. They are reliable, industrious, and good mill operators."

Neither Petersen nor Webb, however, anticipated the precipitous decline in lumber sales in 1937–38. An unstable political climate in Europe, coupled with a slow economic recovery at home, brought housing construction to a standstill. Although they struggled to fulfill their contract obligations, the Arizona lumbermen were hard pressed to succeed. Recurring mechanical failures added to their problems, forcing Webb and Petersen to request a one-year extension on the contract. Donner assured BIA officials in Washington of the honest intentions of the Mormon industrialists. "Failure of Petersen and Webb to meet contract cutting requirements has resulted from market conditions rather than from low mill capacity," he advised. Donner endorsed Webb's intent to move a second portable mill onto the reservation during that summer in order to fulfill the contract obligations. Donner recommended BIA approval based on the excellent record of these contractors.

Despite a gallant effort, Webb and Petersen defaulted on the North Boundary Unit contract in August 1938 after logging all but 50,000 board feet of salable timber. The two lumbermen were not criticized for failing to log out the North Boundary Unit in its entirety. In fact, in July 1941 Fort Apache foresters awarded A. G. Petersen a second, smaller contract known as the Hop Canyon Unit, where he employed his portable sawmill to cut approximately 1 million board feet annually until May 1945. Soon thereafter Petersen activated a small circular-saw facility near Peridot on the southeastern edge of the San Carlos Reservation, processing logs taken from the Nantanes Ridge until the late 1950s. In 1961 Petersen left the Indian reservation for good to build his last sawmill at Overgaard, eighteen miles northwest of Show Low. There, Golden Petersen lived out the final days of a successful industrial career.

Following the fiasco on Fort Apache, Henry Webb and his two sons, Ray and Carl, briefly withdrew to the more attractive surroundings of their Lakeside operation. In 1953 Henry Webb and Sons followed A. G. Petersen's example and moved to San Carlos, where they anticipated a handsome profit from the Turkey Tank Unit #1. Still partners with Southwest Sash and Door, the Webbs foresaw an inexhaustible market for finished lumber among the explosive desert communities of the Salt River Valley.

Years later, Ray Webb reminisced about the Hilltop endeavor with fondness. Using an old '52 Ford pickup engine as its power source, the 56-inch circular sawmill cut logs that, according to Webb, were "bigger than those cut at McNary." Notably, the Webb family employed twenty-eight San Carlos Apaches, half of whom worked in the woods and the remainder of whom found employment at the sawmill. The Webb brothers recalled that volleyball, basketball, horseshoes, and "more than an occasional swimming party in nearby Black River" provided welcome relief from an otherwise arduous cutting schedule. Their willingness to employ local Apaches enabled the Webbs to enjoy moderate financial success on both the San Carlos and the Fort Apache Reservations until resuming their Lakeside base of operations in 1964.[18]

Not all independent operators who logged the Mogollon Rim in the late 1960s and early 1970s recall their experiences as being especially pleasant. Indeed, L. D. "Donny" Porter's story ended in bitter disappointment. Like others of his ilk, Lorin Donald Porter was literally born a lumberman. By all accounts, Porter not only knew the lumber and timber business, but, more importantly, he had an acute sense about where to market the products he and his sons, Dean and Bernie, milled at their Western Pine Sales plant in Snowflake. A modest, soft-spoken man of Mormon heritage, Porter recalled that corporate giants like Phelps Dodge developed quite an appetite for superior-grade mining timbers. In response, the BIA and the San Carlos Tribal Council invited Porter to construct a band saw mill at Cutter on the reservation near State Highway 70. From 1962 to 1974 the Cutter Mill produced a substantial volume of finished lumber products from Apache-owned timber. Referring to the operation as a "very productive little mill," Porter judged San Carlos timber, though generally too poor in grade for housing construction, to be well suited to the mining industry. "It was solid timber," Porter said about the pitch-laden beams his company processed. In 1965, the company's peak year, the copper mines in Bisbee and Douglas alone purchased 10 million board feet of structural timbers from the Cutter facility.

Like the Webb operation, Porter's sawmill proved to be an excellent provider of Apache employment. During periods of high productivity, the Cutter Mill hired as many as forty-five San Carlos tribal members to work "in every facet of the industry." "With the exception of the management, and an experienced planer man from Winslow," Porter recalled, "80 percent of the Cutter sawmill was Indian operated." Unlike his predecessors, however, L. D. Porter was one of a handful of non-Indian independent operators who trained his Apache workforce for promotion to "skilled" labor positions. The experience the Apache people gained while working in privately owned sawmills at San Carlos and Fort Apache proved invaluable to the tribe in future years.

The economic partnership between Western Pine Sales and the San Carlos Apaches was so successful, in fact, that the White Mountain Apache Tribal Council allowed Porter to establish a second sawmill at Cibecue at the western end of the Fort Apache Reservation. With a ten-year agreement penned in 1967, Porter sought to double his production by using the Cibecue sawmill to process rough-cut lumber while employing the planer operation at Cutter to "finish out" the product. "A planing mill can handle more lumber than can be produced through a sawmill," Porter explained. "We had to have access to additional timber for our plants." The BIA's principal charge was to improve living conditions on the reservation through the promotion of timber sales. It appeared to all at the time that the match between the BIA, Western Pine Sales, and the Apache people was a lucrative one.

The symbiotic and seemingly healthy economic union between the federal government, the tribal council, and the independent sawmill operators continued unabated through the early part of the 1970s. In 1974, however, problems ensued that in time spelled disaster for the newly expanded Western Pine Industries. Local lumbermen heralded Porter's company as the "biggest of the small businesses in the Southwest." Nonetheless, the White Mountain Apache Tribal Council terminated Western Pine's ten-year lease three years before its completion date. "We had what we thought was a definite contract with the Indian people," Porter said bitterly. "We spent a lot of money putting in a mill [Cibecue] in the belief that we would have timber to saw."

Unfortunately for Porter, the White Mountain Apache Tribal Council decided that year to terminate agreements with all non-Indian timber contractors in its resolve to reserve future timber harvesting for the Fort Apache Timber Company (FATCO), the tribe's own sawmill. The tribal council offered to buy Porter out for, according to the owner, "one-third of the sawmill's total estimated value." "There was nothing we could do," he lamented. "They [the White Mountain Apaches] were going to build their own sawmill." Inasmuch as the Cibecue operation represented 40 percent of Western Pine Industries' total production, its loss nearly forced the company into bankruptcy.[19]

The problems that befell Henry Webb, A. G. Petersen, and Donny Porter during their cumulative years in the lumber business were in many respects reflective of the industry throughout the Southwest. With few exceptions, small, independent operations rather than well-financed corporations marked the industry in Arizona, New Mexico, Colorado, and Utah. Market fluctuations were the curse of every sawmill regardless of its size or the expertise of its owners. Added to that were staggering interest rates during inflationary years. Finally, a rising environmental consciousness that began in the 1970s, embodied in the regulatory provisions of the Environmental Protection Act and the Occupational Health and Safety Act, further taxed an already beleaguered industry. "Most of us [independents] could not sustain the enormous expenditures required to conform to the new federal regulatory standards," Porter criticized. In the final analysis, however, it was Indian self-determination and corporate consolidation within the industry and not federal restrictions that hastened the demise of independent lumbermen throughout the Southwest.

The shift from independent sawmill operations to corporate production of building and packaging materials, a trend that came to dominate the nation's lumber industry, began in northern Arizona with James G. McNary, former banker and schoolteacher. Out of the remains of the Cady Lumber Corporation, which had endured organizational collapse and financial capitulation in the 1930s, McNary forged a lumbering empire during the postwar decades that would challenge Weyerhaeuser, Georgia Pacific, and other preeminent producers on the Pacific Coast.

When Cady Lumber failed in 1934, the resilient McNary reorganized the defunct Louisiana-based firm under the trademark Southwest Lumber Mills, Inc. McNary aspired not only to maximize Fort Apache timber resources earlier made accessible to Cady Lumber but also to improve the underutilized Flagstaff sawmill. Thus for a third time, the indomitable McNary revamped his lumber interests in November 1935.

With a personal loan of $1.5 million from Phoenix banking acquaintances and a low-interest loan of $300,000 from the Reconstruction Finance Corporation, McNary resumed the long road to financial solvency. He committed $80,000 of his new assets to replace obsolete logging equipment with modern caterpillars and loaders. He applied $76,000 to road construction equipment and new logging trucks, a clear indication that McNary sought an alternative to the costly Apache

Railway. He reconditioned the McNary sawmill for $98,000 to include installation of a molding manufacturing plant. Finally, McNary reactivated the Flagstaff sawmill for $80,000. With one-third of his capital exhausted, the feisty chief executive appealed to the Fort Apache Tribal Council and the BIA for additional timber resources.

In late 1935 the BIA consigned the timber remaining to be cut on the Unit #1 contract to Southwest Lumber Mills. BIA records indicate that Cady Lumber had cleared 430 million board feet of timber since February 1918. Just 40 million board feet remained to complete the contract by the agency's reckoning. The provision to fix the minimum annual cut at 30 million board feet, however, presented McNary with a potentially serious problem. True, the company's expansion program during the spring and summer of 1936 greatly increased its production capacity. But it was also a fact that the McNary sawmill had operated at a loss throughout the Great Depression. While New Deal relief programs such as the Civilian Conservation Corps stimulated demands for building materials in Texas, Arizona, and New Mexico, markets on the whole recovered slowly. In effect, Southwest Lumber Mills had yet to create a wholesale market that could absorb even the minimum annual cut required in the Unit #1 contract.

In an unusual appeal, McNary asked the BIA to reduce the minimum annual cut to 12 million board feet per annum until the contract expired. This measure enabled Southwest Lumber Mills to harvest Apache timber at a more leisurely pace with the hope that lumber markets would soon improve. Concurrently, McNary negotiated additional contracts with the White Mountain Apaches in anticipation of future sales. Again, discussions revolved around the Maverick Mountain Unit as McNary pressed the BIA to open bidding. As the Great Depression lingered, American industrialists observed increasing political instability in Europe. Southwest Lumber Mills, recapitalized and assured that its contractual obligations would be met, planned a corporate strategy that anticipated a post-Depression resurgence of the lumber market.[20]

The U.S. mobilization of its military forces after Pearl Harbor, however, abruptly reversed the nation's economic status. As Congress declared war on the Axis powers in 1941, wartime industrialists assigned maximum priority to the production of lumber and other forest products. Accordingly, requests for timber from both public and private sources increased. In response, Forest Service rangers cooperated with BIA foresters to assure the War Productions Board a generous supply of

Logs felled near the tracks of the Apache Railway Company northwest of the lumber camp of Maverick, 1938. This photo makes a dramatic statement about the negative effects of logging on the natural environment. Courtesy Martha Chilcote Collection, Pinetop, Arizona; photo by Forest Service forester Fred Moffatt

timber. The public's insistence on harvesting government timber did not subside with the surrender of Germany and Japan. After a brief recession, the economic and industrial recovery of the United States that began with World War II continued well into the 1950s and 1960s. The regional appetite for Indian-owned timber—particularly in the booming metropolitan capitals of the postwar American West—increased lumber production on Arizona's Indian reservations to record levels.

Nowhere in the industry was the proposal to step up timber production to meet wartime needs more resolutely supported than in the office of Southwest Lumber Mills President James McNary. In numerous appeals to Superintendent Donner to make more Fort Apache timber available to his company, McNary had underscored the urgency of the wartime effort. "The demand for lumber in connection with the nation's war program is very acute," McNary stressed. "It is estimated that the government's requirements for the current year are about seven billion feet in excess of production." Most of the orders processed in 1942 were from surplus stockpiles accumulated at the nadir of the depressed lumber market. Market projections for 1943 could not be met, in McNary's judgment, without access to virgin stands of ponderosa pine such as those of the often discussed Maverick Mountain Unit.

Cognizant of the importance of lumber to the War Productions Board, the DOI authorized the sale of the Maverick Mountain Unit in the spring of 1943. The announcement proposed the sale of 400 million board feet of ponderosa pine with some Engelmann spruce. The contract required a security deposit of $36,000, thus eliminating most of the smaller independent operators from participation. To no one's surprise, Southwest Lumber Mills was the sole bidder on the Maverick Mountain Unit sale. The BIA awarded the contract on May 20. No other producer in northern Arizona had the additional $1 million to invest in a railroad needed to harvest the remote new stand. With the Apache Railway still in place, McNary had few competitors.[21]

Acquisition of the Maverick Mountain Unit made Southwest Lumber Mills the largest contract producer of timber in northern Arizona, although the honor was short-lived. First, postwar labor unrest, rampant throughout the western lumber and mining industry, especially hurt Southwest Lumber Mills. For example, most of the skilled African American lumbermen who had arrived with Cady and McNary in the 1930s, returned to Louisiana. McNary replaced them with unskilled Hispanic, Navajo, and Apache workers. Second, the tremendous cost to initiate logging operations in 1944 caused Southwest Lumber Mills to default on the contract in its first year as it failed to meet the minimum annual cut of 50 million board feet. Finally, in 1948 fire destroyed the original McNary sawmill. Though he rebuilt and modernized the facility, McNary's patience with the BIA and the attendant restrictions to produce Indian-owned timber had eroded. For these reasons, McNary shifted his energy to the firm's Flagstaff plant.

Curiously, it was not the housing industry that provided the most lucrative market for Flagstaff-produced lumber; rather, it was the crating and packaging industry, which required box shook. During the war, wooden crates were critical to the military mobilization effort. Nearly 60 percent of all lumber milled in the Southwest went into the manufacture of boxes used to transport war matériel, food, and medical supplies. Oscar Z. Brewer, president of the Central Wooden Box Association, which met in Chicago in October 1945, expressed optimism for the postwar conversion of the lumber industry. New markets for packaging furniture, refrigerators, and other household goods suggested a healthy outlook for the nation's wood products industry.

A second promising market for wooden box material was the West

Coast agricultural producers, whose national importance increased with the war effort and continued into the postwar period. A report published in November 1945 noted that the demand for fruit and vegetable crate material west of the Rocky Mountains had grown to "a level sufficient to absorb all available production from southwestern mills." The report predicted 1946 as a banner year in the production of wooden containers.

Encouraged by the good news, National Lumber Manufacturers Association (NLMA) members on the Pacific Coast eagerly awaited the effects of the boom on their industry. West Coast Regional President G. T. Gerlinger of Portland, Oregon, advised Jack Bedford, general manager of one of the Flagstaff sawmills, to prepare for a major upturn in the lumber business. "Because reconversion and postwar construction employment are so greatly dependent upon lumber supplies," Gerlinger wrote, "it is evident that [the] government is finally going to do something about lumber."[22]

In 1945 three lumber mills operated within Flagstaff's city limits. The smallest, which belonged to the Babbitt Brothers Trading Company, specialized in cattle sales with a modest mercantile lumber trade. Saginaw and Manistee Lumber Company, a Chicago-based operation that inherited the facilities of Flagstaff's pioneer venture, Arizona Lumber and Timber Company, was more formidable. Southwest Lumber Mills was the city's largest producer. These three plants, combined with the output of smaller, independent operations, produced 241 million board feet of lumber in 1946.

Lumbermen foresaw an upturn in the postwar industry with good reason. The return of hundreds of thousands of war veterans, many of whom were trained in the West and had immigrated there after 1945 to find jobs, caused a serious housing shortage. President Truman's plans for reconversion, as outlined in his 1946 State of the Union address, closely linked the achievement of a stable, high-level economy to the maintenance of national security. Maximum utilization of national resources was paramount to America's conversion from a wartime to a peacetime economy. For its part, the forest products industry stood poised to make its contribution to national recovery.

On February 7, 1946, President Truman announced the Veterans' Emergency Housing program, which was intended to alleviate the housing shortage through the construction of low-cost, affordable housing. Since most of the nation's lumber supplies had been diverted from civilian construction during the war, the legislation made materials available for new homes under the authority of the Civilian Production Administration (CPA). In a memo to all sawmill producers in the West, CPA President F. H. Brundage stated his expectation of the construction of 200,000 units in most of the principal western cities. Brundage requested that 40 percent of all lumber materials produced be set aside for federal housing projects. As predicted, 1946 was a year of peak production in northern Arizona, as sixty-six sawmills were reportedly in operation.[23]

Increased production both nationally and regionally triggered radical price fluctuations. A Department of Agriculture report entitled *Forests and National Prosperity* predicted a gradual decline in lumber prices beginning in 1948, citing intensified sawmill activity as the primary cause. In effect, accelerated production glutted the market, and lumber prices fell sharply as supplies consistently exceeded demands. The report also noted that timber had become unaffordable because it could no longer be harvested fast enough to keep pace with rising sawmill capacities. While the West held 65 percent of the nation's saw timber reserves, it accounted for only 35 percent of the total national output.

In northern Arizona, producers expressed concern over the steady decline in commercial lumber prices contrasted with rising production costs due, in their judgment, to the unavailability of high-grade saw timber. M. E. Kuhn, president of Saginaw and Manistee, warned forest managers in Albuquerque that dwindling supplies of merchantable timber threatened "the closure of additional small mills and some of the larger plants in Flagstaff." The federal government's solution to this regional problem in 1949 was the establishment of the Flagstaff Federal Sustained-Yield Unit in the Coconino National Forest. Its objective was to guarantee economic prosperity to the Flagstaff community by assuring local forest product manufacturers a dependable supply of timber. Under the provisions of this policy, "eighty-five percent of the saw timber, excluding salvage, sold from the unit [Coconino] must be given primary manufacture within the City of Flagstaff or three miles thereof." The Flagstaff Federal Sustained-Yield Unit clearly illustrated the government's desire to cooperate with private industry toward the maximum exploitation of northern Arizona's prime timber resources.[24]

Such was the economic milieu of the northern Arizona lumber industry when McNary turned his attention to Flagstaff. After the loss of the original mill in 1947, Southwest Lumber Mills built a completely automated facility that doubled the output capacity of the previous operation. Curiously, McNary resigned as president of Southwest Lumber Mills shortly thereafter, selling his interest in 1952 to James Ben Edens, a third-generation lumberman from Corrigan, Texas. Edens promptly vowed to upgrade the company from a regional to a national forest products giant.

Promises of a virtually limitless supply of timber from surrounding national forests assured the Flagstaff corporation an era of unfettered growth and prosperity. In November 1953 Saginaw and Manistee manager Freeman Shultz announced a merger with Southwest Lumber Mills. The integration of the city's two largest producers promised year-round employment to local sawmill workers for decades to come. That same year, the company reorganized under the name Southwest Forest Industries and transferred all of Saginaw and Manistee's inventory to the newly modernized and more efficient Southwest Lumber Mills plant. One year later, Edens purchased 87,000 acres of prime timberland from the Aztec Land and Cattle Company, making his firm the largest private owner of commercial timber in the state.[25]

The conversion of Southwest Lumber Mills into a national corporation in the mid-1950s mirrored a trend in the lumber industry toward the consolidation of single-operator sawmills into modernized, fully integrated operations. Guarantees of availability of federal timber—on and off Indian reservations—to private industry hastened the concentration of sawmill ownership into the hands of a few regional producers. The gradual disappearance of independent operators from the timber-rich environment of the San Carlos and White Mountain Apache Reservations, coupled with congressional resolve to promote Indian economic self-sufficiency, gave rise to northern Arizona's first Indian-owned timber corporation, whose success would exceed the expectations of both the BIA and the White Mountain Apaches.

In November 1952 Americans elected the Republican Party to the executive and legislative branches of the federal government. President Eisenhower and the staff of the DOI ushered in a new but short-lived Indian policy program under the generic name "termination." The program signaled a reduction—if not an end—to federal direction of economic activities on most American Indian reservations. In the case of the San Carlos and White Mountain Apaches, the federal government encouraged Indian initiative, consent, and participation in all programs affecting the social and economic well-being of the tribe. In effect, the long-term goal of the termination policy was to "get the government out of the Indian business." Newly appointed BIA Commissioner Glenn Emmons, an experienced southwestern lawyer, foresaw that the new policy was directly applicable to land and resource issues on the region's Indian reservations. Not surprisingly, the new emphasis embodied in the termination policy was welcome among those who shared long-held hopes for greater Indian self-determination.

The Fort Apache Wholesale Lumber Company (FAWLCO), originally established as the Fort Apache Sawmill Enterprise in September 1951, marked the government's initial experiment with commercial lumber sales on the reservation. Local contractors logged timber, which they cut into rough-cut green lumber for delivery to the planing mill located in Whiteriver, from the East Fork Unit. The Apache Mercantile Company, at that time a separate financial entity, redistributed the finished lumber products for local consumption. In early 1953, however, the BIA advised that the two businesses be merged under a single accounting system.

The ill-fated Fort Apache Sawmill Enterprise suffered numerous financial setbacks beginning in February 1954, when the sawmill and store burned down. Concurrently, the BIA charged the company's first manager with embezzlement, which led to his discharge. The following summer, Superintendent Albert M. Hawley hired a new sawmill manager to oversee the mill's reconstruction. Ross McSpadden, described by BIA forester Harry Kallander as "a real Arkansas hill-billy, but a terrific sawmill man," assumed the managerial position in June 1954 and immediately put the enterprise on a paying basis for the first time. McSpadden acquired his rough-cut green lumber from the Petersen-Osbourne sawmill north of Whiteriver. Tragedy struck again in 1956; this time the entire facility, including supply warehouses, lumber inventory, and the mercantile, were destroyed. The White Mountain Apache Tribal Council called a special meeting to discuss the future of the disaster-ridden enterprise.

A budgetary analysis clearly revealed the causes of the financial failure. While the first year showed potential for profit, the combined sawmill and mercantile endured $19,000 in losses from September 1951 to December 1955. Proponents of the venture, however, pointed to willful mismanagement and criminal misuse of funds as the principal reasons for the company's poor showing. Furthermore, an in-depth accounting of lumber sales for that period revealed that both grading and sale prices during 1955 were consistently lower than those received among comparable sawmills throughout northern Arizona. "It would seem that poor sales management must also have had some part in the loss, since 1953 was a very good lumber market year in which a substantial profit should have been realized instead of a serious loss," remarked area forester Carthon Patrie in his official report on the matter. In defense of McSpadden, both the BIA and the tribe agreed that the company had sustained most of its losses prior to his arrival. Patrie summarized: "In 1954 the tribe employed a new manager who realized a profit of $3,208.80 on the lumber operation compared with losses five times this amount in the preceding year."

On the basis of this information, the tribal council decided on June 14, 1956, to keep the sawmill active but to close the mercantile. They renamed the enterprise the Fort Apache Wholesale Lumber Company (FAWLCO). The company's stated objective—"to promote the economic development of the White Mountain Apache tribe and its members through utilization of the tribal timber resources"—was consistent with earlier BIA goals. The business would be primarily that of a wholesale lumber company, but limited retail sale of lumber to members of the tribe and other local patrons would continue on a strictly cash basis. The plan of operation for FAWLCO took effect on July 1. McSpadden promptly ordered the company to process its first load of surfaced lumber. Inasmuch as FAWLCO evolved from the earlier Fort Apache Sawmill Enterprise, the new venture began operations $36,500 in debt to the tribe. The tribal council allowed McSpadden to repay the sum at the rate of $6,500 in the first year of operation and $6,000 thereafter until 1961.[26]

McSpadden's enthusiasm about FAWLCO's profit potential was such that he outlined an expansion plan to Superintendent Hawley in the spring of 1957. First, he requested a fleet of delivery trucks to market Fort Apache lumber throughout New Mexico. "I can make them pay for themselves, pay the driver, pay the expenses and upkeep, and still make a profit from the freight," he assured his boss. Next, McSpadden asked permission to hire a full-time salesman at 3 percent commission. "This way he will be working for this enterprise only," McSpadden explained. "He will get every dollar possible for our lumber." The new salesperson was to approach mining and other industrial companies as well as wholesale dealers in an effort to increase the sale of rough timber produced from low-grade timber.

McSpadden argued articulately. Still, the BIA Phoenix area office denied his request to expand the operation. Federal officials remained skeptical about the long-term success of a tribal-owned entity. While they applauded McSpadden's optimism, the BIA reminded Superintendent Hawley that FAWLCO was "still on probation." For the moment, FAWLCO, producing an average 8 million board feet per annum, seemed destined to remain a small undertaking.

The decision to reject the proposed FAWLCO expansion was only temporary. In 1956 Southwest Forest Industries (SWFI) announced its intent to build a multi-million-dollar pulp and paper mill near Snowflake, Arizona, just south of Holbrook. The prospect that the White Mountain Apache tribe obligate even larger parcels of Apache-owned timber to sustain the corporation's diversification plans caused tribal council members to reassess the tribe's ability to compete with rather than accommodate SWFI in becoming the region's preeminent lumber producer.

The White Mountain Apaches' reevaluation of their future in forest products manufacture on the reservation coincided with administrative changes in Washington. In November 1960 President-elect John F. Kennedy appointed Stewart Udall, former Arizona congressman and a strident proponent of Indian economic self-sufficiency, as his secretary of the interior. Together, they established a special task force on Indian affairs that called for a movement away from termination toward the implementation of the New Trails program, which promoted improved human and natural resource development on all Indian reservations. Especially promising to the White Mountain Apaches was the task force's call for stronger federal commitments to develop Indian-owned industries, to increase vocational training and placement programs, and to provide low-interest federal loans for such purposes. Significantly, the task force espoused a policy of multiple land use for reservation forested lands with the long-term objective of economic diversification.[27]

Such was the political climate when Apache tribal leaders met with representatives of SWFI in May 1961. The meeting, which took place at the Fort Apache Agency, marked a turning point in the management of FAWLCO, the ongoing, albeit struggling, Indian enterprise. SWFI offered to underwrite the construction of an agency sawmill and, after seven years under SWFI management, sell it to the tribe. At this juncture, the tribal council took a decisive step toward economic self-determination and rejected the proposal. The tribe preferred instead to secure a low-interest loan, now available through the federal government, to construct its own tribal-owned sawmill. Some council members expressed misgivings about the tribe's ability to pay back a one-million-dollar loan in five years, while others questioned the sustainability of reservation timber needed to make the venture successful.

After much discussion, the White Mountain Apaches authorized the formation of the FATCO, concluding a master agreement with the BIA to set aside the Tribal Sawmill Block exclusively for Apache use. The agreement, signed in July 1961, was binding as long as the FATCO remained in existence. The Tribal Sawmill Block comprised 218,684 acres of forested lands, including 166,719 acres of old growth and 52,684 acres of residual growth. Government foresters estimated a net volume of 62 million board feet of timber earmarked for cutting during a twenty-five-year cycle. To comply with the economic ideals of the Udall task force, sawmill managers promised to employ and train registered members of the White Mountain Apache tribe.[28]

In one momentous decision, the White Mountain Apaches turned the ideal of political self-determination into economic reality. "Practical ideas diligently applied transformed the tribe from a dependent ward to the responsible steward of one of the Southwest's unique renewable resources," noted one historian. Under the astute guidance of Hal Butler, general manager and seasoned Arizona lumberman, the FATCO, built with $2.1 million of joint federal and tribal funds, soon became one of two major lumber producers in the American Southwest. With a combined annual output of 200 million board feet of lumber from three working mill sites (Flagstaff, McNary, and Eagar), SWFI remained the dominant manufacturer for the remainder of the 1960s. Ever wary of potential shortages of timber for the tribal sawmill, the tribal council ordered all non-Indian logging contractors—including L. D. Porter's Western Pine Industries—to cease operations and leave in the summer of 1973. Now the sole producer of ponderosa pine on the Apache reservation, the FATCO became SWFI's foremost industrial competitor.

In the interim, SWFI entered the paper products industry when its new thirty-two-million-dollar pulp and paper mill became operational in December 1962. Anticipating that corrugated cardboard boxes would soon replace wooden crates as the nation's primary shipping material, SWFI officials predicted a gradual shift away from wood products to paper production. Indeed, the 1969 annual corporate report revealed that for the first time in the firm's history, the paper products division—not wood products manufacturing—accounted for the majority of all corporate sales. The prospect that paper products would assume more importance than lumber in the future of SWFI no doubt prompted the corporation's decision in the mid-1970s to suspend operations at McNary after the sawmill burned down for the second time.[29]

The federal policy of termination ended officially in 1970 during the Nixon administration. With BIA advisement, the White Mountain Apache Tribal Council discussed its own plans for economic diversification based on multiple land-use management of the forest. Besides the hugely successful FATCO, a perennial producer of lumber products in excess of 100 million board feet per year, the tribe ventured into outdoor recreation and tourism with the creation of the Apache-owned Sunrise Ski Resort. The BIA forest management plans published since 1970 have predicted that the White Mountain Apaches will derive economic benefits for years to come from timber harvesting, milling, and outdoor recreational activities on the reservation.

As for their competitors, tougher federal environmental standards, inflationary costs to manufacture lumber products, and a consumer shift away from wood products to substitute plastics and metal alloys as construction materials have placed strenuous limitations on the industry nationwide. Locally, the steady, high-pitched whine of band saws and planers no longer heralds the industrial vitality of Flagstaff. SWFI closed its doors in the 1970s, leaving only its Eagar sawmill and a smaller facility in Southfork, Colorado, to process lumber. Once the mecca of the forest products industry on the Colorado Plateau, Flagstaff's municipal prosperity has clearly become less reliant upon an extractive economy and increasingly dependent on tourism and outdoor recreation. A recent assessment entitled *Flagstaff 2020* indicated overwhelming community identification with scenic beauty, climate, and clean air—not the forest or the wood products industry—as the city's most marketable assets.

In the interests of Indian self-determination, a dominant precept in federal policy in the 1950s, the BIA yielded its role as decision maker in favor of one as advisor in Indian economic affairs. In partnership with the White Mountain Apache Tribal Council, the BIA can offer years of forestry experience and technical expertise. The tribal council and the Apache tribal membership, however, make the core decisions that determine the outcome of the FATCO and other tribal endeavors. After more than a century of commercial exploitation, the impressive stands of timber that blanket the Mogollon Rim have reverted to their traditional owners. Today, it is the Apache tribe—not the federal government—that ultimately must manage the vast but no longer limitless forest resources for tomorrow.

NOTES

1. Henry P. Walker and Don Bufkin, eds., *Historical Atlas of Arizona*, 2d ed. (Norman: University of Oklahoma Press, 1986), 4–7; Walter E. Hecox with Bradley L. Ack, *Charting the Colorado Plateau: An Economic and Demographic Exploration* (Flagstaff: Grand Canyon Trust Publication, 1996), 7. For an in-depth analysis of the metropolitan development of this region since World War II, see Arthur R. Gómez, *Quest for the Golden Circle: The Four Corners and the Metropolitan West, 1945–1970* (Albuquerque: University of New Mexico Press, 1994; reprint, Lawrence: University Press of Kansas, 2000); Richard E. A. Lyons, ed., "Fort Apache Indian Reservation: Geography" (BIA, Branch of Forestry, Whiteriver, Ariz., n.d.), 1–2. Lieutenant Beale's quote is cited in Edward F. Beale, *Wagon Road from Ft. Defiance to the Colorado River*, 35th Cong., 1st sess., 1858, S. Exec. Doc. 124, 70–71.

2. Greenville Goodwin, *The Social Organization of the Apache* (Tucson: University of Arizona Press, 1942), 63–68; Jack August Jr. and Art Gómez with Elmo Richardson, "From Horseback to Helicopter: A History of Forest Management on the San Carlos Apache Reservation" (BIA, Branch of Forestry, San Carlos Apache Reservation, Globe, Ariz., 1983), 1–5, 12–14; Morris E. Opler, "The Apachean Culture Pattern and Its Origins," in Alfonso Ortiz, ed., *Handbook of North American Indians*, vol. 10 (Washington, D.C.: Smithsonian Institution Press, 1983), 368–92.

3. For a summary of early military sawmill activity on San Carlos, see Letters Sent, San Carlos, Ariz., April 10, 1890, Records of United States Army Continental Commands, 1821–89, Arizona Territory, San Carlos; quotation as cited in Lt. Col. Simon Synder, Commander, San Carlos, to Commissioner of Indian Affairs (CIA), August 20, 1888; Capt. John L. Bullis, Acting Agent, to CIA, August 26, 1889, *Annual Report, 1889*; Bullis to Commanding Officer, San Carlos, June 16, 1890, *Annual Report, 1890*, all in Record Group 393, Box 8, National Archives, Washington, D.C.

4. James H. McClintock, *Mormon Settlements in Arizona: A Record of Peaceful Conquest of the Desert* (Phoenix: Manufacturing Stationers, 1921), 135; Charles S. Peterson, *Take Up Your Mission: Mormon Colonization along the Little Colorado River, 1870–1900* (Tucson: University of Arizona Press, 1973), 24–27; Walker and Bufkin, eds., *Historical Atlas of Arizona*, 28.

5. Robert L. Matheny, "The History of Lumbering in Arizona before World War II" (Ph.D. diss., University of Arizona, 1975), 212–13; J. P. Kinney, "Forestry Administration on Indian Reservations," *Journal of Forestry* (December 1921): 836–43; Hal K. Rothman, "A Regular Ding-Dong Fight: Agency Culture and Evolution in the NPS-USFS Dispute, 1916–37," *Western Historical Quarterly* 20 (May 1989): 140–44; quote cited from Luther S. Kelly, Agent, San Carlos, to CIA, March 13, 1908, Records of the Bureau of Indian Affairs, Record Group 75, Central Classified File 339 (Forestry), Box 1, National Archives, Washington, D.C. (hereafter cited as RG 75, CCF 339).

6. All of Gutches's quotes as cited in "General Report: Care and Protection of

Timber on Fort Apache Indian Reservation," George A. Gutches, District Forester, to CIA, June 18, 1910, RG 75, CCF 339, Box 6; Joe H. Norris, Inspector, Department of the Interior, to CIA, May 4, 1910, RG 75, CCF 339, Box 6.

7. Gutches's quotes taken from "General Report: Care and Protection of Timber on the San Carlos Reservation," George A. Gutches, District Forester, to CIA, June 25, 1910, July 30, 1912, RG 75, CCF 339, Box 1.

8. "Regulations and Instructions for Officers in Charge of Forests on Indian Reservations," February 5, 1918, RG 75, CCF 339; Kinney, "Forestry Administration," 5; Carl Hayden to Secretary of the Interior Cato Sells, October 30, 1919, Box 623, File 25, Carl Hayden Papers, Hayden Library, Arizona State University (hereafter cited as CHP); Sells to Hayden, March 11, 1921, CHP, Box 622, File 11; August, Gómez, and Richardson, "From Horseback to Helicopter," 54–57; Henry T. Getty, "The San Carlos Indian Cattle Industry," *Anthropological Papers*, No. 7 (Tucson: University of Arizona Press, 1963).

9. Robert L. Matheny, "Lumbering in the White Mountains of Arizona, 1919–1942," *Arizona and the West* (Autumn 1976): 237–41; Sells to Henry S. Graves, Regional Forester, January 20, 1916, Records of the United States Forest Service, Record Group 95, Box 24, Timber Sales: Sitgreaves National Forest, National Archives, Washington, D.C. (hereafter cited as RG 95, Sitgreaves); "U.S. Department of Agriculture: Sale Prospectus, August 1917," RG 95, Sitgreaves; Timber Contract: Fort Apache Indian Reservation and the Apache Lumber Company, February 23, 1918, RG 95, Sitgreaves; Frank C. Pooler, District Forester, to Sells, October 23, 1920, RG 75, CCF 339, File 51, Apache Lumber Company.

10. Matheny, "Lumbering in the White Mountains," 242–43; Matheny, "History of Lumbering," 233–34; "Apache Lumber Company: Corporate History," RG 95, Sitgreaves, Box 24, Timber Sale File; E. B. Merritt, Assistant Commissioner, to Sells, August 22, 1917, RG 75, CCF 339, File 51; Sells to Secretary of the Interior, October 31, 1917, RG 75, CCF 339, File 51.

11. Matheny, "Lumbering in the White Mountains," 244–45; J. P. Kinney's quotes as cited in Merritt to Sells, August 22, 1917, RG 75, CCF 339, File 51; Jo Jeffers, "The White Mountain Scenic Railroad," *Arizona Highways* (July 1966): 8–10; Martha McNary Chilcote, *Memories of McNary* (Pinetop, Ariz.: Commercial Graphics, 1985), 1–5.

12. Matheny, "Lumbering in the White Mountains," 245–46; Veronica E. Tiller and Arthur R. Gómez, "Fort Apache Forestry: A History of Timber Management and Forest Protection on the Fort Apache Indian Reservation, 1870–1985" (BIA, Branch of Forestry, Whiteriver, Ariz., n.d.), 28–35; Quincy Raules, "A Reappraisal Report of the Apache Lumber Company, December 21, 1923," in RG 95, Sitgreaves, Box 24; *Holbrook News*, August 15, 1919; H. I. Crane to CIA, July 9, 1918, RG 75, CCF 339, File 51.

13. Alan S. Newell et al., *A Forest in Trust: Three-Quarters of a Century of Indian Forestry, 1910–1986* (Washington, D.C.: U.S. Department of the Interior, Bureau of Indian Affairs, Division of Forestry, 1986), 3–15; Matheny, "History of Lumbering," 273; *What's the Matter with the Lumber Industry?* (Washington, D.C.: National Lumber Manufacturers Association, 1925), 18–19; Forest Service comments as cited in Charles Burke, CIA, to Apache Lumber Company, October 18, 1921, RG 95, Sitgreaves, Box 24; "Reappraisal Report: Apache Lumber Company," December 1923, RG 95, Sitgreaves, Box 24.

14. James G. McNary, *This Is My Life* (Albuquerque: University of New Mexico Press, 1956), 71–73; Lee Muck, "Cady Lumber Corporation History," August 1928, RG 75, CCF 339, File 44408; McNary to Burke, November 16, 1923, RG 95, Sitgreaves, Box 24; Unit #1 Contract Modification Agreement, November 16, 1923, RG 75, CCF 339, Unit #1 File.

15. Matheny, "Lumbering in the White Mountains," 249–50; Muck, "Cady Lumber

Corporation History"; Chilcote, *Memories of McNary*, 1–12; McNary, *This Is My Life*, 158–65. McNary's comments are found in Charles L. Davis, "Fort Apache Reservation, Annual Report, 1925," 4–5, RG 75, CCF 339, Box 50.

16. Matheny, "History of Lumbering," 248, 280–90; Newell et al., *A Forest in Trust*, 3–15; Tiller and Gómez, "Fort Apache Forestry," 35–44; quote in James McNary to CIA, March 28, 1928, RG 75, CCF 339, File 44408, Cady Lumber File; Lee Muck, "Financial and Operating Characteristics of the Cady Lumber Corporation," August 1928, RG 75, CCF 339, File 44408, Cady Lumber File. Although McNary tried to argue that BIA timber in Unit #1 was on a par with substandard timber being cut on the Sitgreaves National Forest, BIA foresters insisted that reservation timber was on the whole superior in grade.

17. Senator Carl Hayden to William Zimmerman Jr., Assistant CIA, May 18, 1935, RG 75, CCF 339, File 44408, Cady Lumber File; quote on Indian New Deal in August, Gómez, and Richardson, "From Horseback to Helicopter," 74; Donner to CIA, April 23, 1935, RG 75, CCF 339, File 10157; "Timber Sale Announcement—North Boundary Unit," June 30, 1936, RG 75, CCF 339, North Boundary Unit File; quote as cited in Donner to CIA, July 22, 1936, June 20, 1938, July 16, 1941, RG 75, CCF 339, North Boundary Unit File; Ray and Carl Webb, interview by Art Gómez, Pinetop, Ariz., July 30, 1987.

18. Virtually all of the personal information concerning Henry Webb and Golden Petersen is derived from the author's interview with Ray and Carl Webb, July 30, 1987; August, Gómez, and Richardson, "From Horseback to Helicopter," 110–13; Turkey Tank #1 contract, in Robert D. Holtz Diaries, especially July 18, 1952, Federal Records Center, Denver, Box 752133, File 63A128.

19. For details on the Point of Pines #1 timber sale and L. D. Porter, see *Area Annual Forestry Report, 1961, 1962* (BIA, Phoenix Area Office Records); August, Gómez, and Richardson, "From Horseback to Helicopter," 134–38; L. D. Porter, "Western Pines, Inc.: Company History," n.d., in author's files; all L. D. Porter quotes are derived from his interview by Art Gómez, Pinedale, Ariz., June 28, 1983; "Western Pine Industries: Annual Report, 1974," in BIA Headquarters files, San Carlos, Ariz., File RF 1107–19; for commentary on early BIA and Western Pine's employment of local Apaches, see Ray Stevens Jr., interview by Elmo Richardson and Jack August Jr., San Carlos, Ariz., October 25, 1982.

20. James G. McNary, Memorandum Regarding Southwest Lumber Mills, Inc., January 25, 1939, Fort Apache Reservation Files, Southwest Lumber Mills, 1935–39 (hereafter cited as FTAP and file name); William Zimmerman Jr. to Southwest Lumber Mills, Inc., June 22, 1936, FTAP, Southwest Lumber Mills, 1935–39; Matheny, "History of Lumbering," 356–58; John Herrick, Assistant CIA, to William Donner, January 4, 1940, FTAP, Southwest Lumber Mills, 1940–44.

21. Statement is cited in McNary to CIA, October 26, 1942, RG 75, CCF 339, File 28017; Robert Patterson, Undersecretary of War, *Chicago Journal of Commerce*, September 25, 1942, 10–12; "Timber Sale Contract: Maverick Logging Unit," May 20, 1943, RG 75, CCF 339, File 28017; Newell et al., *A Forest in Trust*, chap. 4.

22. On the McNary fire, see Chilcote, *Memories of McNary*, 5; William G. Robbins, *American Forestry: A History of National, State, and Private Cooperation* (Lincoln: University of Nebraska Press, 1985), 170–204; Gómez, *Quest for the Golden Circle*, 42–46; quote on markets for box shook taken from Minutes, Central Wooden Box Association Meeting, October 25, 1945, in Saginaw and Manistee Lumber Company Records, Special Collections, University of Northern Arizona, Box 7, Letters 1945 (hereafter cited as SMLR); Gerald D. Nash, *The American West in the Twentieth Century: A Short History of an Urban Oasis* (Albuquerque: University of New Mexico Press, 1977), 236–37; National Wooden

Box Association, Pacific Division, December 18, 1945, SMLR, Box 7, Letters 1945; G. T. Gerlinger to Jack Bedford, July 25, 1945, SMLR, Box 7, Letters 1945.

23. Platt Cline, *They Came to the Mountain* (Northland Press, 1976), 190–93; Frank C. Lockwood, *The Life of Edward E. Ayer* (Chicago, 1929); Kenneth J. LaBoone, "The Arizona Lumber and Timber Company, 1881–1981" (master's thesis, University of Northern Arizona, 1981); Thomas Clifford Moses, "An Institutional and Economic Analysis of the Arizona Timber Industry" (Ph.D. diss., University of Arizona, 1981), 48; Veterans' Emergency Housing Program Order 1, March 26, 1946, in SMLR, Box 8, Letters 1946; F. H. Brundage to Sawmill Operators in Douglas Fir, Pine, and Redwood Areas, August 17, 1946, SMLR, Box 8, Letters 1946.

24. A Civilian Production Administration press release issued in December 1946 reported that the supply of lumber did not yet equal active current demands, SMLR, Box 8, Letters 1946; a memorandum published in May 1949 noted that lumber supplies had overtaken market demands, SMLR, Box 11, Letters 1949; *Forests and National Prosperity: A Reappraisal of the Forest Situation in the United States* (Washington, D.C., 1948), 43–44; Raymond Price, U.S. Forest Service, to Saginaw and Manistee Lumber Co., March 1, 1949, SMLR, Box 11, Letters 1949; quote as cited in M. E. Kuhn to Otto Lindh, Regional Forester, June 2, 1949, SMLR, Box 11, Letters 1949; *Forests and National Prosperity*, 60; quote as cited in *Flagstaff Federal Sustained Yield Unit: Periodic Reanalysis* (Albuquerque, 1970); Moses, "The Arizona Timber Industry," 44–45.

25. Matheny, "History of Lumbering," 360–62. In 1965 there were twenty-five active sawmills in northern Arizona owned by twenty different operators; ten years later that number had been reduced to fourteen, nine of which were owned by four firms producing 96.3 percent of the state's lumber. See *Arizona Landmarks*, vol. 1 (Phoenix, 1971), 49; *Coconino Sun*, January 17, 1947, March 19, 1948, November 11, 1953; "Southwest Forest Industries," *Builder Architect Contractor Engineer* 39 (August 1977): 6–7.

26. August, Gómez, and Richardson, "From Horseback to Helicopter," 109–10; L. L. Nelson, Acting Area Director, to Commissioner, BIA, June 28, 1956, FTAP, Fort Apache Wholesale Lumber Company File; "Plan of Operation of the Fort Apache Wholesale Lumber Company," June 14, 1956, FTAP, Fort Apache Wholesale Lumber Company File; Carthon Patrie, "The Fort Apache Wholesale Lumber Company: A Tribal Lumber Planing Enterprise."

27. "Southwest Forest Industries," 8–10; White Mountain Apache Tribe, Tribal Resolution No. 59-59, November 6, 1959, and Tribal Resolution No. 6-33, June 1, 1960, RG 75, CCF, Fort Apache Agency, Box 80, File 3285; personal information on McSpadden in Tiller and Gómez, "Fort Apache Forestry," 156–58.

28. *Feasibility Study of the White Mountain Apache Sawmill–Planing Mill, 1960*, Fort Apache Agency, Branch of Forestry Files, Whiteriver, Ariz.; quote on Indian self-determination as cited in Tiller and Gómez, "Fort Apache Forestry," 156, 158–62, 56; Master Agreement, White Mountain Apache Tribe and FATCO, July 14, 1961, RG 75, CCF 339, Fort Apache Agency, Box 80, File 3285.

29. Quote cited in Tiller and Gómez, "Fort Apache Forestry," 156, 163–75; Gómez, *Quest for the Golden Circle*, 161–62; "History of Southwest Forest Industries," in *Corporate Fact Book* (Phoenix: Southwest Forest Industries, 1983), 30–32; Moses, "The Arizona Timber Industry," 45–48; *Annual Report 1969* (Phoenix: Southwest Forest Industries, 1969), 2–6.

2

A Social History of McPhee

Colorado's Largest Lumber Town

DUANE A. SMITH

Far down in southwestern Colorado, underneath the usually placid waters of McPhee Reservoir, lies the abandoned site of McPhee, once a lumber town and sawmill. Far deeper underwater than Shakespeare's "full fathom five," McPhee has faded from both scene and memory. Fishermen cast lines above it, boats race over it, and visitors look overhead, admiring the striking variety of vistas of land and water without realizing that they have, in fact, overlooked a grave. For a generation people called McPhee, nestled along the wandering banks of the Dolores River, home. Now, it is no more.

The industry that gave McPhee life is no longer much of an economic factor in the region. Gone, too, is the railroad, the Rio Grande Southern, that transported its lumber products to the outside world. Only abandoned satellite lumber camps, located where vast stands of trees once waited to be converted into timber products by the McPhee mill, are still evident. And the residents who gave the town of McPhee life and vigor? They, too, are gone, scattered like leaves before the fall wind.

McPhee will not resurrect itself to greet the coming century like some water-logged Phoenix. The era, the industry, and the town are dead. All that remains is the story of what occurred there before 1980, when water filled the reservoir and inundated the once-thriving community.

McPhee: The Company

Lumbering became a major business in southwestern Colorado during the 1880s, when the Denver & Rio Grande Railroad (D&RG) extended to Durango and beyond the Animas Valley in order to tap the rich mining regions of the San Juan Mountains. Before the arrival of the D&RG, lumber mills had been small operations, their products sold and consumed locally. As the railroad forged new markets, it profited enormously in those early years by shipping lumber and coal into the bustling mining districts.

In the Dolores River Valley, lumbering developed possibly as early as 1874 with the opening of the mines near Rico and the arrival of the camp's first settlers. With the arrival of the Rio Grande Southern railroad in 1890–91 the lumber industry expanded. From 1874 to 1918, approximately 254,477 board feet of lumber were harvested in the Dolores Valley. By the turn of the century, Dolores, Colorado, one of the towns spawned by the railroad, claimed two of the nation's largest lumberyards.[1]

The lumbering potential of the area soon attracted outside interests, especially from Denver and the adjacent state of New Mexico. The firm of McPhee and McGinnity, formed by Charles and William McPhee and John McGinnity and one of Colorado's oldest lumber firms, expanded into New Mexico in 1892 under the name New Mexico Lumber Company. In 1900 Edgar M. Biggs, one of the company's partners, started to purchase individual cutting rights to timber located north of Dolores. A complicated power struggle in 1913 forced Biggs out of the company, at which time New Mexico Lumber Company assumed his cutting rights in southwestern Colorado.

By the early twentieth century, McPhee and McGinnity had expanded their operations with lumberyards throughout Colorado, Nebraska,

Wyoming, and New Mexico. The firm had at least four subsidiary companies, including the New Mexico Lumber Company. The latter operated in Edith, El Vado, and Chama, New Mexico, in addition to Chromo, Colorado. As timber gave out, sawmill operators moved equipment again and again to the next prospective lumber town. This pattern was repeated when the company arrived at McPhee.[2]

Against this background, local lumber operations in the Dolores region increased until the early 1900s, when mining expansion and building began to decline. That trend reversed after World War I. In January 1924 the U.S. Forest Service announced the sale of 4 million board feet of lumber. Already controlling acreage in the area, the New Mexico Lumber Company successfully bid on the timber and turned to the task of locating its mill. Unfortunately, neither McPhee nor his company advisors realized that most of these reserves, located seven miles north of Dolores, were comprised of what had been described as "overmature" trees of extremely poor quality. This seemingly innocuous oversight haunted McPhee throughout the history of his lumbering operations.[3]

In February 1924 the company finally selected a mill site on the western side of the Dolores River, five miles south of Dolores. Surveyors and crews arrived from the soon-to-be-abandoned lumbering community of El Vado, New Mexico, to start work on the new Colorado community named after William McPhee. McPhee had a daring concept in mind for his most recent project. He would consolidate all of the company's scattered holdings into one site. Moreover, the town would be based upon the concepts of planned obsolescence and rapid production. Thus by its own design, the McPhee sawmill and its community would have a comparatively short existence, only eighteen years.

In the interest of economic expediency, some of the buildings were transported from El Vado, while others were constructed on the site. One crew surveyed and laid tracks connecting McPhee's own narrow-gage railroad system to the Rio Grande Southern at Dolores. When those tracks were completed, work accelerated on the town's buildings, mill, and logging camps, which were quickly established among the tall timber where the rails stretched out to reach them. The idea was to have the mill completed by the time the first log train arrived. But when the locomotive chugged into town on September 29, the mill was not yet in full operation. The company required an additional month to solve all the problems before full operations finally began in October, when construction was finished.[4]

The McPhee railroad, the last narrow gauge built in southwestern Colorado, was unprecedented in the United States. Because of the town's isolation, the New Mexico Lumber Company decided to make McPhee a railroad hub to facilitate operations from the Rio Grande Southern main line to its more remote lumber camps. Railroad spur lines moved with the lumbering operations throughout the lease. At its peak, the company had sixty miles of railroad with five logging locomotives and forty-five flatcars. The company later added a passenger car to transport employees and families to Dolores behind the regular lumber train. Saturday night was particularly popular, as townspeople went to Dolores for dances, movies, and a good time. The faded elegance of red plush seats and flickering gas lamps took the passengers back to the more prosperous era of the 1880s.[5]

This section of the Dolores Valley had never witnessed such a beehive of activity, and the Rio Grande Southern had not been so busy since just before the 1893 crash and the devastating depression that followed. By the end of 1924, McPhee, a company town much like others in the Southwest, had its company-owned store, which when opened advertised itself as "ready to supply your wants" and pay "top prices" for farmers' produce. The impact of the McPhee operations spread beyond its town and camps into the surrounding area to provide a market for local ranchers and farmers that previously had not been available. The store, the largest "department store" in the area, offered clothing, groceries, dry goods, shoes, and other items for sale to the company workers. Those workers, at least in the early years, proved to be a transitory lot, the August 6, 1926, issue of the *Dolores Star* observed, "as there is in a place like this an ever changing population."

Delighted with developments and with typical booster enthusiasm, the *Dolores Star* featured McPhee in its August 13 issue with photographs of the mill, town, coal mine, and logging train. According to the article, people of Dolores "feel McPhee is part of their community and are proud of the mill and town." The New Mexico Lumber Company, the paper noted, had branched out to include a coal mine with a seven-man mining crew. It produced about twenty tons of coal per day, all consumed by the company operations. At this time, the town had an estimated population of 800 people, 350 of whom worked for the company. With a total payroll of $50,000 per month, McPhee ranked as the largest employer in the region.

Denver's *Rocky Mountain News* featured an article on McPhee on April 18, 1927, under the headline "Southwestern Colorado is enjoying thru the patronage of Denver, one of the largest sawmill propositions in the middle west." After praising McGinnity and McPhee, the article went on to say that slightly over 30 million board feet of lumber had come from the McPhee operations the previous year. A large portion of that was sold in Colorado, but it reached an "exceedingly broad market" as far east as Connecticut and New York as well. The *News* claimed a population of "about 1,500" for the town, with an estimated crew of 500 men. It was, the article concluded, "not only a splendid asset for southwestern Colorado but for the whole state."

Both newspapers were right—these represented the best days the mill and logging operations would ever experience. It was a time when the McPhee mill was praised because it "revolutionized the lumber mill business" of the state. Lumber from the mill was used for a variety of purposes. Most of it was roughly cut and then transported to the McPhee and McGinnity Denver plant. Lower grades were made into railroad ties and precut for boxes to ship fruit, vegetables, and paint. Superior grades were used for construction and were shipped to the Midwest to sash and door factories. Federal contracts included, during the New Deal days of the 1930s Depression, a contract for lumber to build Civilian Conservation Corps (CCC) camps.

Even in the promising 1920s, however, not all developments worked to the benefit of the New Mexico Lumber Company. McPhee's logging operations and the mill seemed to have a cloud hanging over them; weather played a vital role in determining profitable operations. The *Dolores Star*, for instance, noted on January 31, 1930, that deep snow had closed the sawmill for the "balance of the winter." Mud, rain, and washouts also slowed or briefly stopped operations, cutting into profits. Accidents, not an uncommon plight for the lumber industry, also bedeviled operations in the woods as well as at the mill.

The good times ended more quickly than the company had originally anticipated. Wishing to retire after decades in the lumber business, McPhee in 1928 sold his stock in the company to John Zalaha, an "experienced sawmill man from Arizona," and "Chicago bankers," who purchased the entire company the next year. Zalaha had ambitious plans for the operation, including more railroad lines, increased production, and expanded sales in the eastern market. Unfortunately, fate

Planer crew for the Montezuma Lumber Company, McPhee, Colorado, ca. 1938. Seated left to right: (1) Roger Oldham, (2) "Lenguita" (nickname), (3) Bill Truett, (4) Jack Ferguson, (5) "Slick" Largent, (6) Pete Gonzáles, (7) Fernando Sosa, (8) Frances Egger, (9) Manuel Aguilar, (10) Celestino Jácquez, (11) unidentified. Standing left to right: (1) Lucas Chávez, (2) Cloro Martínez, (3) Donaciano Aguilar, (4) Joe Maestas, (5) Maximiño Valdéz, (6) Alejandro Herrera, (7) Ray Gómez, (8) Joe Uliberrí, (9) Guillermo Gómez, (10) Juan Martínez, (11) unidentified, (12) Rumaldo Valdéz, (13) Jim Foy, (14) Woody Culp, (15) Clarence Cunningham (planer crew boss). Courtesy Gómez Family Collection, Santa Fe, New Mexico

intervened in the form of the 1929 crash and its resulting depression. The building and lumber trades were among the first and hardest hit. Zalaha defaulted on his payments, and the original owners regained control of the mill in November 1930. Unable to salvage its operation, the New Mexico Lumber Company slipped into receivership and eventually into the hands of the International Trust Company.

The economic morass of the early 1930s proved to be the beginning of the end for McPhee. Other problems hurt the company as well. To its dismay, the New Mexico Lumber Company realized that large stands of timber it had purchased years earlier turned out to be too sparse for profitable cutting. Gradual changes in railroad rates wiped out the advantages McPhee had once enjoyed in shipping, thus constricting much of the mill's market. As the Depression dragged on, the accumulation of these factors became too much for the company to handle. The company, therefore, temporarily shut down and did not resume operations until August 1932, and then only at half capacity. Despite announcements that "old employees" would be given preference in jobs, the desperate Depression days found McPhee flooded with men seeking employment.

It would not be until 1935, after a reorganization under the federal bankruptcy law and a name change to the Montezuma Lumber Company, that the town began to regain a semblance of its former prosperous appearance. The 1935 *Colorado State Business Directory* concisely summarized McPhee as a community with 1,200 residents, an altitude of 6,700 feet, and a postmaster named Frank Smith. From then until World War II, a measure of prosperity returned. An astonished 1940 visitor exclaimed, "A person would have to see the operation with his own eyes to realize how much lumber they really put out in a day's time."[6]

According to John Gómez, whose family moved with New Mexico Lumber Company in the 1920s from northern New Mexico to southwestern Colorado, working in McPhee through its cycles of boom and bust included beneficial as well as unfavorable aspects: "Every one of those people liked working for the company because that was security. Yet, by the same token, they hated it because of no benefits to begin with [and] the treatment; because, generally speaking, you were treated like a bunch of sheep. You are here to do a job, do the job, that's all there was to it." According to Gómez, discrimination was not an uncommon practice. "The 'Spanish' had their no-good, back-breaking jobs," he recalled. On the other hand, "Anglos would go to work one time and then, by God, you would turn around and he's [the Anglo workman] already got a position."

Chris Gómez, John's uncle, also worked at the McPhee mill. He described the experience in a similarly caustic manner. "I hope that I never have to go back to the lumber business. The wages," he bitterly joked, "seemed too low for the twelve hours they gave you to do a day's work." Yet, as he pointed out, the mill was a good place for the uneducated to work. It was, Gómez summarized, "either this or herd sheep." Accidents happened all too frequently; unfortunately, workers had to be "laid up more than ten days" before they were eligible to collect compensation. Any breakdowns longer than fifteen minutes were repaired on the workers' time, and the lost production time had to be made up. "You worked because you had to," Chris Gómez commented, "and you earned your wages."[7]

As the Gómezes and others worked, changes were being made in the McPhee lumbering operations. In 1932–33, the company tore up the railroad tracks, replacing them with less expensive trucks and logging roads. Loggers continued to rely on four-horse teams instead of mechanical loading devices and donkey engines to skid trees as far as the loading station. Not completely out of the railroad business, the company maintained trackage from the mill to the Rio Grande Southern spur at Dolores.

President Franklin Roosevelt's New Deal encouraged labor unions, which appeared even at company-controlled, geographically isolated McPhee. In July 1939 a unionized walkout at the sawmill shut down the entire facility. Nonunion loggers, who deplored their loss of pay, put pressure on the mill workers. The strike lasted only two days. It terminated with a complete management victory.

Despite that victory, the company suffered another setback two years later when the mill burned on June 30, 1941. Fire had long plagued McPhee. In 1934, for instance, a dry kiln was nearly consumed, and 500 acres of prime forest were destroyed. As it continued to operate out of the smaller logging camp mills, the company built a more modern, electricity-driven mill at McPhee. In its effort to complete defense-related contracts, the McPhee sawmill could ill afford a prolonged shutdown. While some worried that cause for the fire was sabotage, there seems little doubt that it started because of faulty wiring in the new mill.[8]

When the Japanese attacked Pearl Harbor on December 7, 1941, and the United States entered World War II, McPhee went to war with the rest of the country. Higher paying defense jobs and military service lured many young men, sometimes entire families, away from the Dolores Valley to the metropolitan capitals of the West. Still, company supervisors did not lack for patriotism, as the *Dolores Star* advertised on October 30, 1942. "Boys did you know McPhee would help you join the Navy? If you have no means of transportation to the recruiting office in Durango call at the filling station and means of transportation will be furnished." As with most small towns in the West, the war irrevocably changed McPhee, bringing to a close an era for the company as well as the town. Despite its myriad problems, residents appear to have enjoyed living there. In fact, as some of them reflected decades later, life at McPhee appears to have been almost idyllic.

McPhee: The Community

McPhee, the community, actually was divided into two sections, one "white," as its residents described it, and the other "Mexican," a section called Chihuahua. Each side of town displayed its own distinct character. Residents of European heritage—Irish, Finnish, Swedish, German, English, and Italian—flourished on one side, and northern New Mexican, or "Spanish," residents prospered on the other. The latter made up the largest percentage of McPhee's population.

The atmosphere between the two groups, old-timers remembered, was neither hostile nor characterized by overt racial animosity. John Gómez expressed it well when he discussed his life at McPhee in the late 1930s and early 1940s: "Probably the best place [McPhee] in the world to grow up in. There was no law enforcement at McPhee, because there was no crime. It was a very tight knit community; I remember that everybody knew everybody. There were, you know, squabbles. There were big dances and there were big fights, but the next morning they [the combatants] got up and went to work." The racial division, however, remained understood. Bernie Torres explained, "They lived up there and we lived down here [in the Hispanic section], and we were rarely in contact with them. If we went to the commissary and saw somebody there, we would talk to them and they would talk to us. That was it."

That was not really "it." Hispanics and non-Hispanics shared many things in common. Their children went to the same school through the ninth grade. If a child wished to go on to high school, he or she traveled to nearby Dolores, an experience that provided quite an adventure because the company allowed passage on the McPhee "luxury" train, as one former resident recalled. In the 1930s, however, many did not attend high school; hard times and individual circumstances caused them instead to enter the labor force at the sawmill. The company not only donated the lumber for construction of the school building but also attempted to hire bilingual teachers, because many Hispanic youngsters did not speak English. Bilingual teachers unfortunately proved hard to come by, and company efforts were soon discontinued.

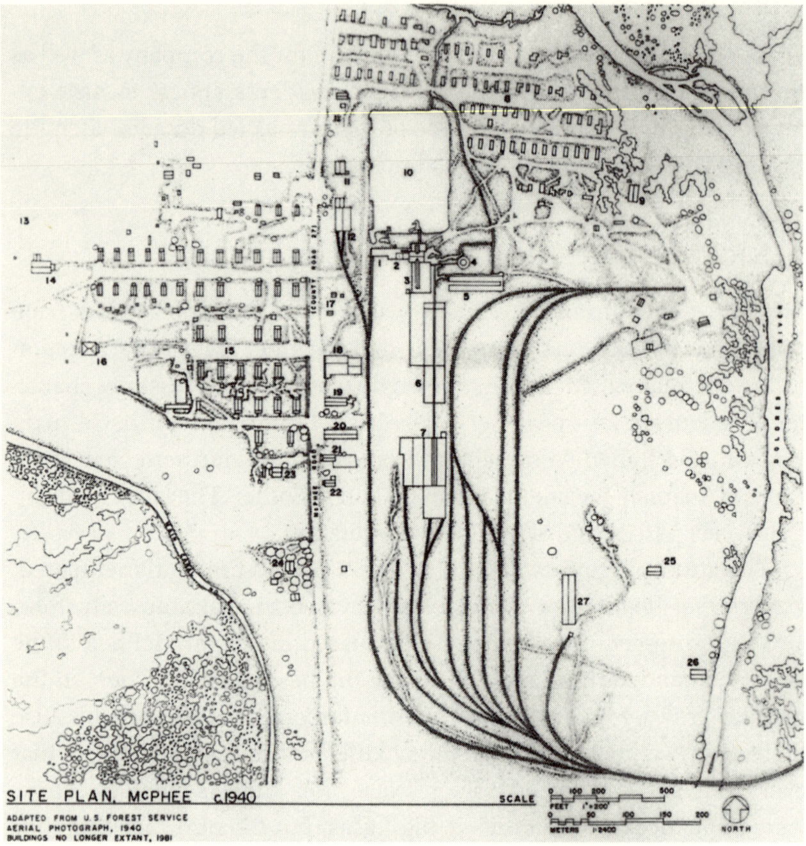

Site plan of McPhee, Colorado, 1940. This Historic American Buildings Survey drawing was adapted from a Forest Service aerial photo taken in 1940. Note the obvious separation between the non-Hispanic housing located at the left side of the photo and the Hispanic housing at the top. The text on the site plan refers to McPhee as "Colorado's largest mill town." Courtesy The Anasazi Heritage Center, BLM, Dolores, Colorado

All McPhee children played on integrated sports teams: the boys ran track and played baseball and basketball, while the girls participated in softball, track, and basketball. Hispanic and Anglo children bonded in other less frivolous ways; Lela Martínez jokingly recalled that the principal "spanked all of the kids the same." "Once, when thirty kids, Anglos and Hispanic alike, played hooky from school for a day, they all learned the same lesson: Don't do it. The next day," Martínez recalled, "each and every one was taken into the office and spanked with a rubber hose."

In the absence of bilingual teachers, all students were required to speak English, leaving those youngsters who had heard only Spanish at home at an early educational disadvantage. Even at recess, if a teacher heard them speak Spanish, they were chided and mandated to speak English. This must have proven especially difficult for the first-year students. Thus, many Hispanic children found themselves at a disadvantage or at the very least with the expectation that they must acculturate—often traumatically—into the unfamiliar Anglo world. The more fortunate students, however, had teachers who at least made an attempt to speak rudimentary Spanish.

According to John Gómez, the teachers did not always encourage cultural harmony. He sadly recalled a racial attitude that he witnessed among some of his teachers. "The teachers that came from outside the area discriminated." He particularly singled out the teachers from Cortez and Dolores. "They were definitely prejudiced," Gómez declared.

Of course, Hispanic youngsters did not have to go to school, nor did they have to learn English. If they hoped to secure a position in the world beyond the McPhee community, however, speaking English quickly became an imperative. Carmen Gallegos, at about age twelve, went to work for a family doing baby sitting and odd jobs. She spoke very little English, but with their help "she learned in no time at all."

The problem Hispanic students experienced in school was in large measure a reflection of their familial upbringing as well as their cultural heritage. Some students received little encouragement to attend school, despite acknowledgment of the importance of education by their parents. This seeming contradiction reflected the fact that during their years in McPhee the parents themselves had not gone very far in school. Moreover, the lingering fear that a formal education might lure youngsters away from the traditional life compounded the issue. Many parents believed that school undermined the traditional family structure, religion, and culture by replacing it with the faster-paced, highly urbanized, English-speaking culture. Although the typical parent-teacher associations were present at McPhee, the extent of Hispanic involvement—if any—remains uncertain.[9]

During the summer months there was plenty of fishing and swimming, as the Dolores River provided the opportunity for youngsters of all ages to play together. Even here, however, incidents that may or may not have been racially motivated got in the way of camaraderie as shouting or shoving often marred a summer's day. As several people remembered, however, children and adults generally chose peers from their own side of town.

Adults, too, found themselves together on occasion. Besides work, manufacturing beer was a common activity among both the Hispanic and Anglo males. Although illegal during the dry twenties, this activity not only endured during Prohibition but continued well into the next decade. Home brew provided a tasty treat and usually warranted soliciting the assistance of friends and relatives to sample it during the brewing process. Some of the local beer became renowned for its flavor and texture.

Hunting was also a universal activity, providing meat for the family table. Occasionally, the meat was sold to neighbors to earn extra money. The latter economic potential naturally encouraged poaching out of season, it being recalled that men "commonly" spent their winter nights hunting deer illegally. Sage hen, blue grouse, and elk graced many a McPhee table. Men of both cultures fished in the Dolores River and nearby streams.

Although adults shared a common enthusiasm for baseball, they rarely formed teams and played together. They jointly cheered the local school teams, however, as many traveled to the surrounding communities of Yellow Jacket, Dolores, Cortez, and Cahone. Local games played in a horse pasture or in the lumberyard by Anglo and Hispanic men were, a resident recalled, "rousing" events. But only at the annual company picnic did the two groups "customarily challenge each other."

The *Dolores Star,* which carried an occasional "McPhee Column," noted on September 3, 1926, that the company picnic featured races and a baseball game with both men and women playing. The next year, in the May 20 issue, it commented that it "looks as if McPhee will have a tip-top baseball club," indicating that there also must have been a town team at that time. A young men's tennis club also played "about every evening." Although the newspaper did not specify this, the members were most likely Anglos, since the Hispanics interviewed years later had no recollection of tennis as being favored among their sports.

A more revealing picture of social life at McPhee appears when we examine the differences between the two groups. Beyond the sport of baseball, most of the remaining recreational activities appeared unique to each particular culture. For example, Hispanic men played a soccer-like game called *chueco,* which involved two teams comprised of twelve men each. The Anglo men enjoyed volleyball in the summer, whereas it appears that the Hispanic men showed no interest in that sport. Both groups enjoyed cards yet preferred different games. Anglo men played poker, and their wives played bridge or pinochle. At ten o'clock, when the electricity was turned off, lanterns lit both sides of town as the card playing continued. Hispanic men and women frequently played poker, though not in mixed company. The women also played pitch. Lela Martínez recalled how popular *píte* became, a game, it appears, that was reserved exclusively for McPhee's Hispanic residents.

One form of entertainment universally enjoyed by McPhee residents was going to the movies in Dolores, a favorite pastime of young and old alike. The fact that the theater would accept scrip facilitated frequent movie attendance. The theater owner turned the scrip back to the company in exchange for cash.

Both groups enjoyed Saturday night dances. Anglos favored the schoolhouse, while Hispanics had their own hall. This activity underscored certain cultural differences regarding dance styles and music. The Anglos' evenings highlighted the latest fads, dances, and popular music, with the standard "oldies" for the senior folks. Hispanic dances, on the other hand, featured popular songs as well as traditional music that accompanied the *varsoliana,* a traditional dance that dated back to the earliest days of territorial New Mexico.

Before World War II, these dances provided memorable moments for the entire Hispanic community. Resident musicians united to form a band (violin, guitar, and accordion, generally), and the dancing lasted late into the night. As John Gómez humorously recalled, "We were born with a guitar." Indeed, music was an intrinsic part of Hispanic culture. The hungry dancers enjoyed lemonade and hamburgers at midnight. Sometime during the evening, announcers called attention to the woman or girl wearing the prettiest dress. Helen Fogel, who lived on the other side of town, recalled how non-Hispanics were welcomed to these dances. In an interview she revealed much about the time and attitudes. "They had a dance down in Mexican Town. You see, the Mexicans didn't live with the white folks then. They had a town of their own down there. They held dances down there too, and sometimes a bunch of us would go to their dances."[10]

While the groups may have intermingled socially, they remained separated in their religious practices. One side of town was predominantly Protestant, the other Catholic. Hispanics worshiped at the Catholic church on a hill to the west overlooking the town. The *Dolores Star* on June 7, 1929, praised the New Mexico Lumber Company for its generosity in furnishing material and supervision during construction of the recently finished building. Nearby was a cemetery for the residents' use. Conversely, Anglos were represented by a denominational mixture. They held church services and Sunday school in the schoolhouse or traveled to nearby communities to worship.

Catholic funerals aroused curiosity among McPhee newcomers such as Bernice Willis, who remembered "the stately, ornate procession in which the mourners paraded on foot from the church to the cemetery." Like the Chinese funerals in western mining towns, however, the novelty wore off for residents who had lived in the community for a while.

The Catholic Church was the center of Hispanic life, essentially a continuation of the tradition they had brought with them from New Mexico. Unique to the Hispanic community were Los Hermanos de Luz (the Brothers of Light), a lay brotherhood commonly known as the Penitentes whose origin dated to medieval Europe. Spanish conquistadors transported their practices to the New World in the late sixteenth century. Juan de Oñate, the colonizer of New Mexico in 1598, is believed to have been a member of this religious fraternity. The members of this secretive organization were devout Catholics who practiced acts of contrition most often through exhaustive prayer meetings and medieval ceremonies but in particular during Holy Week through self-inflicted punishment and religious reenactment.

For some Anglos, Penitente activities represented perhaps the most fascinating aspect of the religious practices of their Hispanic neighbors. Vic Sundquist recalls that as a youngster he became curious about the highly secretive nature of Penitente activities. His friends warned him never to intrude upon these ceremonies, or the Penitentes would "get him." The Penitentes' frequently misunderstood desire for privacy to worship as they had for centuries was too often misconstrued as being either mysterious or at times sadistic. The Penitentes, who as lay religious administered to the spiritual needs of rural Hispanic communities that rarely saw a priest, were very active during the first years after the move from El Vado, New Mexico. After decades of seclusion, the lay brotherhood has reemerged in the remote mountain villages of northern New Mexico as well as the lush San Luis Valley in southern Colorado.

The influence of the Catholic Church and its New Mexico heritage on the Hispanic community evolved in many ways. For example, the Christmas season as a religious and holiday celebration went well beyond anything celebrated in Protestant festivities. It began in November with the butchering of pigs (known as *matanza*) to provide pork for use in making Christmas goodies such as *empanaditas, chicharrones,* and *chile.* Generally, the *matanza* itself served as a coming-of-age ceremony for many teenage Hispanic males.

Another anticipated event was the staging of the community pageant known as Los Pastores. A centuries-old religious folk play, Los Pastores was an annual affair in which the entire community participated. One author who described the pageantry notes that the folk drama is best enjoyed "in the romantic mission atmosphere of an ancient church . . . on a cold winter's night." Indeed, once a year McPhee's Hispanic neighbors stood before a theatrical re-creation of the manger in anticipation of the angelic pronouncement of the birth of Christ to the entire world. The play, in which local residents portrayed all of the principal characters—the Holy Family, angels, devils, shepherds, and kings—opened with the kings' search for the baby Jesus. It ended with a devout procession in which one select individual traveled a considerable distance on his knees for the honor of placing the infant Jesus into his cradle. Christmas Eve mass usually followed, after which the celebrants would go from house to house, indulging in hot chocolate and other treats along the way. En route, *el diablo* relentlessly tried to tempt the crowd, but the angel San Miguel struggled with and eventually triumphed over the mischievous devil. For the adults, a feast and dance ended the evening. On Christmas morning, in a tradition similar to that of the Anglo observance of Halloween, the Hispanic children went door to door with bags in their hands requesting Christmas candy or fruit. Holiday traditions such as these were common among Hispanics, who carried them from lumbering community to lumbering community.[11]

Hispanic weddings, too, had a festive air. Following the marriage ceremony and mass, a wedding feast and dance took place that generally featured *la marcha,* a tradition in which the newly married couple led all the other celebrants around the dance floor. The festivities concluded with the *entriega,* or "giving away," during which the bride and groom, kneeling side by side, received the blessing of the groom's parents as everyone present joined the *trobador* in singing a special, personal tribute to the young couple.

Each side of town had its own physical character and composition, most easily recognized by the differences in the size and the quality of housing. The Anglo homes were framed structures—four or five rooms—that were built on cement foundations. Most had a front porch,

were lighted by electricity, and included running water. Wooden sidewalks stretched along the front of each residence. In contrast, many of the much humbler Hispanic houses included only two or three rooms. Lacking foundations, they were constructed with unfinished lumber. Hispanic laborers were not supplied with either electricity or running water. Instead, a water pump and outhouse furnished their plumbing requirements.

Hispanic residents took great pride in their homes. Lela Martínez observed that "everybody fixed their house real nice." Her father, for example, improved their home in several ways. "My Dad put nice lumber on the inside. We didn't have carpet, but we had linoleum. They [her dad and uncles] built cupboards. My Dad added another room to our house and built a porch with a screen around it. He built a sidewalk

The New Mexico Lumber Company, McPhee, Colorado, ca. 1928. This Forest Service overview photo of the sawmill at McPhee shows the housing for non-Hispanic workers in the foreground, the company store (with company logo) and sawmill operation in the background. At this time, the sawmill had been in operation only four years; note the recently constructed houses. Courtesy Gómez Family Collection, Santa Fe, New Mexico

with lumber to our outhouse."[12] The pride Hispanic residents displayed in their work as well as in their homes contributed greatly to the vibrancy of the McPhee community, even during the Depression.

Not all residents, however, remembered the McPhee mill or its adjacent town as exemplary in either social equality or community solidarity. Lumber worker Chris Gómez recalled his years at McPhee with a less romantic glow than most of his contemporaries. He believed the company more or less imposed a dictatorship over the community and

Housing for Hispanic workers, McPhee, Colorado, ca. 1935. The modest clapboard houses that constituted Chihuahua, or "Mexican" town, as it was commonly referred to, stood in remarkable contrast to the housing that the company provided to its non-Hispanic labor force. Hispanic families made up about 75 percent of McPhee's estimated twelve hundred residents. Courtesy Colorado Railroad Museum, Robert Orr Collection

the lives of its people. For example, the McPhee mill paid in scrip, which could only be traded at the company store, whose prices Gómez criticized as "forty percent higher than anybody else's" rather than less, as the company liked to advertise. On the other hand, long-time plant general manager Charles Artz explained years later that the employees received 40 percent of their wages in scrip and the rest in cash. The coin-shaped, lightweight metal scrip was intended to pay only for the predetermined monthly expenses such as rent and groceries. The rest, according to Artz, was theirs to spend however and wherever they preferred.

Chris Gómez also bitterly reminisced that if an employee tried to quit his job, "the first thing they would tell you was to get out of your house." There were few places to seek employment, so Gómez, like many others, saw no recourse but to remain with the company. Despite working there for many years, Gómez never received a Christmas greeting from the company. If they gave out hams and turkeys at Christmas to the workers, as some recalled, "I never got any," he remarked.

Whether or not the Anglo workers received such gifts of appreciation while the Hispanic laborers did not, Gómez did not indicate. There seems to be little doubt that discrimination existed at McPhee, although perhaps more often subtly implied than blatantly imposed.

John Gómez, however, remembered overt prejudice. "There was a lot of prejudice all right, but you see we were in one school, and we were all together. They would call us 'Mexicans,' and we would call them 'Gringos.'" Nonetheless, Gómez believed that McPhee taught him "how to overcome it [racism] as we went along, and how to succeed." According to him, McPhee folks basically got along, until the outside world crashed in on them. Cortez proved particularly hostile; "they were different people." Gómez recollected going with his father and brother to the Montezuma County fair in Cortez, where they were denied admittance because of their ethnicity. Such racial attitudes were clearly evident in the local news. For example, when an accident or killing occurred in Chihuahua, the Dolores paper simply made reference to a "Mexican" who had been hurt or killed. The victim's first name rarely appeared in the article, and, in most cases, even the last name was not given. Anglo residents of McPhee, on the other hand, received more detailed coverage in newspaper articles.

Other factors separated the two cultural communities at McPhee. The Hispanic families were generally larger than those of their neighbors. The extended family—*la familia*—included grandparents, uncles, aunts, and cousins. It was always nice, one woman observed, to have many children living nearby as the parents aged and needed assistance.

After World War II, social and economic relationships at McPhee changed. The postwar era witnessed the decline of both the town and its lumbering industry. No new contracts appeared, hastening the end of an era. Unemployed Hispanic families were forced to move on, only this time there were no nearby lumbering communities in which to begin their familiar life patterns anew.

Hispanic culture, too, was affected as younger people moved to distant metropolitan communities, where they identified more closely with an urban society that differed radically from that of their hometown. Older McPhee residents sought refuge in the neighboring, predominantly Anglo communities of Dolores, Cortez, and Durango; they no longer took comfort in the security of *la familia* or the familiarity of Hispanic tradition. As the pressure to abandon their rural tradition and cultural identity increased, customs of the past seemed awkward and obsolete. Long-time residents of McPhee remarked how difficult it was to make the transition into a "strange" world from which the once thriving lumber town had for decades shielded them.

As Hispanic families abandoned McPhee, the larger world absorbed them. Erased from memory were the age-old traditions such as Los Pastores as English replaced Spanish to become the dominant language, especially among the younger generation. Postwar-generation Hispanics learned early on that mastery of the English language and acceptance of Anglo customs were the keys to success and social adaptation in their new world. Saddened old-timers, meanwhile, helplessly watched the dilution of their cherished culture. Most died with the hope that their urbanized offspring would in time come to embrace the traditions of their McPhee past as they themselves got older.

For most Hispanics, however, it had been a good life in McPhee, one they would never forget. Bernie Torres spoke for many when she observed, "We loved that. It [McPhee] was freedom. We were there in our little town. We had everything that we wanted that was available at the time." Many of Torres's contemporaries appear to agree with her conclusion: "[Living in McPhee] was one of the happiest times. There have never been times happier than those." Another elderly Hispanic who remembered his days there provided the perfect epitaph for the town during the tumultuous 1930s: "That's [McPhee] all we knew, and we were comfortable. We lived real happy. We never were hungry. We never were cold."[13] Indeed, few Americans could say that about their experiences during the Depression.

While the Hispanic experience at McPhee may appear to have been typical for Colorado lumber towns, insufficient research regarding similar communities precludes making generalizations about the state or the region. McPhee in many respects represented both a beginning and an ending for most of its Hispanic citizens. It provided a point in time during which Hispanic culture flourished in tandem with the Anglo traditions that were common to this remote Colorado subregion. At the same time, McPhee represented a point of departure as Hispanic residents—either willingly or unwillingly—abandoned their traditional life at McPhee in an effort to prepare themselves for entry into a world far beyond the familiar parameters of the Dolores River Valley. Once departed, there could be no return; McPhee the community was, as a consequence, gone forever.

The End of an Era

The war years were not good ones for the company and its town in several respects. Despite the promise of economic revival, fire destroyed the mill, which had not been fully rebuilt before a second fire consumed the machine shop in June 1942. The following October the company's smaller Lost Canyon mill burned. As the *Dolores Star* observed on October 23, the company "seems to be under some kind of curse." While fires were certainly not unusual in the milling and lumbering business, McPhee had experienced an unusually high number of them. A continual headache that had plagued the company since its inception was the difficulty in hiring and retaining skilled workers of any kind. War-related industries lured laborers from McPhee as well as other southwestern Colorado communities to higher paying jobs in Denver, Pueblo, and Colorado Springs. The lack of reliable labor forced the premature curtailment of lumbering operations. To add to the company's frustration, the loss of local manpower coincided with a tremendous demand for all kinds of lumber.

The Montezuma Lumber Company, the new name for the McPhee operation as a result of its financial reorganization in 1935, now ran three small mills near the location of its logging operations from which the company hauled lumber to be finished at McPhee. No sooner did the company announce that the McPhee sawmill would probably never be replaced than the machine shop burned and had to be rebuilt. Local boosters hoped the operation would continue because of "its vital importance to this entire part of the state." As machinery either broke down or wore out, it became harder to replace because of wartime priorities. Thus McPhee limped through the war years, becoming even less profitable. All the while, more people moved out while fewer moved in. Not surprisingly, there was only sketchy commentary in the McPhee news column about lumbering operations during these years.

Then in February 1944 Lewis Mack of Moab, Utah, headed a group of investors who purchased Montezuma Lumber Company. Initially, their plans called for the liquidation of the town of McPhee and its original plant. The new owners preferred instead to build entirely new sawmills in Dolores and other nearby communities. It was even rumored that Mack, who at the same time was also drilling for oil west of Cortez, might convert the McPhee site into "a refinery, if oil of sufficient quantities is encountered." It appears that nothing ever came of his plan.[14]

The new owners, however, sold the houses from McPhee, stipulating that the buyers had to transport them to their new locations. All but fifteen out of approximately one hundred houses were removed from the now abandoned town. Even the Catholic church eventually was transported to Dove Creek, Colorado, near the Utah border.

Although the company faced the serious prospect that it would soon deplete all of its timber reserves, Lewis Mack continued to produce finished lumber. Despite the company's losses, Mack built a new, supposedly fireproof sawmill. Lewis Mack's mill remained in operation until January 1948, when the end finally came. Although his latest building may have been designed to be fireproof, it proved otherwise. During a typical winter storm with temperatures of thirty-eight degrees accompanied by a strong gale, fire erupted in the mill house on the afternoon of January 19. Dolores, Cortez, and McPhee firemen could not control the flames, and witnesses watched the estimated $100,000 worth of machinery and buildings go up in smoke. According to the news story, insurance covered approximately a third of that amount. The *Dolores Star* reported on January 23, 1948: "The loss is a serious blow to Mr. Mack, who is just recovering from a serious illness, to the men who are employed at the mill, and to the community as well."

It was all that and more. Montezuma Lumber Company decided not to rebuild; the end of an era for McPhee had finally arrived. Instead, company owners moved the mill to Dolores. "There was no point in keeping a plant going at McPhee any longer," management told the *Star.* Within weeks, scrap from the burned mill was sold along with the railroad tracks that had once extended from McPhee to Dolores. Salvageable machinery was sold to firms operating in Mexico. The last houses had been moved by the end of March. Only the commissary and a few remaining buildings occupied the land when it was finally sold. The McPhee post office closed in July, just two months short of the town's twenty-fourth anniversary.[15]

The end, when it came, proved merciful, a quick finish to what otherwise would have been a prolonged death struggle. The story of McPhee, Colorado, appears to be typical for lumbering towns in the region. It had developed as older logging areas declined; it drew life from their death. After its usefulness expired, the company abandoned the McPhee operation. William McPhee's "planned obsolescence" of 1924 had finally come to pass, although not quite in the way he had envisioned.

Several factors contributed to the decline of McPhee. The costly series of fires was not only the most spectacular but also the most devastating factor. The early bankruptcy in the 1930s and the continual financial problems that followed weakened the company's operations. In effect, McPhee never completely recovered. No doubt any pessimist might have forecast the ultimate fate of the company and the town as early as 1933. Throughout the company's history managers overinvested in government and private timber purchases, the quality of which proved substandard for meaningful profits. Erratic weather, frequent accidents, and wartime labor and equipment shortages further plagued McPhee and hastened its demise.

The McPhee story in southwestern Colorado in many respects mirrors that of the logging industry throughout the West. The environmental, technological, and market factors determined not only the industry's size but also the extent of its success in the Pacific Northwest, the Midwest, the Southwest as well as elsewhere. That the story occurred in the twentieth century does not mean that the industrial patterns significantly differed from those practiced throughout the nineteenth century. Perhaps because of early direct federal involvement in regulating the industry, northern Arizona's development may constitute a notable exception to this familiar cycle.

As with numerous other lumber towns throughout the country, McPhee altered the western landscape without noticeable concern for the environment. The clear-cut logging practices of the past were a wasteful enterprise driven by a market economy. To its residents, however, it seemed sufficient that the sawmill provided jobs as well as an economic outlet for nearby farmers, ranchers, and merchants. It is, after all, one of the great myths of the West that natural resources are there for the taking.

McPhee as a company town is not very different from others that coexisted in the lumber business. However, McPhee did not display some of the worst aspects of company exploitation practiced in the even more isolated coal communities of Kentucky and West Virginia. Living only a few miles from several other Montezuma and Dolores County municipalities, McPhee residents frequently exercised their choice when shopping for local services. Notably, the company did not insist on paying its employees solely in scrip, which had value only at the company store. This policy, which gave McPhee residents some modicum of freedom, was not a common practice among contemporary company towns across the United States. Sadly, the racism noted within the McPhee community was not unique; it was prevalent throughout southwestern Colorado. Unlike McPhee, racism and ethnic tensions have not disappeared from the region.

McPhee's passing in the early 1990s was scarcely noticed except by its former residents and some of their descendants. Situated in the southwestern corner of Colorado's remote Four Corners subregion, its eventual inundation by the waters of the Dolores River Storage Project was considered hardly newsworthy in the Denver newspapers. While much appreciated by the local citizenry, McPhee was, unfortunately, briefly mourned by those who long ago had abandoned the little industrial town to become absorbed into the frenzy of a vastly urbanized, faster-paced, twenty-first-century West.

NOTES

1. Lisa Mausolf, "McPhee, Colorado," in *The River of Sorrows* (Denver: Government Printing Office, 1982), 57.

2. *Rocky Mountain News,* April 18, 1927.

3. *Dolores Star,* February 15, March 7, April 11, 1924; Mausolf, "McPhee," 57–58.

4. *Dolores Star,* October 3, 14, November 14, 21, 1924; Gordon Chappell, *Logging along the Denver & Rio Grande* (Golden: Colorado Railroad Museum, 1971), 145–47; Mallory Hope Ferrell, *Silver San Juan* (Boulder: Pruett Publishing, 1973), 393–94.

5. Chappell, *Logging,* 162–64; Ferrell, *Silver San Juan,* 395; Mausolf, "McPhee," 62–63.

6. *Dolores Star,* June 7, 1929, January 31, November 14, 1930, January 1, 1931, June 10, August 19, 1932; Chappell, *Logging,* 157, 162–63, 165; *Colorado State Business Directory 1935* (Denver: Denver Gazetteer Publishing, 1935), 773; LeRoy Hafen and James Baker, *History of Colorado,* 2 vols. (Denver: Tinderman, 1927), 2:746.

7. John Gómez, interview, May 14, 1997; Chris Gómez, interview, Cortez Public Library, April 15, 1974.

8. *Durango Herald-Democrat,* June 30, 1941; *Dolores Star,* July 7, 1933, July 14, 1939, July 4, 18, 1941.

9. John Gómez, interview; Julianne Martínez, "McPhee: The Life and Times of a Company Town in Southwestern Colorado" (senior seminar paper, Fort Lewis College, 1994), 19–20; Bernie Torres and Lela Martínez, interviews quoted in Martínez, "McPhee," 10; Dorothy Gómez, "Problems in the Southwest" (independent study paper, Fort Lewis College, 1976), 84; Molly K. Wardrip, *Montezuma's Trails of Time* (Cortez, 1993), 188, Carmen Gallegos quote on 184. See also *Our Past—The Portals to the Future* (Dolores: Dolores Public Library, 1992).

10. Helen Fogel quoted in *Our Past,* 46; Gómez, "Problems," 32, 33; Martínez, "McPhee," 21–22, 23, 25–26; John Gómez, interview.

11. Bernice Willis quoted in Gómez, "Problems," 33; Martínez, "McPhee," 11, 12, 14–15, 17–18, Vic Sundquist and Lela Martínez quotes on 13.

12. Chris Gómez, interview; John Gómez, interview; Bernie Torres, interview, quoted

in Martínez, "McPhee," 26, 28–29, 18, 27; Gómez, "Problems," 75, 79, 80, 81, 85.

13. *Dolores Star,* June 19, July 24, October 23, 1942, March 3, 1944; sampling of the paper, 1942–45.

14. *Montezuma Valley Journal* (Cortez), January 22, 1948; *Dolores Star,* February 6, March 26, 1948.

15. Robert Bunting, *The Pacific Raincoast* (Lawrence: University Press of Kansas, 1997), chap. 7, "Transforming the Forests." This book provides a perceptive and revealing analysis of lumbering's impact in the Northwest and of lumbering in general.

3

The Vallecitos Federal Sustained-Yield Unit

The (All Too) Human Dimension of Forest Management in Northern New Mexico, 1945–1998

SUZANNE S. FORREST

On January 21, 1948, Congress authorized the creation of the Vallecitos Federal Sustained-Yield Unit (VFSYU) on the El Rito Ranger District of the Carson National Forest in northern New Mexico. The unit included approximately 73,400 acres of national forest, about 61,400 acres of which were classed as usable timberland.[1] The unit was designed with two purposes in mind: to protect the forest through sustained-use management and to stabilize three small forest-dependent Hispanic communities within its boundaries.

The legislation that created the unit specified that, rather than putting the timber out for competitive bid, the U.S. Forest Service would sell it to a lumber company at its appraised value in exchange for the company's agreement to employ only local loggers, provide a sawmill, conduct all primary manufacturing on-site, and provide local

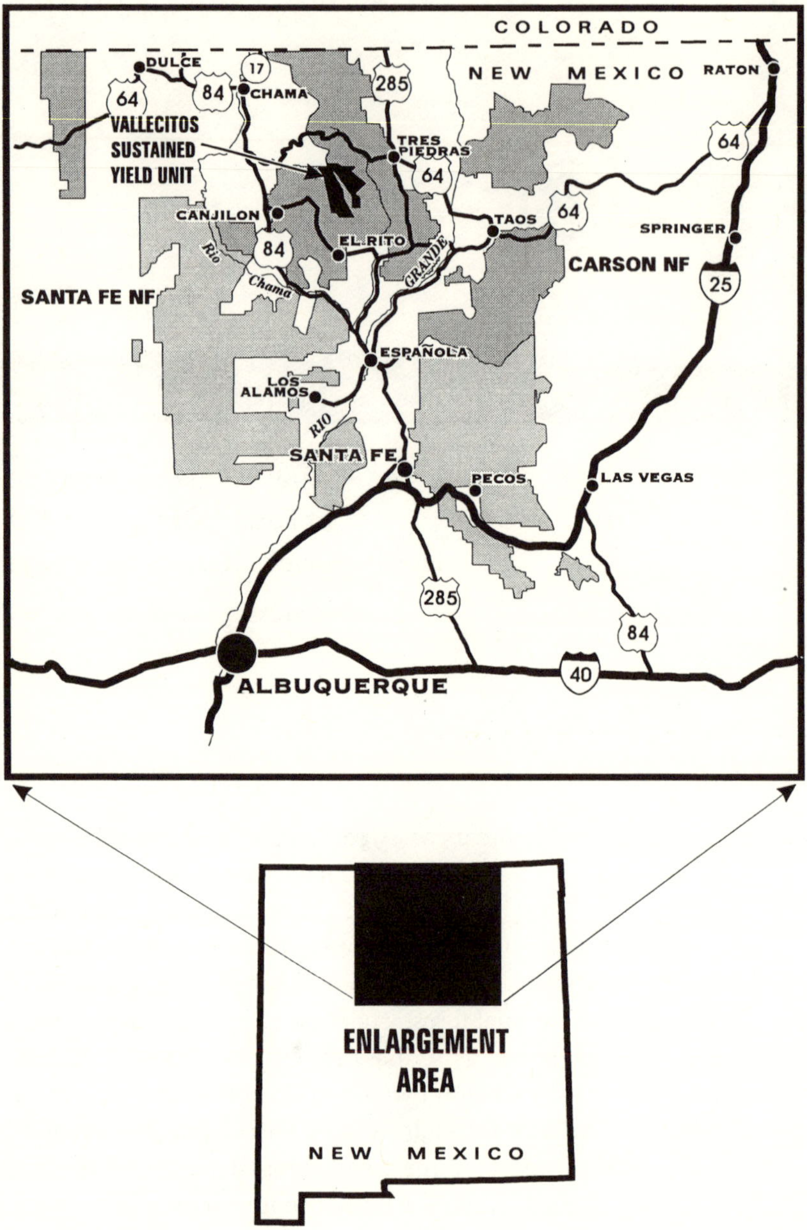

Northern New Mexico.

residents with a supply of timber for domestic needs. The legislation also required the Forest Service to post a notice and call the residents to a public meeting to approve all decisions concerning sales and designated responsible operators as well as any changes in policy governing the unit.[2]

In the fifty years since its founding, the VFSYU has been a source of almost continuous controversy. Forest Service personnel, charged with selecting approved operators and overseeing their compliance with the policies that created the unit, have been caught in a crossfire of demands, often conflicting, from different interest groups at local and regional levels. On numerous occasions the agency has tried to terminate the unit and manage this section of the forest in accordance with general Forest Service regulations. The lumber companies, for whom the unit has been a frustrating obstacle to the cost and labor efficiency of their operations, have repeatedly pressured the Forest Service to terminate it.

The Vallecitos villagers, for whom the unit was created, have not enjoyed the employment security and livelihood promised to them. Logging on the unit has been held up or forbidden for most of the unit's existence because of the Forest Service's inability to find approved operators willing to comply with the necessary regulations. When logging did proceed, the approved operators often subjected the area residents to discriminatory and exploitative labor practices. The loggers protested to the Forest Service but had little choice but to accept the terms of employment offered.

In recent years, the welfare of the Vallecitos villagers and other rural residents of northern New Mexico has been overlooked in the course of a power struggle between the Forest Service and environmental groups that object to Forest Service policies in general and to the logging of stands of old growth timber in particular. The symbol of this power struggle has been the Mexican spotted owl, an endangered species that, according to Forest Service accounts, has never been found in the Vallecitos unit.

With such a long history of wrangling and bitter disputes, one might expect the people of Vallecitos to be eager to terminate the unit. Such is not the case. Their frustrations and disappointments notwithstanding, the villagers have clung tenaciously to the rights promised them in the founding legislation. A review of this history provides an instructive overview of the problems, successes, and challenges of this human dimension of sustainable forest management.

Creation of the Unit

The VFSYU came about through the enactment on March 29, 1944, of Public Law 273, known as the Sustained-Yield Forest Management Act. This act authorized the secretaries of agriculture and interior to create two kinds of sustained-yield units, both of them designed to stabilize forest industries while simultaneously stabilizing permanent communities dependent upon the forests for their livelihood. *Cooperative units* merged the management of public and private lands in order to increase the resource base of local lumber companies. *Federal units* reserved national forest timber in given areas for the exclusive use of local operators. Both kinds of units permitted the sale of timber within their boundaries to private companies on the basis of periodically appraised value rather than the usual competitive bid procedures, in essence giving them a monopoly over the national forest timber in their area.

The framers of the legislation, David T. Mason and Edward T. Allen of the Western Forestry and Conservation Association, feared that the lumbering practices of the late nineteenth and early twentieth centuries were destroying the nation's forests. Concerned that there should be a continuous supply of timber for the future, they reasoned that the most conscientious stewards of the forests would be residents of communities who depended upon these natural resources for a livelihood. If these communities were provided with employment through a sawmill and wood products industries, their members would work willingly with the Forest Service to preserve the future supply of timber. The result would be stability for the forest and prosperity for the communities. Although their concept of sustained yield was consistent with that held by the Forest Service, Mason and Allen received no encouragement from that agency. Support for the passage of their bill came almost entirely from the timber industry.[3]

The concern for the nation's natural resources, combined with a concern for its resource-dependent communities, identifies the Sustained-Yield Forest Management Act as an offshoot of the Community New Deal, a broad-based collection of Depression-era programs aimed at creating or stabilizing small, self-sufficient rural communities on the land. In the decades leading up to the 1930s a multitude of individuals with widely divergent philosophies—industrialists, land economists, agriculturists, politicians, social workers, and city planners—became convinced that the ills of the nation derived from its crowded, crime-ridden, industrialized cities. Driven by a quasi-Jeffersonian idealization of rural values, these individuals believed the nation's strength and moral fiber could be rejuvenated only by restoring life and vigor to its rural populations. This "back to the land" movement collided in the 1930s with the reality that thousands of small-scale farmers, ranchers, and loggers were being displaced from the national economy as a result of a reckless overexploitation of the nation's natural resources. The coalition of leaders who sought and found help from the Roosevelt administration for a variety of experimental programs designed to save the nation's rural communities were the authors of the Community New Deal. Created at the very tail end of the New Deal years, the Sustained-Yield Forest Management Act of 1944 bears many of the imprints of New Deal philosophy and experimental thinking.[4]

Historical Background of the Vallecitos Federal Sustained-Yield Unit

The Forest Service was placed in a unique situation in New Mexico when it fell heir to thousands of acres of old Spanish and Mexican grant lands. The area had been inhabited by Native Americans for thousands of years and by Hispanic settlers since their ancestors moved north from Mexico in the late sixteenth and early seventeenth centuries. The Spanish and, later, Mexican governments granted large tracts of communal agricultural, grazing, and forestlands (known in Spanish as *ejidos*) to both groups to encourage the peaceful settlement of the countryside.

Relations between the native Indian tribes and the Hispanic settlers were tense during the early decades of their coexistence, but after the Pueblo Revolt of 1680 and the Spanish reconquest in 1692–96, the two agricultural peoples—Hispanic villagers and Pueblo Indians—settled into relatively peaceful and remarkably similar ways of life. They became trading partners, helped each other to ward off attacks by marauding nomadic tribes, and intermarried. While each adapted traits from the other, they kept their separate languages and cultural identities.

The United States guaranteed the previous property rights of all Mexican citizens of New Mexico in the 1848 Treaty of Guadalupe Hidalgo, which officially ended the Mexican War. Both Indian and Spanish lands were covered by the treaty because, under Mexican rule,

Indians had been citizens with equal rights under the law. Subsequently, the United States made separate treaties with the Indian tribes, setting aside much, but by no means all, of their lands as federal property.

While the Indians became wards of the government and lost their rights of self-determination, the Hispanic people fared even worse. In the years that followed American conquest, the Hispanic villages lost all or most of their communal lands through litigation, fraud, government seizure, and tax delinquencies.[5] Many of these lands came under Forest Service administration.

The traditional economy of northern New Mexico had been, of necessity, highly diversified. No single resource was sufficient to sustain even a relatively small population. Hispanic and Indian villagers farmed small plots of irrigated bottomlands adjacent to their villages, grazed sheep and goats on the surrounding communally owned highlands, and harvested the forests for fuelwood, domestic construction materials such as *vigas* (roof beams) and *latillas* (roof decking), piñon nuts, and medicinal herbs. They also produced crafts from local materials and traded them, as well as other products and services, with their neighbors.

The introduction of a cash economy in the second half of the nineteenth century, combined with a voracious overuse of the region's natural resources, brought an end to this centuries-old way of life. While Hispanic sheepherders competed with Anglo cattlemen for access to and dominance over the public grasslands, commercial timber operations, largely to provide ties for the advancing railroads, denuded the mountains of miles of ponderosa pine forests. The beginnings of large-scale commercial agriculture in the San Luis and Mesilla Valleys depleted scarce water resources.[6]

The Indians, confined for the most part to limited reservation lands, increased their herds of sheep in their efforts to feed and support a growing population. In the process, they severely overgrazed these lands. The region's Hispanos, equally hard pressed and with fewer land resources and limited access to public lands, turned to wage labor. In the 1930s, when badly needed conservation efforts put strict limits on the use of all public lands, including Indian reservation lands, the Indians had no choice but to become dependent on the Bureau of Indian Affairs. At the same time, the decline in wage labor brought on by the Great Depression resulted in the return of thousands of jobless Hispanos to home villages unable to support them. The extreme poverty among both groups eventually resulted in aid that came in the form of a raft of federal programs, called, respectively, the Indian and Hispanic New Deals.[7]

World War II brought an end to the New Deal and a temporary reprieve from poverty as thousands of New Mexico men and women entered the armed services or found jobs in the defense industry. But when the war was over, the countryside was even poorer than before. The Forest Service, as custodian of the former grant lands, found itself responsible for hundreds of destitute Hispanic families living within or adjacent to forest boundaries.

Early History of the Vallecitos Federal Sustained-Yield Unit

Frank Pooler, New Mexico's regional forester in 1943, took his responsibilities very seriously. In a letter to his superior, Forest Service Associate Chief Earle H. Clapp, he explained that he had been able to increase small permittee grazing on lands already at their maximum carrying capacity by reducing the number of sheep permitted to large, commercial wool growers. Timber in the forests went largely to small mill operations, there was a large volume of free use, and he was making efforts to acquire more land for the dependent populations through purchase or exchange. The Forest Service had provided employment and "desperately needed" transportation facilities for the Spanish American communities through its road program and had helped the Farm Security Administration in its efforts to improve herd quality. Pooler was sure that there were more ways in which the Forest Service could help the rural population, but he believed no complete solution was possible as long as the growing population continued to rely primarily on the region's limited natural resources.[8]

In communities like Vallecitos, the Forest Service encouraged the Hispanic villagers to become loggers. As early as November 1940, the *Management Plan for the Tres Piedras–Vallecitos Working Circle* specified twelve small Hispanic communities as dependent upon the forest for much of their living. The objectives of forest management, according to this plan, were to produce logs, ties, and other forest products on a sustained-yield basis in order to protect watershed health, favor recreational development (though no substantial development of this kind was anticipated in this logging area), and *provide seasonal employment for local labor* at a fairly constant rate in order to stabilize the local communities.[9]

The policy statement prepared for the VFSYU in 1948 left no doubt as to its goals. Using statistics gathered for the *1940 Management Plan,* it called for aggressive action to provide employment for the residents of Vallecitos, Cañon Plaza, and Petaca from forest products industries by providing steady employment throughout each year and from year to year for local labor, lumber sufficient to meet the needs of local residents, and efficient operation and technically up-to-date plant facilities. In order to provide these conditions, the policy statement required that all national forest timber sold from the unit had to be given primary manufacture in or within one mile of the village of Vallecitos (with a few exceptions) and that the timber would be offered for sale only to approved operators maintaining suitable manufacturing facilities at Vallecitos and only in such amounts and at such times as approved by the regional forester.[10] In accordance with regulations set forth in the Sustained-Yield Forest Management Act, Forest Service officials notified local landowners of their intent to create a sustained-yield unit in November 1947. The same letter, written in both Spanish and English, also announced a public hearing to explain the matter in detail.

Pedro Martínez, a local resident living and working for the Kennecott Copper Corporation in Bingham Canyon, Utah, articulated local concerns in a letter to the Forest Service. The Forest Service, he wrote, had already taken away the "rights" of the poor people "who [before] were permitted 25 head [of livestock] on a free permit and . . . 60 free sticks of wood." Now he feared the proposed plan would benefit only strangers from other states who would say, "To hell with the poor people of Vallecitos, Petaca, and Cañon Plaza. We got the money and we are going to drive them out."[11]

The hearing took place on a cold, snowy day in December and was conducted in Spanish with Ranger L. P. Martínez as translator. After a lengthy question-and-answer session, the residents gave the unit their unanimous approval. In a memorandum composed after the meeting, Regional Forester P. V. Woodhead wrote: "It was stimulating, as well as sobering, for Forest officers to sit together with 100 people from local forest communities and realize how eager they are to improve their economic conditions with the only opportunity to do so from surrounding national forest resources. Success is going to depend on top flight leadership and possibly at times local refereeing."[12]

The following March the Forest Service approved O. D. Connery, president of the nearby Vallecitos Lumber Company, as the designated operator of the unit. That same month Forest Service Chief C. Otto Lindh wrote an article for the *West Coast Lumberman* praising the unit. Describing the small Spanish American communities it served, he noted their dependence upon the national forest for water for their small farms, forage for their cattle (at an average of seven per permittee), fuelwood, *vigas*, and employment. The establishment of the unit, he predicted, would result in a fourfold increase in employment opportunities for the area as well as the production of box lumber for the orchardists and farmers of the Española and Velarde communities and the San Luis Valley.[13]

In the meantime, U.S. politics did an about-face. A conservative Republican administration under President Eisenhower succeeded the liberal Democratic administrations of Presidents Roosevelt and Truman. An equally conservative Republican-controlled Congress presided over a nationwide effort to roll back the "socialistic" programs of the New Deal. When the unit was reauthorized in March 1953, the Forest Service, in keeping with the policy of making services pay for themselves, raised the annual sustained-yield cut from the original 1.5 million board feet to 3.5 million board feet.[14]

Despite what appeared to be an auspicious beginning, problems in compliance with the requirements of the unit surfaced immediately. Six months after approving the Vallecitos Lumber Company as designated operator of the unit, the Forest Service canceled the company's contract because it had not moved its sawmill to a new location near Vallecitos and established a remanufacturing plant. Although other lumber companies applied, four years passed before the Forest Service approved the Jackson Lumber Company as the next operator.[15]

Ignoring the requirements clearly stated in its contract with the Forest Service, the Jackson Lumber Company hired workers from beyond the established borders, as well as many non-Hispanic workers who had only recently settled in the area. It also failed to operate the mill for at least nine months of the year. Rather than force compliance, the Forest Service allowed the company to augment the supply of timber for manufacture at Vallecitos with logs from areas outside the unit as well as hire labor from the communities of Ojo Caliente, Servilleta Plaza, La Madera, Las Tablas, and other adjacent rural areas not specified in the original policy statement.[16]

Disgruntled residents of the unit exercised their right to a public hearing on August 29, 1956. Speaking through their representative, attorney Matías Chacón, they demanded a different operator. Forest Service officials attending the meeting were equally disgruntled. They argued that the lack of competitive bidding for national forest timber had resulted in a loss of revenue to federal and county treasuries. If the unit were abolished it would eliminate controversies over employment and guarantee a continuing industry in the Vallecitos area in keeping with the free enterprise system.[17]

The Vallecitos residents wanted nothing to do with these arguments. More determined than ever to keep the unit and its policies intact, they called for help from one of New Mexico's U.S. senators, Dennis Chavez. Chavez obliged and in a meeting held with representatives from the Forest Service, the Jackson Lumber Company, and the Vallecitos labor force on September 7, 1956, negotiated a compromise agreement. The sustained-yield unit was continued with the following adjustments: an advisory board of local citizens was to be appointed by the forest supervisor to review all employee lists and requests for exemptions, and 90 percent of the workers must be selected from a broadly defined Vallecitos employment area. Seventy-six percent of the employees were to be employed from the original three communities of Vallecitos, Petaca, and Cañon Plaza. The remaining employees were to come from the communities of Servilleta Plaza, La Madera, and Las Tablas.

The Jackson Lumber Company agreed in principle to abide by the compromise proposal but discontinued its operations before any of the changes were implemented. In May 1957 the Jackson mill burned amidst rumors that the fire had been deliberately set. Rather than rebuild their facility, the owners shut down the Vallecitos operation. The Forest Service revoked the company's contract the following October.[18]

Although both the Forest Service and the lumber companies operating in the area wanted to abolish the sustained-yield unit, the Forest Service hesitated to call the required public hearing with Vallecitos residents for fear of a repeat of the previous contentious session—or worse. In the meantime, nine more years passed with virtually no activity within the unit.

Civil Rights Activism and the War on Poverty

By November 1966 the political climate had again changed, and there was a new concern for the economic and civil rights of the nation's minorities. On the national level, President Lyndon Johnson had initiated a War on Poverty that, like the New Deal, directed both money and jobs toward the problems of rural America. On the local level, a Chicano activist and utopian visionary named Reies Lopez Tijerina had focused a national spotlight on the problems of New Mexico's rural Hispanos.

The son of Texas sharecroppers, Tijerina came to northern New Mexico in the 1950s and experienced firsthand the poverty of its rural communities. Declaring the Hispanos' problems to be the loss of their *ejido* lands, he incorporated the organization known as La Alianza Federal de Mercedes (the Federal Alliance of Land Grants) in February 1963. Three years later, with a following of twenty thousand rural Hispanos, the charismatic Tijerina made his stand. On October 15, 1966, after the Forest Service had again reduced grazing permits for small farmers, this time by 45 percent, Tijerina and 350 of his supporters staged a mock occupation of a national forest campground a few miles from the village of Vallecitos. Stretching a chain across the entrance and citing the terms of the Treaty of Guadalupe Hidalgo, Tijerina demanded the return of the San Joaquín del Río Chama land grant to its Hispanic owners.[19]

The series of events that followed led, in June 1967, to a well-publicized raid by angry Hispanos on the Rio Arriba County Courthouse. While Tijerina led New Mexico police on an equally well publicized manhunt, his followers protested by burning haystacks, rustling cattle, setting forest fires, vandalizing campgrounds, and cutting fences. In the atmosphere of panic that pervaded the state, rumors flew that a full-scale, Cuban-led revolution was under way.[20]

Tijerina was eventually apprehended and sentenced to three years in a federal penitentiary, but, as he had intended, the movement he started brought the plight of New Mexico's Hispanos to the attention of federal and state authorities. Undersecretary of Agriculture John Schnittker flew to New Mexico to have a firsthand look at Forest Service operations, and Secretary of Agriculture Orville Freeman asked the Forest Service to make a "stronger effort to work in rural development and poverty programs to help the Mexican American people."[21]

In Vallecitos, Forest Service officials resigned themselves temporarily to the retention of the sustained-yield unit. Fearing that any further attempts to eliminate it would bring a deluge of bad publicity to the agency, they suddenly found funds to reclaim and create new pastures. They also defused local tensions with War on Poverty training programs for local workers.[22]

The Vallecitos residents, for their part, were more than willing to accept help. They requested a hearing in September 1967 to ask permission to start their own locally owned lumber operation. Father Robert Garcia of the State Office of Economic Opportunity was present and offered them the assistance of his office in organizing and financing a cooperative. Representatives from another New Mexico poverty program known as HELP (Home Education Livelihood Program) offered to contact a manufacturer of rustic prefabricated fences to determine if there was a market for such a product. In his report of the meeting, District Ranger Jack Miller noted that the Forest Service wanted local leaders to take the initiative in developing action programs. The Vallecitos villagers needed little encouragement. Shortly thereafter, they formed the Vallecitos Federal Sustained-Yield Unit Association.[23]

If the Forest Service was reconciled to the continuation of the unit, the Duke City Lumber Company of Albuquerque, next operator to bid for work on the unit, was not. Its president, Yale Weinstein, argued that the unit's allowable cut of 3.5 million board feet was insufficient to sustain a complete operation. He wanted to transport the lumber to Duke City's central plant in Española, contending that this was more economical and safer than processing the logs in the circular-type sawmill at Vallecitos. Hoping to get the people of Vallecitos to agree to terminate the unit, he offered to help them set up their own logging operation and buy their logs. He further sweetened the deal by offering to help them with the financing if they agreed to set up a joint venture with Duke City and share the profits. Turning on the Forest Service, he accused the agency of abdicating its management responsibilities and yielding to "town meeting" pressures. Privately, he conceded that, "if forced," he would establish a sawmill in Vallecitos.[24]

In a united show of force, the Vallecitos residents refused to approve any development program that threatened to modify or do away with the unit. Duke City gave in after prolonged discussions with the Forest Service and admitted that the problems were not insurmountable. When the lumber company submitted its formal request to be named responsible operator, Tony Jaramillo, VFSYU Association president, called for a public meeting. The local people voted their approval only after being convinced by Duke City that it would comply with the objectives set forth in the sustained-yield unit policy statement. The Forest Service formally awarded the contract in April 1972.[25]

By early 1972, even the Forest Service showed a new concern for civil rights, at least at the local level. In that year, New Mexico Regional Forester William D. Hurst directed Jean Hassell, a member of his staff, to evaluate how the Forest Service might better serve the needs of Hispanic villagers. Hassell's report, "The People of Northern New Mexico and the National Forests," dated March 13, 1972, became the basis for new Forest Service policies in northern New Mexico. In what can only be described as an impassioned memorandum, Hurst sent copies of Hassell's report to all district forest rangers. W. R. Snyder, Carson forest supervisor, endorsed the new policies and added:

> *I firmly believe . . . that our success or failure in Northern New Mexico will be based upon our ability to recognize the unique qualities of both the land and the people that we are dealing with.* The Carson must not be run as a bureaucracy: we must prepare ourselves to be fair and have the time to listen to people when they want to talk. . . . We are not going to give anything away, but . . . we are going to make our resources stretch as far as they possibly will in meeting the needs in this part of the country.[26]

The new policy statement described the poverty of the area and the lack of commercial resources to generate jobs for the Indian and Hispanic residents. Hassell noted that the people had deep roots in the land, since 22 percent of the Carson and Santa Fe National Forests were former Spanish and Mexican land grants. "In the eyes of many," he added, "their rightful owners were unjustly deprived of these properties." There had been periodic uprisings ever since 1848, when the United States gained control over the area. Now these old grievances were being expressed through civil rights organizations. Forest Service policy, Hassell argued, should recognize the Spanish American and Indian cultures as "resources," just as "wilderness" was considered a resource. "There is no middle ground," he concluded. "Failure to meet this challenge will mean conflict, frustration and a loss of prestige to both the Forest Service and the Department of Agriculture."[27]

Despite the official change of heart, problems continued much as before within the VFSYU. In April 1975 Rudy Jaramillo of the VFSYU Association asked that the Forest Service withhold the Lookout timber sale until a public hearing had been conducted concerning Duke City's hiring of nonlandowning "hippies." The association also requested training for local people in the higher paying skilled jobs that were exempt from the local hiring policy and the reduction, if not elimination, of exempt positions.

When the Forest Service offered to help community members secure training through the approved operator, Duke City argued that nothing in its contract required it to do so. Nevertheless, the lumber company agreed to the request because its directors knew, as did the members of the VFSYU Association, that a delay in the sale would mean a reappraisal of the value of the timber and a cost increase to Duke City. While the lumber company's unusual cooperation was due in part to a desire to get on with the sale, it was also because specialized training funds had recently become available through the Employment Security Commission/Manpower Development Training Act enacted in 1962.[28]

New Players, New Concerns

Just as the local people were developing a strong public presence, the political situation took another turn. By 1980 national civil rights activism was fading, and President Ronald Reagan's socially and fiscally conservative administration introduced new Forest Service policies at both the national and local levels. In May 1984, in response to a new assessment of the biological potential of the forest, the Forest Service announced a plan to raise the allowable sale quantity on the VFSYU from 4.2 million board feet to 9.5 million board feet.[29]

The potential for vastly increased timber cutting would have meant more employment for local people, but Vallecitos residents were outraged. As concerned stewards of their land, most considered the earlier cut of 4.2 million board feet to be too high. Convinced that this much pressure on the forest would threaten sustainable management, the watershed, and their entire rural lifestyle, they circulated a petition of protest. The Forest Service scheduled a hearing for June 4, 1985, to review the new forest plan and consider whether the unit should be reauthorized. The Vallecitos residents approved the provision allowing them to develop their own forest products business but continued to protest the enlarged cut. They also protested a plan to build more roads into the unit and the fact that they were paid less for the same type, quantity, and quality of work and received none of the employee benefits of workers at the Duke City sawmills in Española, Cuba, and Albuquerque.

Once again the Forest Service suggested discontinuing the unit. As before, Vallecitos residents remained adamant in its support. It appears that the Forest Service hoped to buy out the protesters. The new plan, when finally published, allowed for the sale of 7.6 million board feet, 5.5 million in saw timber to Duke City Lumber, the designated operator, and 2.1 million in saw timber and timber products to local residents to establish a wood products business.[30]

Dissatisfied with the outcome of the public meeting, Rudy Jaramillo of the VFSYU Association wrote to U.S. Congressman Bill Richardson and asked that the House Agriculture Committee conduct a hearing to review local concerns. Jaramillo put the association on record as opposed to increased timber harvesting and accused the Forest Service of being incapable of believing that the people in the area were vitally interested in the health of the unit. Instead, he wrote, the Forest Service ignored the wishes of the local populace to cater to the needs of the designated operators. Jaramillo recommended a program of small contracts for thinning and harvesting trees for firewood, *vigas,* Christmas trees, landscape trees, and fence posts.

Jaramillo reported that Duke City had told its employees in no uncertain terms that any organizing for collective bargaining was dangerous. The company even frowned upon workers attending meetings of their own association. Whenever people in the unit had asked for better pay, they were fired or laid off. Was not this "economic terrorism," he asked? "When people are so fearful for their jobs that they will not even attend a meeting of vital interest to them, we believe that the above question must be answered in the affirmative."[31]

Two grassroots organizations resulted from the 1985 authorization. Energized by the opportunity to develop their own wood products business, the men of Vallecitos asked for advice from community developer Luis Torres, a frequent visitor in the area. Torres, who was employed by the Southwest Research and Information Center, a nonprofit organization concerned with the needs of indigenous people, began meeting with the Vallecitos men on a weekly basis early in 1988. In April they organized and incorporated La Madera Wood Products Cooperative to cut and sell fuelwood, *vigas, latillas,* and other forest products. The organization, which included virtually every able-bodied man in Vallecitos, elected Antonio "Ikey" DeVargas, Rudy Jaramillo, Delbert DeVargas, and Manuel Gurule as its first officers. Two years later, in May 1990, Ikey DeVargas, Delbert DeVargas, Steve Chavez, Manuel Gurule, Dennis Valdez, Patricio Valdez, Marco Peña, and Rudy Jaramillo split from the cooperative, borrowed approximately $200,000 to buy heavy equipment for a logging operation, and founded their own logging company, La Compañía Ocho.[32]

Duke City Lumber Company contracted to buy timber from La Compañía in the Felipito timber sale, but at the end of the contractual period the Vallecitos lumbermen filed suit on a number of contractual matters. Their principal grievances were that Duke City had measured the number of board feet of lumber processed in an unfair manner and that the subcontractor had prevented them from fulfilling their contract by failing to provide roads into the area's high country before snow made it inaccessible. Arbitration proceedings took place in November 1990 and concluded with La Compañía being awarded $253,500 in damages. The compensation was only a fraction of the amount requested, but the loggers were jubilant with their victory.[33] The victory was short-lived, however. Beginning in 1991, a number of environmental groups, together with the New Mexico Department of Game and Fish, began to pressure the Forest Service to delay timber sales until more was known of the effects of logging and road building on stands of old growth forest, streams and wetlands, and fish and game habitat.

The Vallecitos people received the news of yet another obstacle to their efforts to make a living off the land with a sense of disbelief. When the Forest Service withdrew the Felipito timber sale, on which La Compañía Ocho was logging subcontractor, tempers flared. Hoping to keep the peace and challenge Forest Service management policies, representatives of the Wilderness Society, Forest Guardians, Carson Watch, the National Audubon Society, and the Rio Grande Chapter of the Sierra Club met with Ikey DeVargas of the VFSYU Association and Pat Valdez of La Compañía Ocho in early November 1991.

The Vallecitos-area residents, like the environmentalists, were concerned about the forest resource base, particularly in light of increased logging in the 1980s. They also objected to the fact that Duke City Lumber, now a subsidiary of the multinational corporation Hanson PLC, was draining timber sale profits from the local economy. Finally, the Vallecitos residents welcomed the suggestion that, with financial and technological help from the environmentalists, the two groups could develop a culturally sensitive model of sustainable forestry, one that would preserve the old growth forests, drive Duke City Lumber out of the area, and provide local employment.

As a result of this meeting, the Vallecitos lumbermen agreed to discontinue logging in the northern part of the VFSYU in exchange for help from the environmentalists in expanding a wood products industry that would use only small-diameter trees from thinning on the southern part of the unit. Sam Hitt, director of Forest Guardians, raised $37,500 from various New Mexico foundations to purchase a wood-processing machine that could cut and split logs into firewood.[34]

Unfortunately, the coalition broke down almost immediately as each group perceived the other as violating the agreement. To the dismay of the Vallecitos lumbermen, the environmentalists did not stop protesting timber sales. Neither did the Vallecitos residents end all timber harvesting. The environmentalists had agreed to help enlarge the VFSYU to include much of northern New Mexico in exchange for an end to logging on the Felipito diversity unit. When they failed to follow through with any action, the Vallecitos residents felt betrayed. The fact that the environmentalists' neglect was not deliberate but was due to the issue occupying a very low priority on their agenda did not help matters. The Vallecitans' sense of injustice was only heightened when the Forest Service failed to follow through with any action on the revised timber management policy that the association had submitted at its request.

On March 15, 1993, the U.S. Fish and Wildlife Service officially listed the Mexican spotted owl as a threatened species. Because areas of the Felipito timber sale on the VFSYU were potential owl habitat, virtually all forest activities in the unit were placed on hold. Rather than take the blame for closing the forest, the Forest Service issued a briefing paper that declared that in five years of surveying almost 230,000 acres at a cost of over $500,000, the agency had found only four owls, and those were in the Jicarilla District far from the VFSYU.[35]

The listing of the owl as an endangered species gave environmental groups the legal ammunition they had long desired. Within months, several groups in Arizona and New Mexico joined forces and filed suit against the Forest Service for failing to consult with the Fish and Wildlife Service on actions that could impact owl habitat. The matter came to a head in August 1995, when Judge Carl Muecke ruled in the environmentalists' favor. He ordered an injunction against all logging and most firewood gathering and wood products collecting on national forests in Arizona and New Mexico. The injunction became permanent in October 1995.

In a separate lawsuit filed in New Mexico, many of the same environmental groups sued the Forest Service and challenged its recent La Manga timber sale to Duke City Lumber. In an odd reversal of fortune, La Compañía Ocho found itself on the same side as its erstwhile adversary and intervened in the lawsuit. A few months later, in January 1996, Duke City Lumber announced that, for lack of lumber, it would close the Vallecitos mill.[36]

The Hispanos of northern New Mexico were outraged at Judge Muecke's ruling. It came just when the people of the mountain villages had to begin collecting firewood for their winter cooking and heating needs. Insisting that Mexican spotted owls did not inhabit the Vallecitos forests, the villagers accused the environmentalists of being more concerned with animals than with people. When the injunction became permanent, Leonard Lucero, Carson National Forest supervisor, sent out a letter warning all residents of northern New Mexico that fuelwood gathering would be ended. The shortage of firewood in the village of Truchas, New Mexico, became national news when the story was published in the *New York Times* and the *Los Angeles Times* and aired on *The NBC Nightly News*.[37]

Hispanos and Native Americans, many of them artists and craftspersons who depended on forest resources for their livelihood, united in protest throughout the area. On November 24, 1995, members of La Madera Wood Products Association and La Compañía Ocho logging company marched to Santa Fe and hanged Sam Hitt and John Talberth, two of the best-known environmentalists, in effigy. Representatives of Duke City Lumber, land grant activists, small-scale Anglo loggers, Pueblo Indians, and former New Mexico lieutenant governor and Green Party member Roberto Mondragón joined them.[38]

Shortly before the loggers' rally, the environmentalists held their own demonstration. They accused the Forest Service of putting all its money and effort into timber sales for multinational corporations and declared that it served neither the interests of wildlife nor economic and social justice. By this time, however, the positions of loggers and environmentalists had hardened virtually beyond repair. Each side accused the other of racism and extremism.[39]

Although the specter of an injunction against grazing on public lands had yet to appear, the conflict evoked great sympathy among the sheepherders of Ganados del Valle, a neighboring Hispanic grazing cooperative with a national reputation for innovative economic development initiatives. Looking for a less-threatening form of protest, the Ganados sheepherders suggested and helped organize a candlelight prayer vigil in Santa Fe on December 20, 1995. The event drew an ethnically diverse crowd of nearly four hundred people from all over the state. During the politically charged aftermath, Santa Fe Mayor Debbie Jaramillo, Roberto Mondragón, María Varela of Ganados del Valle, Ikey DeVargas of La Compañía Ocho, and Santa Fe artist Erwin Rivera addressed the audience. Speaking of the hardships faced by rural Hispanos and the continuing erosion of their access to resources, the protesters called the Endangered Species Act of 1973 just another way of abrogating Hispanic and Native American rights to moral and economic justice. The native peoples of New Mexico were the "true environmentalists," they declared, with centuries of experience in living close to nature. The Treaty of Guadalupe Hidalgo plus the Hispanos' mixed racial and cultural heritage made Hispanos, as much as Indians, the land's legitimate owners and managers. Erwin Rivera expressed these sentiments when he proclaimed, "Tenemos sangre de Indios. Tenemos raíces en la tierra. Somos indígenas." [We have Indian blood. We have roots in the land. We are indigenous people.][40]

In January 1996 Duke City Lumber carried out its threat and closed the Vallecitos mill. Although both the company and the loggers blamed the environmentalists, the environmentalists claimed the real reason behind the sawmill's demise was deflated lumber prices brought on by the North American Free Trade Agreement of 1995. The loggers also charged that the Duke City sawmill was outdated and could only handle old growth trees. The Forest Service put the blame squarely on the rules, regulations, and restrictions that stood between loggers and a forest that "has more trees per acre than in recent history."[41]

Summary and Conclusions

What can be learned from this long and continuing history of conflict, confrontation, and occasional compromise? First, no definitive conclusion is possible because it is an evolving story, with new actors entering the scene and old ones dropping out.

At first glance, it would seem that the unit was and continues to be a dismal failure since it has pleased no one. The historical record, however, shows that the philosophy and principles behind the founding legislation were sound. Although the Vallecitos residents have been disappointed that they have not benefited from the secure employment promised them in the initial legislation, they have been concerned stewards of their land. With little more than a weak veto power in determining its fate, they have steadfastly resisted exploitative use by the timber industry.

Both the Forest Service and the various approved operators have complained that the requirements written into the original legislation were inefficient and costly. However, Forest Service policies have been aimed at commercial timber exploitation rather than sustained-yield management. Formulated at the regional and federal levels, the policies have too often been inappropriate when applied locally. Even when economically or ecologically justified, they have been inadequately explained and excessively harsh. It is hardly surprising that Forest Service directives have led to serious misunderstandings and bitter wrangling. Therefore, the requirement written into the VFSYU that Forest Service officials listen to the needs and concerns of the local residents and work out compromises with them has been salutary. An educational experience for all involved, the requirement of open discussion has kept the bureaucracy in touch with the human dimensions of natural resource management.

If there is a major flaw in the legislation creating the VFSYU, it is that a bureaucracy like the Forest Service is ill equipped to assure social and ecological justice. It has not protected the Vallecitos lumbermen from exploitation by various approved operators, and it has not protected the forest from commercial overuse. It is highly unlikely that the Forest Service, an agency not known for its compassion, would have required the approved operators to provide training and employment to residents of the local communities had it not been forced to do so.

A continuing problem—one that has existed since the initial legislation in 1944—is the paternalistic assumption on the part of Forest Service officials that Hispanic villagers cannot be trusted to make responsible decisions concerning the management of their land. The historical record reveals, however, that they have repeatedly fought both the Forest Service and the various lumber companies to protect their land from overuse and abuse.[42] Given technical assistance and the opportunity to develop alternative, low-impact ways of harvesting forest resources, they might well have been better forest managers than their counterparts in the Forest Service.

In part because of this paternalistic bias, the policies written when the unit was created were not allowed to evolve as the local population acquired the skills and knowledge needed to run the logging and wood products industries themselves. In the words of community developer Luis Torres:

> The Vallecitos Unit is a perfect example of government missing a tremendous opportunity . . . for local development. They even had the right kind of language to use the community as a resource base, but then they put it in the hands of the Forest Service. That might have been the right thing in 1951 and probably through the '50s, but there came a time when . . . the Forest Service . . . should have been thinking about how the agency could make it possible for the local people to be the managers, and eventually the owners, and then upgrading the ownership from just making rough sawn lumber to . . . making something like picnic tables and work benches. And they could have saved the resources. I have done a study that shows you could employ the whole work force using as little as one million board feet with added value activities.[43]

The ongoing controversy between the Hispanos of northern New Mexico, their Native American neighbors, and environmentalists hinges primarily on the definition of sustainability. Neither Hispanos nor Indians deny the vital importance of preserving the nation's endangered natural resources. However, the indigenous cultures insist that the practical application of sustainability must include their traditional uses of the land. Hispanos like María Varela of the sheep-and-wool-based cooperative Ganados del Valle argue that traditional management practices will sustain and protect both the rural economy and the environmental health of public and private lands. "The environmentalists ought to want our kind of economy instead of a resort economy," she insists. "Our native cultures must be treasured and protected for the good of the planet just as much as any endangered species."[44]

Environmentalists dispute such claims as totally irrelevant and insist that Hispanos and Native Americans are just as capable of destroying the forests with their chain saws and pick-up trucks as any other group. They base their definition of sustainability on the findings of ecological science. Unfortunately for the Native Americans and Hispanos, there have been no studies of the beneficial effects of traditional uses of the land with which to counter the environmentalists' scientific evidence.[45]

Sam Hitt and John Talberth, among others, do agree, however, that the native peoples should not be displaced from their ancestral homes. They recommend the Forest Service implement an economic diversification study made for the VFSYU Association that would pay members of the Vallecitos community to thin the forest and restore it by obliterating roads and replanting native vegetation. One environmental organization, the Santa Fe–based Forest Trust, devotes a substantial amount of its annual budget to a community forestry and grazing program focusing on resource management training and business development in northern New Mexico.

The most bitter areas of disagreement relate to the use of the land. Hispanos and Native Americans are thoroughly frustrated by their lack of control over the resources they must depend on for a living. They argue adamantly that they can never be economically self-sufficient unless they are given priority access to the lands guaranteed to them by the Treaty of Guadalupe Hidalgo. Environmentalists and most other Americans counter this claim by insisting that national forest and other public lands belong to all the people of the United States. They contend that a dangerous precedent would be established if the residents of the VFSYU and other land-based communities were granted a greater say in the management of their lands than other interested parties.[46]

The problem will not be easily resolved. A growing local population is putting an ever greater strain on limited resources, but this pressure is nothing compared to that exerted by newcomers from outside the region. While ski resorts and golf courses increase the demand for scarce water resources already depleted by industry and the state's growing cities, wilderness advocates yearn for a romanticized state of nature devoid of human presence. Those with abundant resources seek to enlarge on the experience by buying scenic mountain land for second homes. These newcomers, though largely unaware of the impact of their presence on land values, make it increasingly difficult for long-time residents with few financial options to buy land or even pay their increased property taxes.

The native cultures have depended from earliest times on a diversified economy for survival in this resource-poor land. Specialization in one industry or crop, such as that envisioned for the VFSYU by the New Deal planners, was impractical from the beginning. If there is a place for a sustainable rural lifestyle in northern New Mexico, it will be within the context of a similarly diversified economy. Yet the multiple pressures and intense competition for the region's water supply, its forests, and its grasslands make the practice of such a diversified economy increasingly difficult.

There are some positive notes. More people are coming to agree with the Hispanos and Native Americans that small-scale sustainable agriculture and forestry actually buffer public lands against the incursions of commercial development. Mainstream tastes are also changing, and a small but significant segment of society is turning to a simpler, more "earth conscious" lifestyle. There is a growing market for organically grown products from small farms and a renewed appreciation for natural fibers, hand-produced, one-of-a-kind items, and the personal quality of small-scale service operations. Ganados del Valle is successfully exploiting this market niche. Given the opportunity, the residents of Vallecitos might be able to do the same with a wood products industry that would not deplete the forest.

Half a century ago the VFSYU came into existence as the result of an Anglo American urban middle class desirous of preserving rural communities and traditional lifestyles. In the years since, succeeding generations of middle-class society have made that survival virtually impossible. They have not only put enormous commercial pressures on the nation's forests and its rural communities through unrestrained development but have competed directly with the native populations for use of and access to the natural resources essential to their traditional rural lifestyle. It would be ironic, indeed, if the present generation of this same Anglo American middle-class society were to deliver the coup de grâce to Vallecitos and other northern New Mexico rural communities simply by writing them out of its definition of a sustainable biodiversity.

NOTES

1. The Vallecitos Unit contains a mix of piñon-juniper and encinal woodlands with ponderosa pine and some spruce at the higher elevations.

2. Sustained-Yield Forest Management Act of 1944, 58 Stat. 132, Carson National Forest Files, Taos, N.M., Vallecitos Federal Sustained Unit File 2410 (hereafter referred to as USFS Vallecitos File 2410).

3. David A. Clary, "What Price Sustained Yield? The Forest Service, Community Stability, and Timber Monopoly under the 1944 Sustained-Yield Act," *Journal of Forest History* 31, no. 1 (1987): 4.

4. For an extensive discussion of the origins and implementation of America's Community New Deal, see Paul Conkin, *Tomorrow a New World: The New Deal Community Program* (Ithaca, N.Y.: Cornell University Press, 1959).

5. The loss of the Hispanic land grants is discussed in detail in Malcolm Ebright, *Landgrants and Lawsuits in Northern New Mexico* (Albuquerque: University of New Mexico Press, 1994), and Victor Westphall, *Mercedes Reales: Hispanic Land Grants of the Upper Rio Grande Region* (Albuquerque: University of New Mexico Press, 1983). William deBuys, *Enchantment and Exploitation: The Life and Hard Times of a New Mexico Mountain Range* (Albuquerque: University of New Mexico Press, 1985) describes in detail the loss of the Las Trampas land grant (171–92).

6. Suzanne Forrest, *The Preservation of the Village: New Mexico's Hispanics and the New Deal* (Albuquerque: University of New Mexico Press, 1989; reprint, 1998), 17–31. DeBuys, *Enchantment,* provides an extensive historical overview of the effects of a changing economy on the degradation of New Mexico's grasslands and forestlands (196–98, 215–25).

7. Forrest, *Preservation,* vi, xi, 15, 16, 78, 80, 129–50, 159–66, 178–80.

8. Ibid., 159–60.

9. Walter G. Thomsen, *1940 Management Plan for the Tres Piedras–Vallecitos Working Circle,* Carson National Forest, Region 3, USFS Vallecitos File 2410.

10. Policy Statement for Vallecitos Federal Sustained-Yield Unit, Carson National Forest, Southwestern Region, April 7, 1947, USFS Vallecitos File 2410.

11. Pedro Martínez, Bingham Canyon, Utah, to Regional Forester, November 21, 1947, translated from Spanish, USFS Vallecitos File 2410.

12. Letter sent in English and Spanish to all landowners within VFSYU, November 1947; "Public Hearing on Forest Management," news release, November 7, 1947; Pedro Martínez to Regional Forester, November 21, 1947; Report on Hearing on the Proposed Establishment of the Vallecitos Federal Sustained-Yield Unit Held December 9, 1947, Title 36 Parks and Forest; Region 3 Memorandum on the Hearing and the Hearing Officer's Report from P. V. Woodhead, Regional Forester; Declaration of Sustained Yield Unit, Vallecitos Federal Sustained Yield Unit, Sec. 221.33, January 21, 1948, USFS Vallecitos File 2410.

13. Lyle F. Watts, Forest Service Chief, to O. D. Connery, President, Vallecitos Lumber Co., March 31, 1948; Forest Service Press Release, March 24, 1948, USFS Vallecitos File 2410.

14. Statement of Regional Forester, August 29, 1956, USFS Vallecitos File 2410.

15. R. E. McArdle, Forest Service Chief, to O. D. Connery, Vallecitos Lumber Co., Revocation of Designation of Approved Operator, Registered Mail, September 17, 1948; Statement of the Regional Forester at Hearing to Consider the Vallecitos Federal Sustained-Yield Unit to Be Held at Vallecitos, N.M., August 29, 1956, USFS Vallecitos File 2410.

16. Statement of Regional Forester, August 29, 1956, USFS Vallecitos File 2410.

17. Memorandum, Fred H. Kennedy, Regional Forester, to R. E. McArdle, Chief, Forest Service, September 4, 1956; Statement of Regional Forester, August 29, 1956, USFS Vallecitos File 2410.

18. Richard E. McArdle, Forest Service Chief, to Jackson Lumber Company, Vallecitos, N.M., January 8, 1957; Fred H. Kennedy, Regional Forester, to McArdle, No-

vember 25, 1958; History of the Vallecitos Federal Sustained Yield Unit, n.d., USFS Vallecitos File 2410.

19. Peter Nabokov, *Tijerina and the Courthouse Raid* (Albuquerque: University of New Mexico Press, 1969); Regional Forester to National Forest System, August 10, 1967, USFS Vallecitos File 2410.

20. Luis Torres, community organizer, taped interview, Santa Fe, N.M., August 3, 1993; DeBuys, *Enchantment,* 261–63.

21. Orville Freeman to Edward P. Cliff, Chief, U.S. Forest Service, October 28, 1967, U.S. Forest Service Regional Office, Albuquerque, N.M.; DeBuys, *Enchantment,* 263.

22. Forrest, *Preservation,* 190; Torres, interview, 1993.

23. Memorandum, F. LeRoy Bond, Assistant Regional Forester, to W. D. Hurst, Regional Forester, March 1, 1967; Jack R. Miller, District Forest Ranger, to Carson Forest Supervisor, January 26, 1968, USFS Vallecitos File 2410. Information on the efforts of the Vallecitos residents to organize comes from interviews with Antonio "Ikey" DeVargas, member of La Compañía Ocho (Vallecitos, March 15, 1993); Luis Torres, a former employee of the Southwest Research and Information Center and now an independent community organizer living in Española (1993); Steve Chavez, president, and Delbert DeVargas, board member, of La Madera Wood Products Cooperative (Vallecitos, March 15, 1993).

24. Yale Weinstein, President, Duke City Lumber, to District Forest Ranger, January 29, 1968; Weinstein to M. J. Hassell, Carson National Forest Supervisor, April 18, 1968, USFS Vallecitos File 2410.

25. T. W. Koskella, Deputy Regional Forester, to Forest Supervisor, May 27, 1969; Weinstein to M. J. Hassell, September 16, 1969; Hassell to Weinstein, April 7, 1971; Weinstein to Hassell, August 6, 1971; Koskella to File, January 11, 1972; John F. Hutt, Timber Staff Officer, to File, February 23, 1972; W. D. Hurst, Regional Forester, to Weinstein, April 4, 1972, USFS Vallecitos File 2410.

26. W. R. Snyder, Forest Supervisor, to Forest Rangers, March 13, 1972, USFS Vallecitos File 2410, emphasis in original.

27. Hurst to Forest Supervisors and District Rangers, March 6, 1972; "Region 3 Policy on Managing National Forest Land in Northern New Mexico," USFS Vallecitos File 2410.

28. John S. Crellin, Deputy for Resource Management, to Tony Jaramillo, VFSYU Association, March 4, 1975; W. R. Snyder, Forest Supervisor, to Regional Forester, April 9, 1975; Gordon R. Struthers, Forester, to File, April 24, 1975; John F. Hutt, Timber Staff Officer, to Forest Supervisor, May 5, 1975, USFS Vallecitos File 2410.

29. Carson Forest Plan of May 1984, USFS Vallecitos File 2410.

30. Ibid.

31. Notice of Public Hearing on June 4, 1985; John C. Bedell, Forest Supervisor, to Regional Forester, May 5, 1986; Bill Richardson, N.M. Delegate to U.S. House of Representatives, to Kika de la Garza, Chair, House Agriculture Committee, Washington, D.C., May 7, 1986; Rudy J. Jaramillo, President, VFSYU Association, to Richardson, May 1, 1986, USFS Vallecitos File 2410.

32. Torres, interview, 1993; Luis Torres, telephone interview from his home in Española, September 29, 1997.

33. Interview with Ikey DeVargas, board member of La Compañía Ocho, Vallecitos, April 11, 1993. The author of this report was a witness for the prosecution in the arbitration proceedings and testified concerning the origins and purposes of the Vallecitos Federal Sustained-Yield Unit.

34. Torres, interview, 1993; Carl Wilmsen, "Fighting for the Forest: Sustainability and Ethnicity in Vallecitos, New Mexico" (Ph.D. diss. in preparation for submission to the faculty of Clark University, Worcester, Mass.), 237–39, 241. My thanks to Mr. Wilmsen for providing me with this preview of his dissertation.

35. Wilmsen, "Fighting for the Forest," 241–43; Carson National Forest Briefing Paper No. 3.

36. Malcolm Ebright, "Current Litigation in New Mexico and Colorado Relating to Common Land Use Rights," *Common Property Resource Digest, Yale School of Forestry and Environmental Studies* 38 (June 1996): 9; Doug McClelland, "Rare Jobs vs. Rare Spotted Owls," *Albuquerque Journal,* August 26, 1995. The environmentalists argue that the Mexican spotted owl, which is dependent upon old growth forests, must be protected in order to preserve biodiversity. They disagree with the Forest Service about the existence of spotted owls in New Mexico's forests.

37. Jeffrey St. Clair, "Fear and Firewood in New Mexico," *Wild Forest Review* (October–November 1995): 32; Wilmsen, "Fighting for the Forest," n. 245.

38. María Varela, "On the Road to Healthy Environments," *Noticias Nuevas de Ganados del Valle* (Spring–Summer 1996): 2; Wilmsen, "Fighting for the Forest," 248.

39. Wilmsen, "Fighting for the Forest," 249.

40. Varela, "On the Road," 2; Wilmsen, "Fighting for the Forest," 249–50.

41. Tom Ragan, "Sawmill Closed for Good," *Albuquerque Journal North,* January 10, 1996, 1.

42. G. Emlen Hall, "Community Land Grants and the Forest Service as Watershed Managers: The Example of Santo Domingo de Cundiyo" (paper presented at Making Sustainability Operational: Fourth Mexico/U.S. Symposium, Santa Fe, N.M., April 19–23, 1993), reprinted as USDA General Technical Report RM-240 (Fort Collins, Colo.: Rocky Mountain Forest and Range Experiment Station, December 1993), 93–101, describes the efforts of another traditional Hispanic community to manage its land and water resources responsibly in spite of a divided authority over watershed management.

43. Torres, interview, 1993.

44. María Varela, taped interview, Los Ojos, N.M., September 16, 1993; Varela, telephone interview, April 14, 1997; María Varela, "Ganados del Valle 1981–1996: Creating Economically Sustainable Ecologies" (position paper, n.d.).

45. Dena Leibman and Anna El-Eini, "The Sustainability Movement: Rhetoric or Reality?" *Friends of the Earth* (September–October 1996): 7–11, 14; Wilmsen, "Fighting for the Forest," 251; full-page ad commemorating the 149th anniversary of the Treaty of Guadalupe, *Santa Fe New Mexican,* February 2, 1997.

46. Wilmsen, "Fighting for the Forest," 253.

4

Grazing the Southwest Borderlands

The Peloncillo-Animas District of the Coronado National Forest in Arizona and New Mexico, 1906–1996

DIANA HADLEY

This chapter presents a case study of grazing in the Peloncillo-Animas District of the Coronado National Forest in southeastern Cochise County, Arizona, and southwestern Hidalgo County, New Mexico. Named for the Peloncillo and Animas Ranges, the district contained two noncontiguous divisions until the Animas portion was eliminated from national forest status during the 1940s. This district formed the easternmost extension of the Coronado National Forest, including the only portion of the forest outside of Arizona. The original southern boundary for both divisions was the international border with Mexico.

The Peloncillo-Animas District provides an excellent case study for national forest grazing because neither the Peloncillos nor the Animas supported alternative economic activities. Commercial logging never occurred because deep canyons and

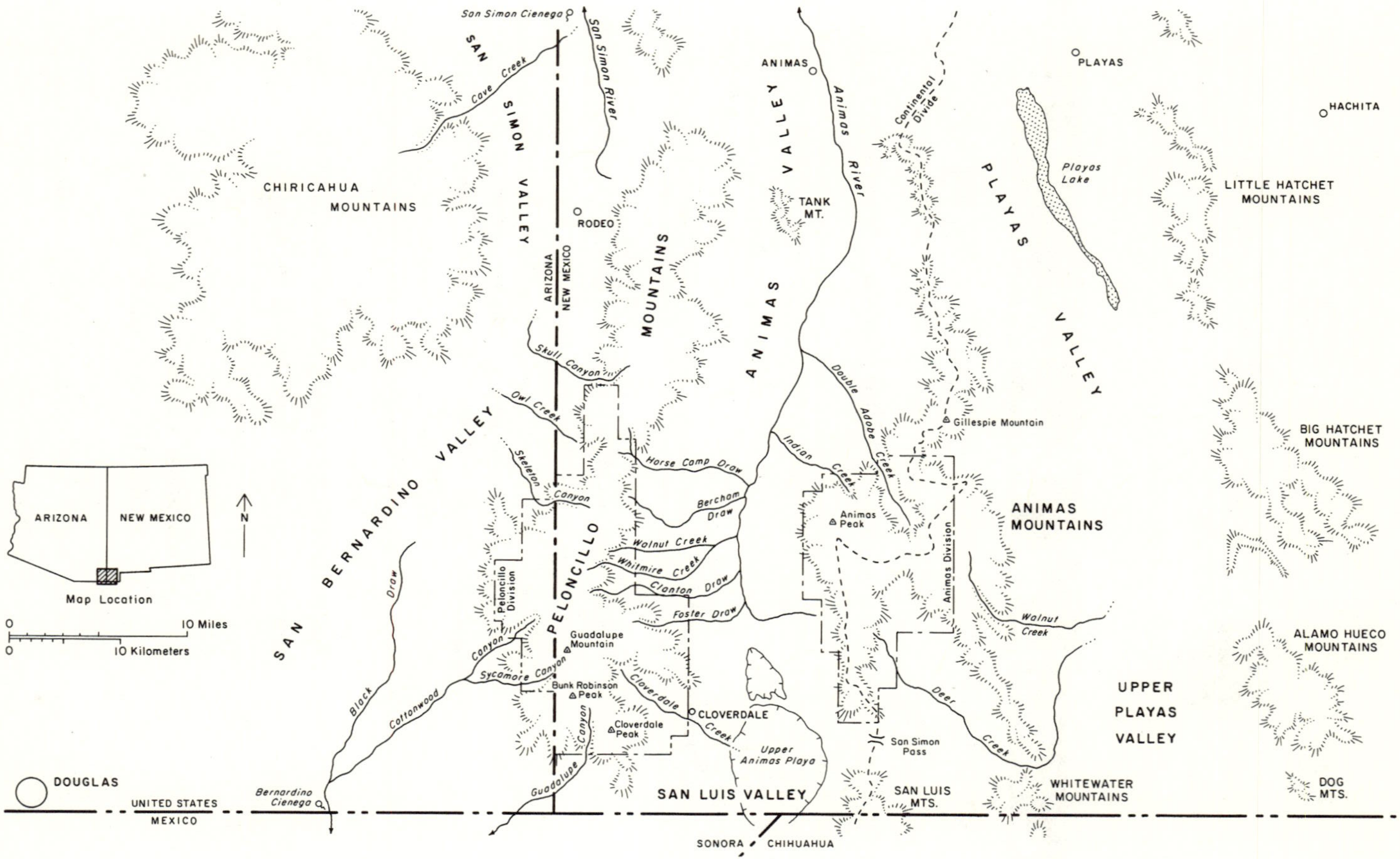

The Peloncillo-Animas District of the Coronado National Forest and the surrounding area.

high, broken ridges made the district's timber resources inaccessible, limiting timber values to fuelwood and fence posts. No significant mineral deposits exist, and few prospects or mines were developed. Recreational values consist of hunting and camping and, in recent years, bird watching. Minor economic activities like harvesting of bear grass for export to Mexico or rockhounding have had little impact. With no significant population centers nearby, the isolated area has few settlers. One graded dirt road crosses the mountains connecting the Arizona and New Mexico sides of the Peloncillo division, and primitive dirt roads penetrate both the Peloncillo and Animas divisions for short distances. Because livestock raising has been the area's single significant extractive activity, the history of Forest Service grazing is largely uncomplicated by competing multiple uses. The present Peloncillo District has fourteen grazing allotments, nine in New Mexico and five in Arizona; one is equally divided between the two states. When the district included the Animas, that division had ten additional grazing allotments, with three allottees leasing multiple allotments.

Since 1906 Forest Service officials have attempted to improve the forage resources of the Peloncillo-Animas District through implementation of a continually evolving series of national regulatory measures and scientific management programs. This chapter describes how regulations and programs influenced local activities within the district and how they were modified to meet local socioeconomic and ethical values. The chapter also evaluates the degree to which these regulations have achieved their goals.

The Setting

Both the Peloncillos and the Animas are small, rugged mountain ranges emerging abruptly from the surrounding grasslands in a series of stacked biotic communities that change from Chihuahuan desert grasslands at the lowest elevations to coniferous forest at the highest elevations. The understanding that the biotic communities of the mountain ranges are isolated by "seas of grass" has given rise to the term "sky island." The Peloncillos and the Animas are sky islands in the Madrean Archipelago, a biogeographic area of exceptional biodiversity and endemism that extends from central Arizona and New Mexico to the Sierra Madre Occidental in northern Mexico. Many species common in Mexico occur only in this portion of the United States.

The Peloncillo Mountains extend for approximately one hundred miles north from Mexico, straddling the boundaries between the states of Chihuahua and Sonora in Mexico and New Mexico and Arizona in the United States. They have an average elevation of 5,800 feet, with peaks of 6,600 and 8,000 feet at Owl Mountain and Mount Baldy. To the west of the Peloncillos in Arizona are the San Simon and San Bernardino Valleys, with average elevations of 4,400 feet. The Animas Valley, approximately thirty miles in width with an average elevation of 4,800 feet, separates the two mountain ranges. East of the Animas Valley, the Animas Mountains parallel the Peloncillos, extending northward from the Mexican boundary for approximately thirty miles into New Mexico. The Animas Range has an average elevation of 6,500 feet, with its highest elevation being the summit of Animas Peak (8,519 feet). The topography of the Peloncillo Range varies from gently rolling to moderately rough on the higher bluffs. The majority of its surface area is accessible to livestock. Although alluvial fans are present on both slopes of the Peloncillos, the western slope is steeper and rougher than the more rolling eastern slope.

The Animas Range has high, well-defined ridges, breaking off sharply to the level Animas Valley on the west and to very broken hills and canyons to the east. The Animas ridge is higher and steeper than the Peloncillo ridge, with a greater proportion of range inaccessible to cattle. Throughout the complex basin-and-range topography found in the two mountain ranges, approximately twenty-four USDA soil types and terrains occur. Few soils are deep, however, and bedrock is visible in many places.

Although rainfall can be highly unpredictable and drought frequent, during "normal" years rainfall has a biseasonal pattern, with the majority of precipitation occurring in the form of thunderstorms during the monsoonal season from July to September. A second rainy season from December through March typically has lighter, gentler rainfalls. Within the district, average annual rainfall varies significantly according to location and topography. Annual averages may differ from ten to twenty-two inches per year within a few miles. Great climatic variability exists within subareas. High and low average annual temperatures vary significantly between valleys and higher elevations. The wind chill factor is frequently severe in the Animas Valley.

In both ranges, the majority of drainages flow east or west. They are typical intermittent desert creeks that have the appearance of dry arroyos but carry large volumes of water during rainy seasons. No drainage has a perennial flow for its full length, although there are short stretches of permanent water in areas below springs. In the Peloncillos, Guadalupe Canyon and Cloverdale Creek flow southward into Mexico; Cottonwood, Skeleton, Deer, Owl, and Skull Canyons flow in a westerly direction toward the San Simon Channel; Foster, Clanton, Whitmire, and Bercham Draws drain in an easterly direction into closed basins in the Animas Valley. In the Animas Mountains, the major drainage, Deer Canyon, runs southward toward Mexico.

Although the forage resources primarily consist of desert bunchgrasses, both mountain ranges present an intricate mosaic of vegetation, reflecting the rough, broken terrain. Vegetation consists of seven general types: piedmont desert grassland, interior chaparral, Madrean oak woodland, piñon-juniper woodland, Madrean-evergreen woodland, riparian oak-sycamore, and rock outcrop vegetation. Common grasses include bluestem, curly mesquite, tobosa, and several varieties of grama, interspersed in some soils by yucca, agave, sotol, *Nolina,* and several species of prickly pear. In the interior chaparral, Toumey oak, manzanita, evergreen sumac, and border piñon are common; in the chaparral woodland, piñon, oak, manzanita, sumac, silk-tassel, and desert buckthorn are most common. In the woodlands, common trees include juniper, Mexican piñon, and several varieties of evergreen oak, with small stands of Chihuahua and ponderosa pine at the higher elevations and very small stands of Douglas fir and a few Arizona cypress at the highest elevations of the Animas Mountains.

Characteristic mammals include Coues white-tailed deer, Chihuahuan pronghorn antelope, mule deer, javelina, coati, ringtail, hog-nosed skunk, hooded skunk, striped skunk, desert shrew, coyote, gray fox, mountain lion, bobcat, and black bear. The grizzly bear and Mexican gray wolf were extirpated during the early twentieth century. A few black-tailed prairie dogs still have colonies near the Mexican border, and, in 1996, a jaguar, thought to be extinct in the United States, crossed through the Peloncillos approximately ten miles north of the international boundary. The jaguar, Arizona shrew, Sanborn's long-nosed bat, southern yellow bat, long-tongued bat, white-sided jackrabbit, and southern pocket gopher are threatened or endangered mammals in the area. Grassland-dependent bird species like the mountain plover are no longer present, and only occasional sightings of Aplomado falcons have been reported. Ferruginous hawks have suffered a sharp decline. The thick-billed parrot was extirpated in the area, and reintroduction attempts in the nearby Chiricahua Mountains failed. Both mountain ranges form direct corridors to the Sierra Madre Occidental in Mexico, making them significant routes for wildlife migration to the Mogollon Plateau in Arizona and the mountains of the Gila Wilderness Area in New Mexico.[1]

Pre–National Forest Grazing History

The earliest grazing in the region dates from the periods of Spanish (1695–1821) and Mexican (1821–54) control. The mountain ranges received their names at this time: "sierra de las Animas" (mountain of the spirits) most likely derives from the existence of many mirages in the "valle de las Animas"; Peloncillo, "little baldy," reflects the appearance of that range and refers to the hard, cone-shaped chunks of brown sugar sold in Mexico. During the Spanish and Mexican periods, continual warfare in this portion of the frontier discouraged Euro-American settlement and prevented the development of a viable cattle industry. Apaches, who had effective control of the borderlands, did not engage in cattle breeding. Instead, Apache bands raided ranches farther south in Chihuahua and Sonora and herded the stolen cattle and horses over well-worn cattle trails through the San Bernardino, Animas, and Playas Valleys to their *rancherías* in the Peloncillo and Animas Ranges, where they held the animals for slaughter in the natural corrals formed by the box canyons. Apaches also consumed the offspring of cattle and horse herds left at abandoned Spanish and Mexican ranches. Impacts from captured livestock occurred only at holding areas and at springs and cienegas (marshes). Intensive grazing did not begin until the 1880s, several decades after Mexico transferred the area to the United States with the 1853 Gadsden Purchase.[2]

In 1881 the construction of the Southern Pacific Railroad across southern Arizona and New Mexico provided the means for quick importation of livestock, precipitating an open range cattle boom. During most of the Territorial Period (1854–1912), cattle were sold by the animal unit rather than by weight, encouraging ranchers to focus on the quantity rather than on the quality of their livestock. Between 1881 and 1906, the desert grasslands of the Peloncillos and Animas supported excessive numbers of livestock. Two large cattle companies, the San Simon Cattle and Canal Company, with headquarters near the San

Simon Cienega west of the Peloncillo Mountains, and the Victorio Land and Cattle Company, headquartered in the Animas Valley, dominated grazing in the area, each with herds of more than 20,000 head. In addition, homesteaders with small herds settled at springs throughout the valleys. Gangs of brazen cattle rustlers, operating from headquarters in the Chiricahua, Peloncillo, and Animas Mountains, moved entire herds of stolen cattle across both international and territorial borders. Livestock numbers peaked during the early 1890s, despite the disastrous drought of 1892–93, when between 50 and 70 percent of the area's cattle died of starvation on the range. By 1900 overgrazing had led to irreparable damage to the grasslands and timbered areas. In the oak woodland and piñon-juniper zones of the Peloncillo and Animas Ranges, damage consisted of heavy trampling near springs and water holes, the permanent drying of many springs, the depletion of native bunchgrasses (where dying cattle had consumed the roots), and an invasion of ruminant-proof brushy plants and toxic weeds. Without the usual grass cover, cattle trails etched deeply into the soil, augmenting hillslope erosion. When the rains returned, runoff on the denuded soil increased, intensifying flood events and causing severe downcutting and erosion in watercourses. Both ranchers and scientists recognized the damage that had been done by the disorder of the open range cattle industry on the border.

In 1891 Congress passed the General Land Law Revision Act, which gave the president the authority to set aside timber reserves on the public domain. Many of the new reserves contained unforested areas of prime meadowland where grazing was an obvious commercial use. The future of grazing on the reserves became the focus of a heated controversy, pitting commercial livestock interests, who supported unrestricted free grazing, against antigrazing agriculturalists and conservationists, who wanted to limit grazing and protect watersheds. When Congress transferred the forest reserves from the Department of the Interior to the Department of Agriculture (USDA) in 1905, the pro-grazing interests won. Three of the top officials of the USDA, Chief Forester Gifford Pinchot, Chief Botanist Frederick Coville, and Director of the Division of Grazing Albert F. Potter (an Arizona rancher), favored utilization of the public forest through regulated livestock grazing, provided that grazing was not harmful to vegetation regeneration. In 1907 the USDA renamed the reserves national forests to reflect the change in policy from one in which resources were reserved to one in

Top: erosion of stream-built slope in the Upper Animas Valley, Grant County, New Mexico, looking east toward Animas Peak from the western slopes of the Peloncillos. The Animas River is visible at the center of the photo. When Schwennesen visited the area in 1913, the river had little or no riparian vegetation, indicating overgrazing. Administrators in the Forest Service noted that in 1906 the Animas Mountains were incorporated into the Peloncillo Forest Reserve because reserve status was the only way to protect the area from overgrazing. Bottom: as a result of "green" ranching practices, riparian vegetation has returned to the Animas River. Top photograph courtesy U. S. Geological Survey, Denver, Colorado, photo by A. T. Schwennesen, photo no. 23; bottom photograph courtesy Ray Turner

which resources were "scientifically" managed by the federal government for the benefit of the nation.[3]

The Peloncillo-Animas District

On November 3, 1906, President Roosevelt established the Peloncillo Forest Reserve, containing approximately 320 square miles in the Peloncillo and Animas Ranges. The reserve had two geographically separated divisions—Grazing District Number 1, Animas Creek, with approximately 55,700 acres, and Grazing District Number 2, the Peloncillo, with approximately 88,000 acres. Newspapers reported that the reserve had been established to protect the Animas watershed, rather than its timber values, because it was "fast becoming denuded." Through a complicated process of consolidations occurring over a fifty-year period, the Peloncillo and fourteen other small reserves in southeastern Arizona were reconfigured to become the present Coronado National Forest. In 1908 the nearby Chiricahua National Forest absorbed the Peloncillo Reserve, and the former reserve was downgraded to district status, still subdivided into two divisions. Indicating the importance of grazing, the forest supervisor's office for the Chiricahua was located in Rodeo, New Mexico, the largest nearby cattle-shipping station. This administrative unit lasted until 1916, when the Chiricahua was incorporated into the Coronado National Forest and the supervisor's office moved to Tucson.

Ranger stations for the Peloncillo-Animas District moved with population shifts. They were located at Skeleton Canyon on the road from Douglas to the Animas Valley (1908–20), the small agricultural settlement at Cloverdale, New Mexico (1920–33), and Douglas, Arizona (1933–present). The Peloncillo-Animas District experienced several reductions in size. In 1910 President Taft removed the southern portion of the Animas division along the Mexican boundary from national forest status. In 1948 the remaining 50,000-plus acres of the division were privatized in a land exchange.[4]

The Allotment System

Regulations for management of all national forests, including the Peloncillo-Animas District, appeared in the *Use Book* (*Use of the National Forest Reserves*), the USDA handbook (or bible of grazing policy, as

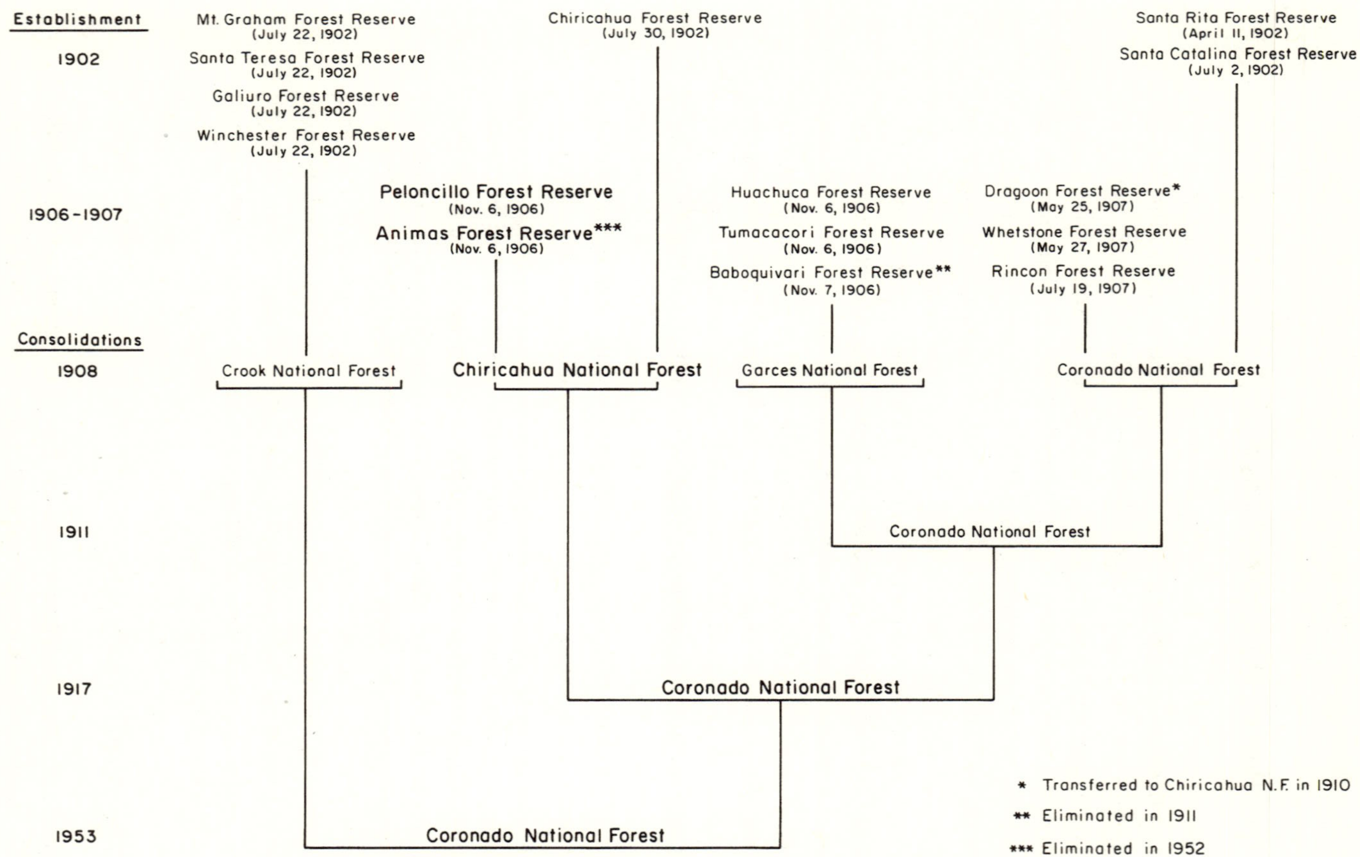

Chronology of consolidations and eliminations in the Coronado National Forest.

it was known), which reflected Pinchot's philosophy of multiple use. Each year, the Grazing Branch of the USDA determined the length of the grazing season and the number of livestock in each category (cattle, sheep, goats, horses, or swine) to be permitted on each national forest. Supervisors for individual forests divided grazing ranges into allotments, specific portions of the forest reserved for specific allottees (permit holders) through a process of negotiation with prior users or with owners of adjacent private property. Initially, the district contained several community allotments shared by several permittees, but as fencing was constructed, individual allotments replaced the unsatisfactory shared allotments. Supervisors issued permits for the use of specific allotments. Class A permits went to livestock grazers who owned private ranch property adjacent to the forest, Class B to those who owned nonadjacent property but had traditionally used the forest range, and Class C to transient herders. Permits were usually issued for one year (occasionally for five years) for cattle, horses, sheep, swine, or goats. On January 1, 1906, Pinchot set grazing fees at 20 to 35 cents per animal unit for the grazing season or for the year. An animal unit consisted of one bull or a cow with her calf, one horse or a mare with her foal, or the equivalent thereof in smaller herbivores. Five sheep, goats, or hogs were considered equivalent to one cow or horse unit (see figure and table).[5]

Supervisors established a maximum limit or total number for all classes of livestock to be permitted on the forest as a whole, as well as a maximum limit for the individual ranchers or cattle companies with permits. They also set a "protective," or minimum, limit, which represented the number of livestock the Forest Service believed necessary to support a family or to make a cattle company economically viable. Ranchers grazing fewer animals than the protective limit were allowed to increase their numbers during good years, while those whose numbers were at or above the protective limit could not obtain increases. Each forest was expected to form a grazing advisory board with representation from local stock associations. Working with forest officials, the local grazing board advised on forestwide grazing plans, assisted in dispute settlement, and made recommendations to the Forest Service concerning questions of local use. Additional *Use Book* regulations encouraged financial cooperation on range improvements, including the construction of roads, trails, and stock waters. The Forest Service supplied materials and permittees provided labor for the construction of

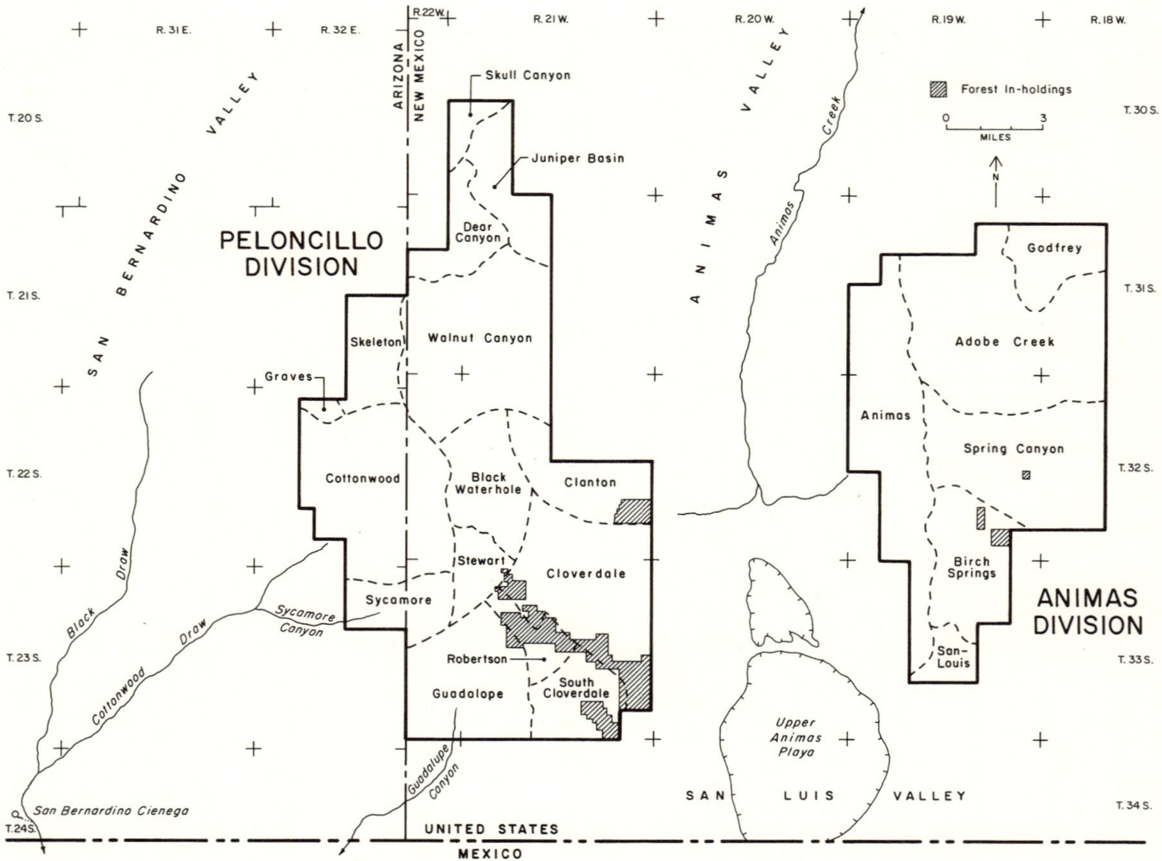

Allotments in the Peloncillo-Animas District of the Coronado National Forest, 1937.

Summary of Allotments and Permits, 1923–1940

Allotment	Permittee	Stock Numbers	Permit Type
PELONCILLO DIVISION			
Clanton	Eicks, J.H.	107 to 157	Yearlong
Cloverdale	Victorio L & C C	356–2100	Yearlong
	Hughes	84 (1923 only)	Yearlong
	Stewart, J.W.	113–158	Yearlong
	Robertson, W.B.	30–268	Yearlong
	Pascoe, W.W.	25–50	Yearlong
	Sanford, J.N.	disapproved	
	Greenhaw	20–(1933 only)	Yearlong
South Cloverdale	Victorio L & C C	120	Yearlong
	Sanford, J.N.	25	Yearlong
	Robertson, W	34–75	Yearlong
Cottonwood	McDonald, W.	95–188–238	Seasonal
	Fairchild, Mrs. F.	46–170	Seasonal
	Taylor, Mrs. L	170	Yearlong
	Fairchild, M.A.	46–92	Seasonal
Deer Creek	Darnell, C.	90–200 (153)	Yearlong
Guadalupe	McDonald, D.	230–321	Yearlong
	Taylor, D.H.	41–261 (1932 on)	Seasonal
Horse Camp	Wamel (Lee, Leslie)	50–200	Yearlong
	Victorio L & C C	105	Yearlong
	Kimble, F.C.		
Noland	Graves, W.S.	20	Yearlong
Pine Canyon	Darnell, C.	70	Yearlong
Skeleton Canyon	Sloan, R	100	Yearlong
Skull Canyon	Wamel, R	50	Yearlong
	Noland, C.	60	Yearlong
	Graves, W.	20	Yearlong
	Kembitch, W.	7–10	Yearlong
Sycamore	Swaggart, W.E.	75	Yearlong
	McDonald, L (after 1940)	75	Yearlong
Walnut Canyon	Fields, A	15–200 (60 sheep)	Seasonal
	McCarthy, C	125–144	Seasonal
	Eaton, F	disapproved (lack water)	
	Taylor, J.L.	disapproved	
	Taylor, J.L.	disapproved	

Allotment	Permittee	Stock Numbers	Permit Type
Allotments added after 1941			
Blackwater Hole	Henderson, L.	45–71	Yearlong
Fairchild			
Graves	Snure, Ben	14	Yearlong
Juniper Basin	Miller, W.	85	Seasonal
Outlaw Mountain			
Robertson	Robertson, M.	54	Yearlong
Stewart	Henderson, L & V	113	Yearlong
ANIMAS DIVISION			
Adobe Creek	Victorio L & C C Godfrey, G.	600–619* 140	Seasonal Seasonal
Animas	Victorio L & C C	175–300*	Yearlong
Birch Springs	Victorio L & C C	188–225*	Yearlong
Horse Camp	Victorio L & C C		Yearlong
Red Hill	Victorio L & C C	70*	Yearlong
San Luis Pass	Victorio L & C C	99–200*	Seasonal
Spring Canyon	Timberlake & Adams Smith, Sim Timberlake, E.	350–400* 50–100 (1936)250–368	Yearlong Yearlong Yearlong
Allotments added after 1941			
Godfrey	Godfrey, G.	191	Seasonal

Source: Records in the office of the U.S. Department of the Interior, Forest Service, Coronado National Forest, Tucson, Arizona.

division fences to separate grazing districts or allotments and "drift" fences, partial fences placed in key locations to obstruct the natural drift of unpermitted cattle and horses onto the forest.[6]

Stocking Reductions and Carrying Capacity

During the period prior to World War I, Forest Service management policy focused on reducing the number of livestock on the forests and on preventing access (trespass) of unpermitted livestock. Reduction and regulation depended on fencing, which became the first priority of rangers in the Peloncillo-Animas District. Chiricahua Supervisor Arthur Ringland instructed permittees on the Peloncillo-Animas District to refer requests for construction of drift fences to him directly. He wanted to assure that new fences would not prevent "full utilization" of the forest by "equitably entitled" permittees or allow large cattle companies to gain exclusive control of any portion of the forest. Rangers and permittees cooperated in gathering wild, unbranded cattle and herds of wild horses that migrated from winter range in the Animas Valley to summer range on the Animas division. To decrease stocking rates with the least amount of conflict, officials imposed transfer reductions. These automatic 10 percent decreases in permitted numbers occurred each time ranchers sold their base land and the Forest Service transferred permits to new buyers.

Forest officials used the concept of carrying capacity as the basis for determining the number of livestock and length of stay for each allotment, attempting to maintain numbers within what they considered to be the natural limit of available range resources. The concept of carrying capacity—the amount of forage needed to support one head of livestock for a one-month period without damage to the grassland—first appeared in the United States in nineteenth-century grazing manuals. In a now-famous 1908 directive, Grazing Division Chief Albert Potter set formal guidelines for determining and applying a range capacity formula. The nation's top range manager instructed forest officials throughout the country to estimate capacity by means of a close, on-the-ground field survey of the ranges in their jurisdictions. They were to evaluate the condition and types of forage, the distances livestock would have to travel from forage to water, the normal climatic conditions, the physiographic nature of the land, and the currently applied livestock management practices. Estimates of carrying capacity varied with forage values (the value of particular types of forage to convert to calories) from approximately twelve acres per head (fifty-three head per section) for summer range on rich meadowland in higher mountains to approximately forty acres per head (sixteen head per section) on more arid desert grasslands. For the Peloncillo-Animas District, no pre–World War I estimates of carrying capacity have been preserved, although rangers made efforts to reduce stocking rates and distribute cattle more evenly throughout the district. Despite these efforts, prior to World War I livestock numbers actually increased throughout the National Forest system.[7]

Forest Homesteads

While local Forest Service officials in the Peloncillo-Animas District attempted to implement stocking reductions, policy on the national level inadvertently encouraged increased use of forest resources by small ranchers and homesteaders. The publicly professed democratic, antimonopoly objectives of the Forest Service stressed protection of "bona fide settlers and homebuilders" from the unfair competition of large owners, a reflection of Progressive Era opposition to business monopoly. In 1905 President Theodore Roosevelt endorsed a grazing policy that gave priority to smallholders, calling the small rancher "the kind of man upon whom the foundations of our citizenship rest." Echoing this opinion, the Forest Homestead Act of June 11, 1906, opened those portions of national forests suitable for agriculture to homestead applications. Homesteads of up to 160 acres could be obtained for a minimal charge after a five-year residence. On August 12, 1912, Congress authorized the Forest Service to reclassify all national forest land to determine which areas would be returned to the public domain for homesteading or other forms of alienation. As the reclassified farm sites came up for entry, announcements appeared in local newspapers, advertising the acreage, location, and advantages of the sites. Land selected under the forest homesteading process did not have to be rectangular in shape and could follow the physiographic conformation of better soils or irrigable areas along floodplains.[8]

In the Peloncillo-Animas District, homesteaders submitted approximately sixteen applications, the majority of which were made during the 1920s. Approval usually came three to four years after application and depended upon a Forest Service survey and inspection of the homestead for evidence of "a genuine attempt to settle and farm." Inspection reports reveal that inspectors often sympathized with the bare-bones existence of homestead applicants. As one official remarked, "the government ought to pay a man to live on one of these God-forsaken places." A review of the homestead inspections made in the Peloncillo division indicates that farming was successful to some degree on the majority of homesteads, many of which had small vegetable gardens and up to forty acres planted in corn, milo, Sudan, millet, or pinto beans. Only one farm plot was enclosed by a hog-proof fence. Farm implements included walking or riding plows, Fresno scrapers, mowers, riding or walking cultivators, and harrows. Livestock consisted of work horses, dairy cattle, chickens, hogs, and a few beef cattle. Two notable exceptions, however, were homesteaders with 150 and 600 head of beef cattle, respectively.[9]

Although the intention of the Forest Homestead Act was to provide land for dry farmers and agricultural settlers, in the Peloncillo-Animas District it actually encouraged additional exploitation of forage resources. Since the reclassification process carved homesteads out of established grazing allotments, new homesteads were entitled to application for Class A grazing permits. Ranchers with prior permits considered the homesteaders intruders, little better than the "nesters" or "squatters" who had preempted public domain lands twenty years earlier. To accommodate the homesteaders, forest officials reduced the permit limits given to existing allottees. Established ranchers objected to the intrusion of the homesteaders' inferior breeds of cattle and to the conversion of prime stands of native grasses along well-watered creek beds into fenced farmland inaccessible to cattle.[10]

The objective of many homesteaders was to abandon dry farming as soon as possible and get started in the livestock business. In the Peloncillo division, forest homesteader Henry Eicks became a successful rancher. In 1926 the inspector who examined his homestead for proof of genuine settlement reported that Henry and his wife had resided on the eastern slope of the Peloncillos since 1919. Henry worked intermittently as a cowboy for the Victorio Land and Cattle Company, and when company roundups took him away from home, Mrs. Eicks remained on the homestead. Eicks and his wife had a two-room lumber house twenty-four by twelve feet in size, a chicken house, a corn crib, a well, a fifty-foot-square wire corral, and nearly two miles of fencing. The combined value of all improvements was $735. The inventory listed farm implements and "all the necessary" household equipment: a wood-burning kitchen stove, bedsteads and bedding, two lamps, two trunks, two chairs, and one phonograph. Indicative of the simple living conditions of many settlers, neither the Eicks family nor any other forest homesteader in the Peloncillos possessed a sofa.

Despite the prevailing opinion that dry farming was doomed to failure, the inspector who visited the Eicks homestead recorded considerable success. During the preceding five years, Eicks had farmed between 20 and 37 acres, producing as much as 40 tons of maize, 2,500 pounds of beans, and 150 bushels of corn in good years. More important to Eicks than farming were his 169 head of range cattle, 16 horses and mules, and 12 swine. Eicks had obtained a permit for 150 head on the Peloncillo division; the remainder of his herd stayed on "outside" private pasture. Officials approved the application, and although the Eicks family sold their homestead to a large ranch long ago, the house and corrals are still standing, and the homestead is known today as the "Eicks place."[11]

Reactions to Federal Regulation

Many ranchers reacted initially with anger to Forest Service grazing regulations, the first ever imposed on any public lands. They particularly objected to the imposition of grazing fees. Smallholders believed that fees would jeopardize their expansion and lead to monopolies by large stock owners. Fees negated the main objective of the homestead laws, which was to stimulate settlement and create communities by dividing up land among many small settlers. Large ranchers, on the other hand, argued that protective limits favoring smallholders ran contrary to the economy of scale needed in the cattle industry. To operate profitably in arid areas, large holdings were necessary. Both large and small stock raisers, however, agreed on one point: short-term leases made ranching an unstable economic activity. One-year leases, subject to cancellation during the annual review process, provided no economic security, discouraged permittees from making long-term investments in range improvements, and made it more difficult to obtain loans.[12]

Reaction to the regulations, however, was not uniformly negative and became more positive with time. Many livestock associations had called for a permitting system on public lands before the forest reserves were established. Only two years after the imposition of grazing fees, the Arizona Cattle Growers Association passed a resolution favoring federal regulation. The 1907 resolution noted that in comparison to cattle on the open range, cattle on national forests could be handled with greater ease and brought higher prices. In 1909 the *Southwestern Stockman,* the voice of southern Arizona stock raisers, reversed its opposition to regulation and stated that since controls had been initiated on national forests, productivity had increased by 30 percent through prevention of misuse. In 1914 the Arizona Cattle Growers Association passed a resolution urging extension of national forest grazing controls to other public lands. In 1915 Arizona rancher Dwight B. Heard, president of the American National Live Stock Association, published an article in the *American Review of Reviews* that used national forest grazing regulations as a model to be emulated. Regulation, he stated, had improved forest range conditions and carrying capacity, increasing stocking rates on some forests up to 50 percent without damage to the range.[13] No organized opposition to the new regulations emerged in the Peloncillo-Animas District, and no negative comments appeared in any of the local newspapers.

By 1916, when the Coronado National Forest absorbed the Chiricahua, a new trend in the cattle industry had become apparent. Large cattle companies like the Victorio were fencing vast areas of public domain. The companies purchased large tracts of land from the General Land Office, from the new state land offices established in 1912 with statehood, and from railroad companies that sold unwanted lieu lands granted by the federal government for railroad construction. Small ranchers found it difficult to gain access to the diminishing public domain and consequently increased their requests for forest allotments. During World War I, to accommodate smallholder and patriotic wartime food production demands, the Forest Service increased the maximum numbers for the Peloncillo-Animas District.[14]

The First Phase of "Scientific" Management

In 1910 the Forest Service established the Office of Grazing Studies to coordinate studies and implement a more scientific system of management on national forests. Proper utilization and correct seasonal use replaced stocking reduction as the Forest Service's primary management strategy. Rangers imposed seasonal use adjustments to implement deferred and rotational grazing, protecting rangeland by means of shorter grazing seasons adjusted to reseeding requirements. They encouraged subdivision of grazing ranges into units of approximately equal capacity, with deferred grazing on one subunit until grass seed had matured and disseminated and rotation of subunits on an annual basis. In the Peloncillo-Animas District, however, range examiners considered the forage strong enough to withstand yearlong grazing. Some allotments did not have deferred or rotational grazing for another three decades.[15]

The two most important tools for making a scientific determination of range condition were forestwide range reconnaissances and district-level range surveys. Reconnaissances were major range inspections conducted for entire national forests at approximate ten-year intervals by personnel from the Office of Grazing Studies. On reconnaissances, teams of three or four trained field inspectors rode through the forest, following township lines and attempting to cover the equivalent of three adjacent sections per person each day. When the field research was completed, personnel in the regional office produced reports that rated range condition for allotments on a scale of 1 to 10 and contained maps of vegetation types, soils, water sources, topography, erosion, and improvements. More limited in scope, range surveys were conducted on the district level to create management plans for individual allotments or grazing plans for grazing districts.[16]

The first major reconnaissance for the Peloncillo-Animas District took place in 1912 as part of the reclassification of agricultural lands for the Forest Homestead Act. At that time, the Peloncillo-Animas District, one of six districts in the Chiricahua National Forest, had a maximum limit forestwide of 12,000 cattle and horses, 2,000 goats, and 300 swine. In 1913 the seventeen allottees in the Peloncillo division received permits for 1,900 head of cattle and horses and the three allottees in the Animas division for 1,400. The maximum limit for individuals or cattle companies was set at 1,000 head and the minimum protective limit at 157 head. In 1914 Acting Assistant Forester Will C. Barnes approved a double grazing season (April 1–July 15 and November 1–March 31) in some portions of the forest in order to rest range from July through October when reseeding took place. If conditions warranted, the season could be extended—with a fee increase, adjusted from the base of 34 cents per head for cattle and 42 cents for horses. In 1916 increased rainfall combined with the World War I beef shortage and high cattle prices led to a forestwide increase in permitted maximum numbers for the Chiricahua to 12,600 head for cattle and 3,000 head for goats. Swine, however, were decreased. Then in 1917–18 the maximum total number for Peloncillo allotments increased to 3,800 and for Animas allotments to 3,000, despite below average rainfall. By the end of the war, the Peloncillo maximum limit had doubled and the Animas limit had more than doubled, resulting in considerable range deterioration.[17]

A second important component of scientific management was cooperation between permittees and federal and state agencies in predator and rodent control. During the 1890s, both Arizona and New Mexico territories enacted bounties on predators. Ranchers and livestock associations, claiming losses of thousands of dollars annually from wolf depredations, paid bounties for predator scalps. The South Chiricahua Cattle Raisers Association paid bounties of $50 for wolves and lions, while the nearby Victorio Land and Cattle Company paid separate bounties. This type of random trapping was found to be ineffective. Walter Birchfield, superintendent of the Victorio, complained that trappers brought him "scalps" of wolves caught in Mexico or other areas off the ranch, and he decided to terminate the payments. After June 1914 the U.S. Bureau of Biological Survey stationed professional trappers in key locations in the district, where they made a concerted effort to exterminate the resident wolf population and to prevent migrant wolves from moving north from Mexico. Between 1917 and 1925 government trappers killed several dozen wolves annually, most of them within or near the southern boundaries of the Peloncillo and Animas divisions. Biologists and local ranchers recognized the Animas Valley as a major corridor for wolf migration from the Sierra Madre into the United States. Cowboys who worked for the Victorio Land and Cattle Company recall roping wolves from horseback whenever they chanced upon them. The famous Biological Survey trapper Clarence Echols, who appears in Cormack McCarthy's 1995 novel *The Crossing*, homesteaded directly on the wolf corridor at the southern end of the Animas Mountains. He lived on his homestead part of the year, using it as a station for trapping and poisoning wolves and other predators, although he was active throughout the region as well. The Biological Survey estimated calf crop losses from predators at 3 percent, while Birchfield claimed the Victorio's losses were 10 percent as late as 1921, despite the Biological Survey's efforts. By 1926 many fewer wolves had been trapped, indicating a steep decline in the wolf population. After that time, agents focused on coyotes, mountain lions, and bobcats.

The Biological Survey also conducted a rodent control program, designed to eliminate rodents that competed with cattle for forage. Agents poisoned kangaroo rats and extirpated entire colonies of pocket gophers and black-tailed prairie dogs (*Arizonensis*) in the San Simon and Animas Valleys adjacent to both forest divisions. Working with local "cooperators," agents also organized jackrabbit drives to herd rabbits into wire mesh traps, where they were clubbed. Old-timers on the Animas Valley recall participating in these drives, which proved very effective in reducing rodent numbers. In the 1940s the Fish and Wildlife Service's Predator and Rodent Control (PARC) took over the predator programs. PARC used the predacide Compound 1080, an indiscriminate method of coyote control that caused secondary deaths of many other species until Congress banned its use in 1972.[18]

Forest Service Monitoring and Policy: Range Inspections during the 1920s

In 1924 inspector Frank Grubb conducted another major range survey of the Peloncillo-Animas District for Ranger George Galer. His detailed report describes the district's agricultural, economic, and environmental conditions approximately a decade after the Forest Service's grazing system had been accepted by local ranchers. By the time of Grubb's observations, a pattern of grazing had been established that changed little during the succeeding two decades. The district had twenty permittees on thirteen fenced allotments. The majority of permittees were smallholders with between 150 and 500 head of cattle, and two of them were forest homesteaders. The only exception was the Victorio Land and Cattle Company, the area's remaining large company headquartered in the Animas Valley, which had a total of 20,000 head both on and off the forest. The other large company, the San Simon, had sold most of its cattle a few years prior to the inspection.

The maximum limit for individuals or companies in both divisions had increased from 1,000 (in 1913) to 1,200 head of cattle and horses. The protective (or minimum) limit on the Animas division was 200 head and on the Peloncillo 150 head. No protective limits had been established for swine. Throughout both divisions, Grubb graded the majority of the cattle as "fair," with the exception of two allotments with high-grade Herefords. Many permittees ran a few head of branded swine, allowing them to free range with the cattle. The Victorio Land and Cattle Company owned most of the swine and distributed them evenly throughout its seven allotments in the Animas division. Although the district was considered too rough for sheep, in past years ranchers had run a few small bands of sheep on the northern limits of the Peloncillo division. Prior to 1920, when the Forest Service terminated goat grazing, several permittees practiced a dual stocking strategy, running bands of mohair-producing Angora goats in conjunction with their cattle and varying the numbers of goats and cattle with fluctuations in the respective markets. Many ranchers claimed that goats, who are primarily browsers, eliminated brush, making grass more accessible to cattle. By 1920, however, cattle and swine were the only classes of livestock in the district.

The majority of permittees owned outside range, or nonforest grazing land, that could be used in connection with the forest allotment. Although most permittees had permits for yearlong use, a few had seasonal permits for double seasons (April 1–June 30 and/or October 1–March 31) that could be used by moving their cattle short distances to their outside pasture. The only requests for additional grazing range came from existing permittees and from a few "640-acre entrymen," homesteaders who had proved up on homesteads adjacent to the forest under the Stock Raising Homestead Act of 1916. The majority of the permittees were exclusively stockmen, with the exception of the few forest homesteaders who had farms along Animas and Cloverdale Creeks, practicing what Grubb considered to be unsuccessful farming. He warned, however, that if all the forest homesteaders along the creeks were to apply for grazing permits, it would seriously disrupt the existing permittees' "acquired preferences" (i.e., their existing stocking rates).

Grubb classified the entire district as combined grass-browse range. Grass forage included native bunchgrasses and varied from curly mesquite/grama types on lower elevations to fescue/mountain sacaton types in the pine belts. Browse provided an important part of the forage supply on all allotments and included oak and mesquite brush, mountain mahogany, and manzanita. "Weeds," undesirable species unpalatable to cattle, comprised a minor part of the forage. Locoweed and other poisonous plants, a common indication of overgrazing, were rare, and little or no stock losses from poisonous plants had occurred. Grubb classified most of the district as yearlong range, except for "waste range" at the higher elevations of the Animas Mountains that could not be used during winter. He classified range condition in both divisions as "lightly to evenly grazed," with the exception of three heavily grazed allotments that had "more stock under permit than should have been allowed in view of current range development." Although both divisions were remarkably free of fires of any magnitude, Grubb classified a few areas as fire hazards. He suggested "proper utilization," or increased grazing in the underutilized areas, to reduce the existing minimal fire hazard even further.

Natural water sources were inadequate throughout both divisions. The district had an approximately equal number of "improved" natural springs and "developed" water sources, the latter consisting of wells with windmills or dirt stock tanks in the form of excavated retention basins. Grubb was optimistic that the construction of additional water sources and the resulting more equitable distribution of cattle would prevent overgrazing near stock tanks, and forage would be sufficient to carry the present stocking rate.

Almost all of the permittees ran Hereford or Hereford-cross cattle, had cow-calf operations, and sold cattle as yearling feeders. Only the Victorio commonly kept steers until age two or three. Since the end of World War I, however, market prices had been so low that other permittees had retained their two- and three-year-old steers, hoping that prices would go up. Ranchers commonly shipped their sale cattle out of the region, trail herding them to the railroad along easy routes through private ranches that did not charge the trail herders for feed or water. Five shipping stations on the El Paso Southwestern Railroad line at an average distance of twenty miles from the boundaries of either division provided large shipping pens and water cars at all shipping stations.

Drought conditions altered management practices in the district. In response to the drought of 1920–21, some permittees provided starving cattle with purchased feed, although it was not a common practice to supplement range forage with supplemental feed. The Victorio had removed its livestock by railroad to the company's range in Kern County, California, as was its practice during times of short forage. Many smaller permittees sold calves, others sold portions of their breeding stock, and others moved their herds to leased pasture in northern Mexico, where the drought was less severe.[19]

By the time of the next range inspection, conducted by Assistant Range Examiner W. G. Koogler in 1929, the patterns of ranching had changed little. Koogler's assessment echoed many of Grubb's opinions.

The Peloncillo-Animas District carried the next-to-lowest recommendation for stocking reduction in the Coronado. Some permits carried stocking rates of eight forage acres per head on a seasonal basis, a much higher than normal stocking rate justified by the district's longer-than-average growing season. Koogler classified range condition as satisfactory throughout the district. He recommended moving cattle to the higher foothills during cool weather in late winter and early spring, when temporary water was available, to utilize forage more thoroughly. Ranchers were cooperating with the Forest Service to remove the remaining wild cattle.

Koogler noted that water facilities were more widespread, but salt distribution was still careless, with excessive placement of salt next to water sources. Erosion was a problem in only two locations in the Peloncillo division—Cowboy Flat and Clanton Draw—and along Deer Creek in the Animas division. Despite what Koogler considered "heavy stands" of grasses, the district carried the lowest fire hazard rating for the Coronado National Forest. Like Grubb, Koogler recommended "full proper utilization" of forage to reduce the fire hazard, which existed only in the association of tall grasses of low palatability. Koogler found three areas of underutilization: Bunk Robinson Peak and the high ridge divide in the Peloncillo division and the south end of the Animas ridge in the Animas division, all locations with rough, coarse grasses of low palatability. Demands for increases in grazing permits came only from the owners of big ranches who had large outside range in the valleys nearby. Several permittees were running fewer cattle than allowed in their permits. Ranchers and Forest Service personnel were cooperating on trapping the remaining wild cattle.[20]

Both Grubb and Koogler found little evidence of overgrazing, with grass coverage adequate throughout the district. They found erosion only in areas where cattle were held for sorting during roundups. The acknowledgment that some allottees ran fewer cattle than permitted in their lease agreements was a tacit recognition that these ranchers were practicing stocking strategies more conservative than those recommended by forest officials. Neither higher officials in the Forest Service nor outside observers challenged either range assessment.

Sources of Conflict

The 1920s, when the inspections were made, were a time of increased acrimony between the Forest Service and ranchers. The Peloncillo Cattle Growers Association, one of three associations formed in the former Chiricahua National Forest, represented the permittees on the Peloncillo-Animas District. Rangers complained that in the early period the associations were not "progressive." Members frequently broke state and federal regulations and even their own resolutions. By the mid-1920s, however, the Peloncillo Cattle Growers Association was holding regular meetings to discuss range improvements, like cooperative construction of a drift fence along the ridgeline of the Peloncillos. By the late-1920s permittees had begun to join county cattle growers associations affiliated with more progressive state organizations, which had a stronger voice with Congress.

From 1906 to 1933 grazing fees on national forests remained approximately 50 percent lower than fees for leased private rangeland. Yet they were a major source of friction. After a small increase in 1910, each subsequent attempt to raise fees engendered vociferous opposition. World War I emergency food programs stressed greater production as a patriotic duty, a sentiment that paved the way for higher stocking rates and prevented an attempted fee increase in 1916. After the war, a severe local drought in 1920–21 combined with a collapse in the oversupplied cattle market left the Peloncillo-Animas District overstocked with undernourished, low-value cattle. In 1920, when the Forest Service unsuccessfully attempted to return to prewar stocking rates and implement slight fee increases, the Arizona Cattle Growers Association passed a resolution endorsing transfer of the national forests to individual states, a strategy repeated during the sagebrush rebellion fifty years later. Then in 1924 officials proposed conducting market-price appraisals of forage crops similar to the national forest timber appraisals done prior to sales. Market appraisals for forage crops would have doubled many permit fees.[21]

Livestock interests responded by pressuring Congress to create a subcommittee of the Public Lands Commission to investigate grazing practices on national forests. Chaired by Oregon Senator Robert N. Stanfield, a sheep rancher, the committee included two Arizona ranchers, Senators Ralph Cameron and Henry Ashurst, whose brother was a permittee in the Chiricahua division. For three days in June 1925 Coronado permittees testified in Douglas, Arizona. Permittees objected to automatic transfer cuts, requested government payment for drift fencing and other improvements, and urged the adoption of ten-year permits to eliminate insecure tenure. Two testimonies illuminate the dilemma faced by the Forest Service in attempting to accommodate the widely divergent objectives of its permittees. J. H. Snyder, representing a large cattle company, complained that the Forest Service had cut his company's permit from 400 head yearlong in 1906 to 175 head yearlong in 1925. Fees had increased from 25 cents to $1 per head, and the company allotment was converted into a community allotment with other permittees. Further cuts would put his company out of business. In contrast, Thomas Mabry, a financially stressed forest homesteader with a wife and six children, wanted to exchange his twenty-five-head permit from a waterless allotment several miles away from his base land to an allotment adjacent to his homestead that had water. He paid fees for two years but could not use the allotment and had to sell his cattle.[22]

Had the Stanfield bill (Senate Bill 2584) passed, it would have given the stock-raising industry vested, or prescriptive, rights on national forests. A temporary solution was negotiated for the 1926 grazing season. After a drought and temporary depression in the cattle market stressed the cattle industry, Arizona Senator Carl Hayden was able to persuade the Forest Service to suspend grazing fees for the following season. Late in 1926 the U.S. Department of Agriculture appointed Arizona rancher Dan D. Casement to review the proposed fee increases. Casement's report recommended incremental fee increases over a four-year period, adjustment of grazing fees to national livestock prices, and expenditure of fee income on range improvements. In 1927 the Department of Agriculture adopted his recommendations and announced incremental fee increases for the years 1928–31, not to exceed an overall increase of 45 percent, still well below commercial grazing leases. The Great Depression, however, intervened in this plan. In 1932 grazing fees, including those in the Peloncillo-Animas District, were reduced by 50 percent.[23]

The 1930s: Depression and Drought

The Stanfield hearings contributed to the polarization of livestock interests and proponents of more conservative stocking rates. Yet the

1930s became a period of new cooperation in range reform. The extreme drop in cattle prices that accompanied the Great Depression and the severe drought of 1933–34 increased public awareness of the plight of ranchers and brought national attention to the poor condition of ranges. In 1934 the Taylor Grazing Act extended Forest Service range management principles to the rest of the unapportioned public domain and incorporated many elements of both the Stanfield bill and the Casement report. The act adopted ten-year permits, tied grazing fees to current livestock prices, and provided for permittee participation in decision making.

In conjunction with the implementation of the Taylor Grazing Act, Forest Service personnel cooperated with local ranchers in gathering the remaining "wild," unbranded cattle and horses. During the worst period of the drought, officials shot all cattle too weak to walk to shipping pens on the railroad. As a result, in the years following the 1933–34 drought, stocking rates on the Peloncillo-Animas District and surrounding areas reached their lowest levels.[24] Between 1933 and 1942 the Civilian Conservation Corps (CCC), a national public relief program for unemployed men, provided labor for range improvements in the district. Enrollees from Camp F-10 (Camp Harold C. Riley) on Cave Creek in Arizona and from two temporary "fly" camps at Pratt and Clanton Draw in New Mexico built miles of drift and allotment fencing, installed water developments, riprapped eroded streambeds, worked on road crews, manned fire lookouts, and fought forest fires.[25]

The 1940s and 1950s: Privatization and Drought

After stocking rates gradually increased during the late 1930s, Forest Service personnel again attempted to implement reductions, now proposing that stocking rates should not exceed the numbers appropriate for dry years. After 1941 rangers began estimating grazing capacity in terms of animal unit months (the amount of forage required per month by a mature, one-thousand-pound cow or its equivalent) and writing grazing plans accordingly. Rangers monitored range condition, frequently using the Parker 3-Step Transect method, in which range examiners counted plants and their forage values along a one-hundred-foot transect line that began at a permanently marked monitoring plot. Allotment records for the Peloncillos from this period commonly contain photographs of monitoring plots and descriptions of the plants found along transect lines.

During World War II administrators attempted to avoid the "patriotic" overstocking of the previous war and recommended sale of cattle at advantageous wartime prices. After the war, Coronado personnel prepared the first forestwide range capacity estimate, which led them to recommend cuts in permitted numbers. The recommendation of thirty-one acres per animal unit resulted in another demand for the transfer of all nontimbered lands out of the Forest Service and into the Grazing Service of the Department of the Interior (which became the BLM). The State of Arizona Land Department supported this proposal.[26]

In 1948 the Forest Service did remove a portion of the Coronado National Forest from Forest Service status, setting what could have been a dangerous precedent. In September 1948 the Forest Service finalized a long-planned exchange of approximately 135,000 acres in several New Mexico national forests for approximately 82,000 acres of severely depleted, overcut, overgrazed timber land in the Zuni Mountains of New Mexico owned by the George E. Breece Lumber Company. The exchange lands in the national forest included all of the Animas division. Prior to the exchange, the Breece Lumber Company made arrangements to sell the 55,000 acres it would acquire in the Animas division to local ranchers, including the existing permittees, the Victorio Land and Cattle Company, Edgar Timberlake, and George Godfrey. Forest Service officials considered the acquisition of Breece's lands critically important for the protection of portions of the Rio Grande watershed. They justified privatization of the Animas division because it had little timber value and its entirely internal drainage into Animas and Playas dry lakes did not contribute to a river basin. According to one Forest Service administrator, the Animas division had been incorporated into the Peloncillo Reserve in 1906 because at the time forest reserve status was the only way to protect rangeland from the destructive competition of the disorganized, speculative open range cattle industry. Although the unequal valuations and acreages involved in the exchange generated some criticism, no organized opposition arose. The privatization of the 55,000-acre Animas division left the 88,000-acre Peloncillo division as the entire district.[27]

During the early 1950s, the Forest Service promoted "range renewal," encouraging permittees to cooperate in erosion control, remove brush and noxious weeds through herbiciding and grubbing, and construct cross-fencing for increased livestock rotation. The severe drought that devastated the cattle industry in southern Arizona and New Mexico during the mid-1950s brought about extreme reductions in carrying capacity and stocking rates. During the drought, many allottees in the Peloncillo District chose to take voluntary nonuse permits. Although stocking rates continued to decline throughout the 1950s, Peloncillo Range surveys reported that utilization continued to exceed capacity estimates, accelerating loss of forage productivity.

In conjunction with drought adjustments, rangers called for the removal of feral hogs from the Peloncillo District, just as the "wild" cattle had been removed in previous decades. The feral hogs were descended from pigs that homesteaders and the Victorio Land and Cattle Company had branded and turned out to fatten on the oak mast of the forest allotments. Rangers allowed sons of local ranchers to gather the hogs, some of which weighed five hundred pounds and had fierce tusks. In true Wild West rodeo style, the young cowboys roped and gathered approximately six hundred large hogs for sale at local livestock auctions to unwitting farmers, who later regretted their purchases.[28] Despite the efforts of the Forest Service, wild hogs still roam the Animas Valley and the foothills of the Peloncillos today.

Recreation and Conservation of Biodiversity, 1960–1997

By the 1960s, a new public perception of the purposes of the national forest system had begun to influence grazing management in the Peloncillo District. Supported by rapidly expanding groups of environmental activists, hikers, campers, bird-watchers, rock hounds, and naturalists, the New Mexico and Arizona outdoor recreation and conservation lobbies began to compete with grazing lobbies. Prior to this period, the borderlands region had received little biological investigation. From the early 1970s on, the number of scientific studies undertaken in the Peloncillos and the Animas Valley exploded. As the new scientific information gradually reached the general public, appreciation for the large number of unique species and the complex borderland ecology increased.

Biological investigations coincided with the passage of federal laws that affected national forest grazing policy on the Peloncillo District. The Multiple Use Sustained Yield Act of 1960 affirmed the Forest Service's philosophy of multiple resource use and imposed the principle of sustained yields. The Wilderness Act of 1964 provided for the creation of wilderness areas in which operation of mechanical equipment was prohibited. Although grazing under prior permit could continue, the act prohibited new road construction and severely limited grazing improvements. The National Environmental Policy Act (NEPA) of 1969 provided for administrative procedures to assure that new developments or extractive activities on national forests would not compromise environmental quality, including biodiversity. For new forest plans or projects, NEPA required the writing of an Environmental Assessment and/or an Environmental Impact Statement (EIS), documents that provided multiple alternative options with public commentary during the selection process. The Endangered Species Act (ESA) of 1973 provided for the identification and protection of endangered species of plants and animals. The Freedom of Information Act of 1974 gave the public access to all agency information. The National Historic Preservation Act of 1966 and the American Indian Religious Freedom Act of 1978 provided for protection of historic sites and access for Native Americans to sacred sites on national forests.

The 1974 Forest and Rangeland Renewable Resources Planning Act required a nationwide inventory of the potential productivity of all federal lands. The 1976 National Forest Management Act (NFMA) set requirements for more sophisticated and informed forestwide multiple-use plans with public appeal. The 1978 Public Rangelands Improvement Act (PRIA) required an inventory of rangeland conditions, established a formula for annual adjustment of grazing fees, and stipulated that management practices must benefit wildlife habitat, recreation, forage, and water and soil conservation. The PRIA fee formula used the fee base of $1.23 established by the 1966 Western Livestock Grazing Survey and took into consideration the Beef Cattle Price Index and the Forage Value Index. It limited increases or decreases to 25 percent of the previous year's fee.

The combined effect of these acts was to create greater public input into national forest policy, management practices, and regulation. The acts led to greater consideration of the multiple-use concept, particularly with regard to wildlife and water resources, and increased consultation between the Forest Service and other federal agencies (e.g., the Fish and Wildlife Service). But they made agency decision making more cumbersome and slowed management implementation. From the permittees' point of view, the new regulations increased their frustration with the Forest Service, since new range improvements on allotments without approved management plans were delayed.[29]

The 1975 Coronado National Forest plan placed greater emphasis on improved management practices than on stocking reduction. Nevertheless, administrators continued to cut permitted stocking rates to achieve a greater balance with the estimated carrying capacity. By 1980 administrators calculated stocking rates at fifty acres per animal unit. To avoid permanent cuts, administrators allowed Peloncillo permittees to bring actual stocking into compliance with estimated carrying capacity by taking voluntary nonuse permits of up to 15 to 20 percent. The 1986 Coronado National Forest management plan, the first written under guidelines established by the NFMA, rated the condition of 84 percent of the grazing land throughout the Coronado National Forest as "satisfactory or better." Nonetheless, the plan included a temporary reduction in permitted numbers from 400,000 animal unit months (1986) forestwide to 338,000 animal unit months forestwide. The plan anticipated that each allotment would be evaluated on its own merits and then adjusted accordingly. It stated that for the 16 percent of grazing ranges in unsatisfactory condition, ongoing efforts were being made to create a balance between forage production and permitted livestock through "range development, improved management, and reduction in numbers." Both the Draft EIS (1982) and the Final EIS (1986) envisioned that improvement would occur gradually over a fifty-year period. During this time, "intensive efforts would be required" to bring permits and capacity into balance. The 1986 plan stated, "As the range responds to improved management, coupled with decreased numbers, use will be allowed to follow the rising capacity." The forestwide temporary reduction on all rangeland (both satisfactory and unsatisfactory) was intended to create an eventual long-term increase in forage capacity, which would be followed by a return to the former higher stocking rates.[30]

In the 1986 Coronado National Forest management plan, two roadless areas in the Peloncillo Mountains, the joint Bunk Robinson/Baker Canyon/Guadalupe Canyon area (15,960 acres) and the Whitmire Canyon area (12,840 acres), came under consideration for wilderness status. Although the wilderness acts for both New Mexico (1980) and Arizona (1984) specifically sanctioned continued grazing where it existed prior to wilderness designation (with many allotments crossing boundaries between wilderness and nonwilderness areas), the plan did not recommend wilderness status for either area. Instead, both areas were designated Wilderness Study Areas, a status the Forest Service believed would provide greater flexibility in management of rare species (e.g., Gould's turkey and desert bighorn sheep), including habitat improvement through methods like prescribed burning. The plan excluded grazing only in "extra sensitive" riparian areas.[31]

During the commentary period for the 1986 plan, several respondents criticized grazing permits as taxpayer-supported subsidies and urged elimination of all grazing on the proposed Peloncillo Wilderness Study Areas. The Forest Service responded that if grazing constituted a subsidized use, then all activities on national forests, including bird watching, fuelwood harvesting, picnicking, and so on, were similarly subsidized, even though they were not extractive activities. In recognition of the area's large number of rare species, Earth First! recommended creation of the Geronimo Wilderness Area to include the entire Peloncillo Range and suggested closure of all roads, elimination of grazing, and reintroduction of extirpated species, including the grizzly, jaguarundi, jaguar, Mexican wolf, ocelot, and thick-billed parrot.[32]

During the 1990s, public awareness of the nonextractive values of sensitive areas in national forests increased pressure for more ecologically aware management. In the remote Peloncillo District, two responses to these pressures have had national consequences. In 1990, partly as a reaction to the increasing polarization of opinions concerning grazing, many permittees in the Peloncillo District participated in the formation of the Malpai Borderlands Group. This nonprofit organization of working ranchers is dedicated to preserving rangeland health and preventing landscape fragmentation on more than a million acres of ranch land in Hidalgo and Cochise Counties. The Peloncillo District is the center of its area of concern. The Malpai Group cooperates with federal and state agencies in coordinated ecosystem management, research projects, Geographic Information System mapping, comprehensive fire planning, brush removal, and native reseeding projects. The group also works with the Animas Foundation, the nonprofit "green-ranching" environmental organization that purchased the former Victorio Land and Cattle Company lands in the Animas Valley from the Nature Conservancy. Together the two organizations have pioneered several "grass bank" exchanges. In these exchanges ranch owners granted easements that forfeited the development rights on their private property to the Malpai Group as payment for leased forage on rangelands formerly owned by the Victorio. In 1996 and 1997 the Malpai Group facilitated prescriptive burns aimed at grassland restoration in the area of Baker and Guadalupe Canyons and below Maverick Springs. The prescriptive burns, the most recent of which covered almost 9,000 acres, required cooperation from many government agencies, including the Forest Service, the BLM, the Fish and Wildlife Service, and the Arizona and New Mexico state land departments and game agencies. Several Peloncillo permittees have voluntarily deferred grazing to allow more fuel to accumulate to support future controlled burns. The Malpai Group has assisted biologists, botanists, and the Forest Service in compiling a list of ninety-three species of concern in the Peloncillo District.[33]

Beginning in 1989, organizations concerned about species and habitat protection initiated a series of ongoing lawsuits to assure that national forest plans and amendments include environmental protection and that the specified protections are implemented. Designed to achieve preservation of riparian species and habitat, many of these lawsuits and appeals challenge failure to comply with procedural aspects of the NEPA. Lawsuits filed by the Southwest Center for Biological Diversity and Forest Guardians in 1996 and 1997 could jeopardize approximately half of the grazing leases on eleven national forests in the Southwest. These lawsuits have made it clear that compliance with habitat protection, threatened and endangered species protection, and clean water regulations presents significant challenges to the continuation of the type of grazing practiced in the past in the Peloncillo District and on all national forest land.[34]

Conclusion

Until the progressive departures of the past decade, the grazing history of the Peloncillo-Animas District has been typical of Forest Service grazing on the national level. With pressures from other extractive multiple-use demands largely absent, the Peloncillo-Animas District developed a pattern of grazing shortly after its creation that remained constant during the subsequent eight decades, although stocking rates steadily declined. The district experienced few significant conflicts, grazing remained the single major economic activity, and permittee continuity was remarkably stable. Three of the present permittees are descendants of the original allotment holders, and the remaining allotments have been transferred infrequently. Most of the permittees reside at ranch headquarters nearby, ride their forest allotments frequently, and provide careful attention to their livestock operations. Variations in range condition have not been as extreme as those in many other forest districts, a result of permittees' on-the-ground management and the presence of the nearby railroad, which provided a means for implementing an off-take or stock removal strategy during dry years.

Throughout the ninety-year history of the Peloncillo District, range examiners and administrators for the Coronado expressed concerns and attitudes consistent with those prevalent in the national forest system as a whole. Over the years, the series of programs and measures designed to improve forage have included stocking reduction and fencing; scientific management with rotation, deferral, and improved distribution; range renewal through brush removal and erosion control; and, most recently, removal of livestock from riparian areas and the restoration of a natural fire cycle to encourage growth of native vegetation. A persistent sense of optimism that current management measures would improve forage has pervaded both official reports and informal statements of district administrators.

During the initial phase of forest management, roughly two decades (1906–29), administrative reports for the district expressed an overriding concern with range improvement through reduction in stocking rates, fencing, livestock distribution, and prevention of trespass. During the period immediately preceding the establishment of the reserves, range conditions had reached their lowest point, a result of rampant overgrazing and the third of three disastrous droughts during the open range period. The initial phase of fencing and regulation happened to coincide with an extended period of high rainfall in southern Arizona and New Mexico. With rain and fencing, forest officials proudly observed a rapid improvement in range condition on the reserves. In 1910 the initiation of "scientific" range management policy also fell within the cycle of wet years. Officials credited scientific management for increased forage. Administrators continued the effort to improve range condition by reducing competition for forage through elimination of predators and rodents and decreasing stocking rates by means of transfer reductions. Optimism that improved management would allow permittees to return to higher stocking rates persisted, despite official acknowledgment that some ranchers were not stocking their allotments with as many animal units as their forest permits allowed.

Administrators dutifully attempted to serve the needs of opposing interests and enforce the democratic goals of equitable allotment, although they questioned the economic wisdom of dry farming and small-scale ranch operations. Despite this continued espousal of Progressive Era democratic goals, the single largest permittee in the district was the absentee-owned Victorio Land and Cattle Company.

During the next three decades (1930–60), the overriding concerns expressed in reports were improved forage production through range renewal, full utilization of available resources to prevent fire, more even livestock distribution, erosion control through construction of retention structures, deferred grazing to allow natural reseeding, and mechanical removal of undesirable plants to accommodate regeneration of native bunchgrasses. Despite these efforts, forage capacity continued to decline, erosion increased, and brush invaded. By the end of this period, managers were calling for further stocking reductions to bring stocking rates into balance with repeated downward adjustments of estimated carrying capacity. Yet despite the evidence of their own reports, range examiners clung to the optimistic vision that scientific management would restore rangeland. Throughout this period, the livestock lobby managed to maintain fees at a politically acceptable level and keep them well below market level.

During the most recent four-decade period (1960–99), a new awareness of the complexity, uniqueness, and limits of the Peloncillo grassland ecosystem developed. The focus of management shifted from production for extractive purposes toward preservation of biological resources. Range managers and allottees incorporated new scientific understandings into management, with a greater focus on habitat protection, wildlife corridors, and prevention of landscape fragmentation. These managers and allottees believe that sustainable grazing is not incompatible with biological diversity, that scientific management will eventually lead to restored rangeland, and that limited grazing will not prevent the use of the Peloncillo District as a "living laboratory," available to natural science investigators from around the world. The future will tell if this new optimism can be implemented with success.

NOTES

1. More information on the biological resources of the Peloncillo-Animas District can be found in *Biodiversity and Management of the Madrean Archipelago: The Sky Islands of the Southwestern United States and Northwestern Mexico,* U.S. Department of Agricul-

tural General Technical Report RM-GTR-264 (Fort Collins, Colo.: Rocky Mountain Forest and Range Experiment Station, September 1995).

2. John P. Wilson, *Islands in the Desert* (Albuquerque: University of New Mexico Press, 1995), 10. The name Peloncillo dates from the mid–nineteenth century, but during the Spanish period four segments of the mountain range bore separate names: from south to north, the Sierra Sarampeón, the Sierra de Nochebuena, the Sierra San Vicente, and the Sierra San Marcial. Apaches resided in encampments to which they returned on a seasonal basis. *Ranchería,* the Spanish name for indigenous seasonal camps, is used by borderlands scholars.

3. Diana Hadley, "Cattle and Drought, the Ecological Impacts of Late Nineteenth Century Grazing" (paper presented at the conference "Biodiversity and Management of the Madrean Archipelago," Tucson, 1993); Samuel P. Hays, *Conservation and the Gospel of Efficiency: The Progressive Conservation Movement 1890–1920* (New York: Athenaeum, 1959), 49, 471; Fred S. Breen, "The Forest Reserves as Seen at Close Range," *Forestry and Irrigation* 13 (1907): 180–83; Will C. Barnes, *The Story of the Range* (Washington, D.C., 1926), 36; and Hearings, Subcommittee of the Senate Committee on Public Lands and Surveys, 69th Cong., 1st sess., 1926, pt. 6.

4. Larry S. Allen, "Livestock and the Coronado National Forest," *Rangelands* 11, no. 1 (1989): 14–19; *Santa Fe New Mexican,* November 21, 1906; Sidney R. Albert, "History of the Douglas District, Coronado National Forest" (manuscript, Douglas Ranger District, Douglas, Ariz., 1967).

5. Allen, "Livestock and the Coronado National Forest," 15; William D. Rowley, *U.S. Forest Service Grazing and Rangelands: A History* (College Station: Texas A&M University Press, 1985), 58–59; *The Western Range,* written by the secretary of agriculture in response to *A Report on the Western Range—A Great but Neglected Natural Resource,* 74th Cong., 2d sess., 1936, S.R. 289, doc. 199, 257–61.

6. *Use of the National Forest Reserves* (Washington, D.C.: Government Printing Office, 1905), 20–21; Rowley, *U.S. Forest Service Grazing,* 72; Paul H. Roberts, *Hoof Prints on Forest Ranges: The Early Years of National Forest Range Administration* (San Antonio: Naylor, 1963), 100–101.

7. Albert F. Potter, "A Few Pointers on Making Estimates of the Grazing Capacity of Lands" (ca. 1908), Pinchot Papers, Box 2873, quoted in Rowley, *U.S. Forest Service Grazing,* 72; Frederick O. Knipe in Charles R. Ames, "A History of the Forest Service," *Smoke Signal* (Tucson Corral of the Westerners) 16 (Fall 1967): 134–36; Albert F. Potter, "How the Forest Service Has Helped the Stockmen," *American Forestry* 24 (1918): 165–69.

8. Rowley, *U.S. Forest Service Grazing,* 94; President Theodore Roosevelt to Secretary James Wilson, December 26, 1905, quoted in Roberts, *Hoof Prints,* 45–46. On March 10, 1921, a typical forest homestead advertisement appeared in the *Lordsburg Liberal,* announcing that 457 acres of land adjacent to the Cloverdale Ranger Station had been restored to entry. Ranger A. W. Piggott of the Cloverdale Ranger Station noted that the land was subirrigated, had several springs, and was in an ideal location for dry farming.

9. Final Reports, Forest Homestead Claims, Supervisor's Office, Coronado National Forest, Tucson. The Forest Homestead files contain over a hundred reports on applications for homesteads in the Peloncillo and Chiricahua Mountains.

10. Rowley, *U.S. Forest Service Grazing,* 81–83, 137; Robert R. Hill, "Grazing Administration of the National Forests in Arizona," *American Forestry* 19 (September 1913): 581.

11. U.S. Department of Agriculture, Forest Service, Homestead Claim, J. Henry Eicks, Homestead Entry 0919332, Las Cruces, N.M., report submitted March 7, 1925,

George E. Galer, Forest Ranger; Gertrude Gardner Eicks, personal communication, Douglas, Ariz., December 1996.

12. *Southwestern Stockman, Farmer and Feeder* (Phoenix), August 2, 1895, October 11, 1901; Barnes, *The Story of the Range*; Rowley, *U.S. Forest Service Grazing*, 90. The earliest fee schedules were one-third less than commercial prices for forage: cattle or horses, 20 to 35 cents per head for the summer grazing season and 35 to 50 cents per head for the whole year for cattle; sheep, 5 to 8 cents per head; and goats, 8 to 10 cents per head (yearlong).

13. *Proceedings of the Annual Meeting, Arizona Cattle Growers Association* (1907); *Southwestern Stockman, Farmer and Feeder* (Phoenix), March 5, 1909; *Proceedings of the Annual Meeting, Arizona Cattle Growers Association* (1914); Dwight B. Heard, "Public Grazing Lands: The Range Homestead," *American Review of Reviews* 42 (September 1915): 333–37.

14. Records of allowances, Chiricahua, permits for 1917, 1918, Supervisor's Office, Coronado National Forest, Tucson.

15. Rowley, *U.S. Forest Service Grazing*, 98–99; records of grazing allotments for the Peloncillo-Animas District, Coronado National Forest, Tucson, and Douglas Ranger District, Douglas, Ariz.

16. Records of grazing allotments; Larry S. Allen, personal communication, Supervisor's Office, Coronado National Forest, Tucson, October 1997.

17. Rowley, *U.S. Forest Service Grazing*, 99–101; records of grazing allotments (2210s and 2230s), including range reconnaissances and range surveys.

18. *A Report on the Western Range*, 259–61; Rowley, *U.S. Forest Service Grazing*, 77, 162; Dan Miles Gish, "An Historical Look at the Mexican Gray Wolf (*Canis lupus baileyi*), Early Arizona Territory and since Statehood" (manuscript, Supervisor's Office, Coronado National Forest, Tucson, n.d. [ca. 1977]); Gary Lee Nunley, "The Mexican Gray Wolf in New Mexico" (manuscript, Division of Animal Damage Control, Albuquerque, N.M., 1977); Bureau of Biological Survey, "Annual Report of Work Accomplished in the State of Arizona/New Mexico against Predatory Wild Animals by the United States Department of Agriculture, Bureau of Biological Survey, Arizona and New Mexico Livestock Sanitary Boards and Cooperating Stockmen," various reports; U.S. Department of Agriculture, Bureau of Biological Survey, Annual Reports, various years, Arizona and New Mexico Districts, Predatory Animal Control, various years. Both groups of records are in Record Group 22, Boxes 7–44, National Archives, Beltsville, Md.

19. C. K. Cooperrider and R. W. Hussey, Grazing Examiners, "Range Appraisal Report for the Coronado National Forest," Field Data by H. G. Calkins, Forest Supervisor, and Frank L. W. Grubb, Deputy Forest Supervisor, September 20, 1924, Supervisor's Office, Coronado National Forest, Tucson.

20. W. G. Koogler, "Reconnaissance Discussion Extensive, Range Surveys for the Coronado National Forest," April 4, 1930, Supervisor's Office, Coronado National Forest, Tucson.

21. *Proceedings of the Annual Meeting, Arizona Cattle Growers Association* (1920), 6; Kenneth B. Platt, "Public Land Policies in Relation to the Range Industry," Part 3: "Land Grants and Reservations," and Part 4: "Immediate Background of the Taylor Act" (manuscript, WPA Files, New Mexico State Archives, 1940); Rowley, *U.S. Forest Service Grazing*, 95; Roberts, *Hoof Prints*, 130–32; Barnes, *The Story of the Range*, 265.

22. U.S. Congress, Senate, Committee on Public Lands, Hearings before a Subcommittee of the Committee on Public Lands and Surveys, 79th Cong., 1st sess., 1926, S.R. 347, 371–429.

23. Dan D. Casement, "The Public Lands from a Cowman's Viewpoint," *American*

Forests 36, no. 1 (January 1930): 83–84; Rowley, *U.S. Forest Service Grazing*, 137–40.

24. George Pendleton, personal communication, Reserve, N.M., February 1997; Margaret García Dunagan, personal communication, Animas, N.M., July, August 1996.

25. Alison T. Otis, William D. Honey, Thomas C. Hogg, and Kimberley K. Lakin, *The Forest Service and the Civilian Conservation Corps: 1933–1942* (Washington, D.C.: U.S. Department of Agriculture, August 1986), 29–33, 155–75.

26. Allen, "Livestock and the Coronado National Forest," 18–19; Rowley, *U.S. Forest Service Grazing*, 201–5.

27. U.S. Department of Agriculture, Forest Service, Record Group 22, various boxes, National Archives, Beltsville, Md. These records include U.S. Department of Agriculture, Forest Service, Tract Valuation Report: Acquisition Cibola (Coronado), Coronado Selections, 55,510.98 acres (October 5, 1944); Forest Service R-3 Office Memorandum, January 3, 1945; U.S. Department of the Interior General Land Office, Decision and Conditional Approval for Forest Exchange Land for Land 2028365 "K" (Las Cruces), July 6, 1945; Edgar Timberlake to U.S. Forest Service, December 5, 1945; Acquisition, R-3, Cibola, Exchange George E. Breece Lumber Company (Consummated), September 7, 1948.

28. Allen, "Livestock and the Coronado National Forest," 18–19; Edward Elbrock, personal communication, Animas Valley, August 1997.

29. *Regulations for Implementing the Procedural Provisions of the National Environmental Policy Act*, Reprint 40 CRF, Parts 1500–1508 (as of July 1, 1986); Allen, personal communication. Although by the 1960s hunting and bird watching were more common recreational uses of the Peloncillo District, it is clear from early documents that recreation had not been a common use in the past. See Harrison Burrall, *Coronado National Forest Recreation Plan* (Albuquerque, N.M.: USDA, Forest Service, 1930), Supervisor's Office, Coronado National Forest, Tucson. The 1930 plan mentioned only one tourist camp and filling station (Project 60) for any area near the Peloncillos.

30. The *Proposed Coronado National Forest Plan* (May 1985) included BLM lands in the planning process and gave numerical land management classifications (1 through 14) to all portions of the Peloncillo District. See maps in the *Draft Environmental Impact Statement, Proposed Coronado National Forest Plan*, EIS 03-05-85-1 (Albuquerque, N.M.: USDA, Forest Service, May 1985). See also the two EIS documents: *Coronado National Forest, Draft Environmental Impact Statement Summary* (Albuquerque, N.M.: USDA, Forest Service, September 1982), 20–24; and *Summary of the Coronado National Forest Environmental Impact Statement and Forest Plan* (Albuquerque, N.M.: USDA, Forest Service, July 1986), 12–13, 16–17.

31. *Summary of the Coronado National Forest Environmental Impact Statement and Forest Plan*; Allen, personal communication.

32. *Public Comments and Forest Service Response to the DEIS, Proposed Coronado National Forest Plan* (Albuquerque, N.M.: USDA, Forest Service, July 1986), 1: 248–54.

33. Bill McDonald, "The Formation and History of the Malpai Borderlands Group," in *Biodiversity and Management of the Madrean Archipelago*, 483–86. The spelling of Malpai (*malpaís*, Spanish for rough, broken land with volcanic intrusions) is intentional, intended to reflect the American pronunciation of the word used throughout the Southwest.

34. Mike Taugher, "Ranchers Feel Heat of Greens," *Albuquerque Journal*, September 14, 1997; Keith Bagwell, "Clarification Sought on Leasing Ruling," *Arizona Daily Star*, August 3, 1997; Peter Galvin, personal communication, Southwest Center for Biological Diversity, Tucson, November 1997.

5

America's First Wilderness Area

Aldo Leopold, the Forest Service, and the Gila of New Mexico, 1924–1980

CHRISTOPHER J. HUGGARD

Lightning was igniting blazes throughout the Southwest during the dry and windy summer of 1922 when Aldo Leopold made his way to the Gila country of New Mexico. The assistant district forester in the Forest Service's District 3 noted the aroma of charred and burning timber as he approached one of the American West's most remote regions that hot June day.[1] For Leopold and his colleagues in the Forest Service, the destructive fires threatened the very core of that federal agency's being: timber as commodity. As a progressive conservationist, he fully accepted the Forest Service's doctrine of fire suppression. The young forester's anthropocentric concern for protection of natural resources was a motivating force at this early juncture in his career.

Ironically, Leopold was becoming a pioneer thinker on the cusp of a shifting conservation paradigm that would question the economic argument for protecting timber and range from fires and other natural forces. His ideas and actions in the Southwest in the 1920s, however, belied his future fame as the author of the "land ethic," an environmental philosophy that served as the core of an emerging biocentric voice in the post–World War II era. The history of the Gila (pronounced "heela") reveals Leopold's preservationist contribution during a period of his life when he did not fully understand the ecology of the natural world despite his intimate relationship with and growing understanding of the wilderness where he worked as a forester.

As chief of operations in the southwestern district, Leopold was visiting the Gila National Forest in 1922 to conduct a special inspection. Since 1920, as part of his job he had been inspecting District 3 forests to consider new management strategies and to evaluate personnel, fire control, roads and trails, recreation, public relations, range, timber, and watershed policies. A first-generation, Yale-trained forester, Leopold had been plying his trade in the Southwest since 1909 and was keenly aware of the issues surrounding timber, range, and wildlife management. A slender man of medium height, Leopold espoused scientific management of the Southwest's forests. His idealism often manifested itself as arrogance and alienation from the non–college trained foresters. Yet he instilled in them a healthy respect for the wilderness idea. Like his university-trained colleagues, Leopold accepted the new tenets of the science of forestry and combined them with his love of the land as a hunter-naturalist.

By the time of the Gila inspection, Leopold had already confronted numerous problems—overgrazing, conflagrations, erosion—that plagued the nation's southwestern forests. These issues, of course, concerned him on his journey into the Gila. Yet there was much more at stake on this special inspection. He was there to test his idea—the "Gila idea"—with the local citizenry for setting aside a wilderness area.

But the sea of flames in the Gila country in the late spring of 1922 forced Leopold to postpone the public discussion of his wilderness proposal. Instead, he and the other firefighters attacked the wildfires that were threatening expansive stands of ponderosa pine and Douglas and white fir intermixed with Engelmann spruce at elevations of from 6,000 to 10,500 feet; likewise, they fought to protect box elder, piñon pine, quaking aspen, Gambel oak, mountain mahogany, and a host of other species that thrived on the rolling terrain, in box canyons, and in river valleys from 5,000 to about 8,000 feet. While Forest Supervisor Fred Winn, his rangers, Navajo firefighters, ranch hands, and other local volunteers fought to suppress the tempestuous fires, Leopold coordinated the teams to insure that enough men arrived at the most threatening infernos with axes, shovels, mattocks, and rakes. With dogged persistence, the crews finally contained the fires on June 19.

The fires under control, Leopold headed to Silver City to test his wilderness area idea. During his visit to Silver City, he, Winn, and other Forest Service officers drew the boundary of the wilderness area to encompass 755,000 acres at the center of the Gila National Forest. This collective act marked a key moment in the national wilderness movement, because the Forest Service had now imposed on itself an informal wilderness policy.

Wilderness preservation also called for a reevaluation of the Forest Service's commodity-based management scheme. Just as former Chief Forester Gifford Pinchot's sustained-yield policy marked a shift from nineteenth-century laissez-faire exploitation of the national forests, forest managers were now initiating another shift from a conservationist to an environmentalist worldview later manifested in the Wilderness Act of 1964. In essence, the modernist paradigm of human separation from and mastery over nature was giving way, ever so subtly, to the realization that humans are part of nature.[2]

In his inspection report to District Forester Frank C. W. Pooler, Leopold warned that the Gila wilderness was unhealthy, making preservation essential. Except for some of the high elevation areas, erosion had devastated the hillsides and riparian zones. The forester blamed overgrazing by cattle and forest fires as the principal culprits for the flooding and erosion damage. Overgrazing, clear-cutting, and trampling of the vegetation by ranchers, loggers, and miners, in fact, had convinced President William McKinley to create the Gila River Forest Reserve in 1899 under the authorization of the Forest Reserve Act of 1891.[3] In 1905 the new reserve was renamed the Gila National Forest and placed in District 3 of the newly formed U.S. Forest Service, a branch of the Department of Agriculture.

Under Pinchot's leadership, the Forest Service became a prime example of progressive conservation designed to allow for efficient grazing, logging, mining, water-power development, and watershed protection. Despite this federally administered conservation mandate, however, the Forest Service had done a poor job of reducing destructive frontier practices on the more than 18 million acres of national forest in the Southwest.[4] Leopold wanted to reverse that trend, at least in a portion of the national forest soon to be known as the Gila Wilderness Area.

Leopold began to define wilderness and its potential role in American life in 1921. In a seminal essay published in the *Journal of Forestry,* Leopold wrote: "By 'wilderness' I mean a continuous stretch of country preserved in its natural state, open to lawful hunting and fishing, big enough to absorb a two weeks' pack trip, and kept devoid of roads, artificial trails, cottages, or other works of man."[5] Wilderness areas were to be set aside mainly for recreational use. He went on to argue that wilderness preserves would not threaten the market in national forest timber, stating that "such wilderness areas should occupy only a small fraction of the total National Forest area"; moreover, "only areas naturally difficult of ordinary industrial development should be chosen" for this use.

Later in the article, he further reassured the capitalists, stating that "no net economic loss would result from the policy of withholding further industrial development, except that timber would remain inaccessible and available only for limited local consumption." Grazing would not be prohibited but would be "an asset from the recreational standpoint because of the interest which attaches to cattle grazing operations under frontier conditions." Ranchers, in fact, would be protected from "new settlers" and from the "hordes of motorists who will invade the region the minute it is opened up" with roads and private resorts.[6] Leopold was remaining true to his utilitarian background by suggesting that only a portion of the national forests should be preserved as wilderness and that grazing did not threaten the integrity of that wilderness; the remaining forestlands should continue to be utilized as Pinchot had mandated: to produce "the greatest good to the greatest number for the longest time."[7]

The "highest use" of some national forest lands, although "a very varied use," Leopold argued, was recreational use. In an obvious response to recent National Park Service (NPS) strategies for development of national parks, he believed that "the old conflict between preservation and use" was reemerging. Looking into the future, he warned that vacation spots were being "spoiled by tourists." And even though democratic principles should determine natural resource management strategies, "it is our duty to vary our recreational development policy, in some places, to meet the needs and desires of the minority also. The majority undoubtedly want all the automobile roads, summer hotels, graded trails, and other modern conveniences. . . . But a very substantial minority, I think, want just the opposite."[8]

With this latter point, Leopold was treading on shaky political ground. The argument to set aside wilderness for "minority wants" provided ammunition for preservation opponents who for the remainder of the century would claim this was an elitist argument for wilderness and contrary to the Forest Service's democratic mandate. Probably realizing his mistake, Leopold limited use of this terminology in his future pleas for wilderness and, indicative of his evolving beliefs, called for a "poor man's right" to hunt in the wilderness as early as 1925.[9]

In 1922, then, Leopold's ideas for defining a wilderness area were still in the formative stage. District Forester Pooler, who accepted the report in October, other Forest Service officers, and state and local game protection associations believed Leopold was on to something important and lasting. Later that year Leopold realized that national forest and wilderness area policies were in the embryonic phase: "the only thing I am sure of is that policy is being written much faster than research can illumine the path, that huge values are at stake, and that there are neglected possibilities in personnel management that might be useful in telling us what to do until research forges abreast of the times."[10] Leopold viewed wilderness as something greater than the subliminal playground envisioned by nineteenth-century naturalists like Henry David Thoreau and John Muir. Rather, he believed that scientific examination via ecological inquiry would ultimately have to be applied to the wilderness to understand its biotic relationships. When Pooler approved the Gila Wilderness Area on June 3, 1924, the Forest Service set in motion a process that would not only preserve an untrammeled reserve but also set aside a recreational playground that would eventually serve as a grand natural laboratory for studying the impact of humans on wilderness areas in the United States.

In 1925 Leopold, although no longer working in the Southwest, clarified his ideas for wilderness and the policy tactics that should be taken to insure its protection in four published essays: "The Last Stand of Wilderness," "Wilderness as a Form of Land Use," "A Plea for Wilderness Hunting Grounds," and "Conserving the Covered Wagon."[11] In these think pieces, the maturing Leopold was fortifying his argument for preservation over development in wilderness areas, clarifying the importance of wilderness in building the American character, and justifying the public's need for recreation, especially hunting.

In addressing these first two themes, Leopold took on as well as borrowed from Frederick Jackson Turner's frontier thesis. He called for a reevaluation of the frontier mindset while at the same time revealing his faith in Turner's nationalist hypothesis. "The question . . . is whether the benefits of wilderness-conquest will extend to ultimate wilderness-elimination," he contended in the "Land Use" essay. Because of the "four centuries of wilderness-conquest the possibility of disappearance has been too remote to register in the national consciousness . . . . [W]e have no mental language in which to discuss the matter."[12] Leopold argued that American culture itself was the motivating force behind the conquest of wilderness, and, even though born of the wilderness experience, if allowed to go unchecked, that conquest would completely destroy the wildlands where that unique way of life originated.

In the "Last Stand" article, he postulated that the answer to this dilemma was within a collective consciousness, a realm that Anglo Americans had not yet properly tapped into: "we are *unconscious* of what the disappearance of wild places would mean, just as we are unconscious of what the disappearance of winds or sunsets would mean." By emphasizing the term *unconscious* in this quote, Leopold suggested two things in this sentence. The first is obvious; he feared losing what he believed made the nation great, the wilderness, because Americans were unaware of its importance due to their unfailing confidence in the "operation of economic laws." The second suggestion, however, is much more subtle and basic. By using a Freudian psychoanalytical term, the *unconscious*, a concept that had become part of the popular lexicon by the 1920s, Leopold implied that deep within the human psyche existed a real need to connect with the primordial self, a truth that transcends nationalism and economics.

His arguments in favor of hunting as perhaps the "highest use" of wilderness support this contention. Hunting as "sport" became for Leopold the next logical expression of wilderness values. "Physical combat between men and beasts . . . was economic fact, but when it disappeared as such, the instinct of the race led us to hunt and fish for sport. . . . Hunting for sport in its highest form is an improvement on hunting for food in that there has been added . . . an ethical code which the hunter formulates for himself and must often execute without the moral support of bystanders."[13]

Leopold concluded that hunting connected humanity to the primitive as well as contributing to continuity in the national character. "Public wilderness areas are essentially a means for allowing the more virile and primitive forms of outdoor recreation to survive the receding economic fact of pioneering. These forms should survive because they likewise are an improvement on pioneering itself." Such endeavors also demonstrated the important "impress of wilderness" on Americans and perpetuated "distinguishing marks" such as "individualism combined with ability to organize, a certain intellectual curiosity bent to practical ends, a lack of subservience to stiff social forms."[14] Leopold argued that to be an American, to engage wilderness, and to hunt required a masculine kind of relationship with the natural world. To this end, Leopold suggested at one point that the Gila Wilderness be named the "Gila National Hunting Ground."[15]

Leopold finalized his plea to save wilderness within a modified Turnerian framework. "In all the category of outdoor vocations and outdoor sports there is not one, save only the tilling of the soil, that bends and molds the human character like wilderness travel. Shall this fundamental instrument for building citizens be allowed to disappear from America, simply because we lack the vision to see its value?" Leopold also interpreted the preservation of wilderness as the logical extension of the frontier process: "it will be everyone's fault if they do not serve as a warning against delaying the immediate inauguration of a comprehensive system of wilderness areas in the West, where there is still a relatively unimpeded field for action." In this classic restatement of the frontier thesis, Leopold inserted the words "unimpeded field" to serve as a metaphor for Turner's "free land," while also attaching regional identity to wilderness preservation as a process that takes place "in the West." His use of "relatively" before "unimpeded field" also reveals his fear that few wildlands remained because the frontier had closed or was in the process of closing.[16]

In his call to preserve wilderness, Leopold was getting at the heart of Americanism as most Americans viewed it then and now: democracy and individualism are born of the land-human relationship. And the best way to perpetuate that symbiotic relationship, he pleaded, was to "act at once." Democratic action or "public demand" through "a definite expression of public opinion" was needed to salvage the wilderness origins of democracy. Leopold was also moving away from his 1921 argument for minority rights and calling on a larger public to demand the salvation of wilderness. The insightful forester was attempting to broaden the base of support for wilderness: "I believe that both the Forest Service and the Park Service would be receptive to the wilderness idea, but neither can be expected to execute it with the vigor and despatch necessary to save the situation, unless they can point to a definite crystallized public demand for such action."[17] He was appealing to Forest Service and Park Service officials, hunters, backpackers, and motor tourists alike to unite in the fight to preserve wilderness areas in national forests *and* national parks.

By calling for special interest advocacy, Leopold was also turning the Turner thesis on its head. Rather than the human-wilderness relationship giving birth to democracy as the frontier theory demanded, the forester was pleading for democracy to spawn the preservation of wilderness. This redefinition of the democratic process, whether Leopold knew it or not, was ironic and, once put to use, greatly contributed to the passage of an avalanche of environmental protection laws in the United States in the generations after his death in 1948.

For Leopold, the Gila Wilderness Area was the starting point for a national wilderness policy. The "last roadless area of any size in the Southwest containing all the best types of mountain wild life and scenery," the Gila was the "logical location for a wilderness playground." This southwestern treasure was unique even in the American West. In "A Plea for Wilderness Hunting Grounds," Leopold explained why five of six candidate wilderness areas were inferior to the Gila. The White Mountain–Blue Range country, the "cream of the Southwest," for example, was spoiled by the Clifton-Springerville Road. The Jemez Division of the Santa Fe National Forest, also a "splendid wilderness," had too much "valuable timber and range" to justify preservation status. "Spanish-American villagers" had long-standing settlement and grazing rights in the Pecos National Forest, and the Tonto Basin in Arizona was also too peopled. These eliminations left the Gila and the Kaibab forests of the Grand Canyon country as the finalists.

In the end, the Gila won out because the overpopulation of deer in the Kaibab was forcing the Forest Service to construct numerous access roads for hunters there, a no-no for purists. Perhaps the Almighty had intervened: "The Creator must have foreseen" that the Gila was "a precious remnant of the old frontier. . . . Every point where roads might enter is set a rugged mountain."[18] Here, Leopold was restating the manifest destiny argument for American expansion. Rather than serving as a call to conquer new peoples and new lands, however, his nationalist plea was being used to justify preservation of a peopleless land.[19]

If it had been a decade earlier, Leopold went on, any of the six "big wilderness areas" could have been chosen for special designation. And even though he favored the Gila for candidacy as America's first wilderness area, Leopold realized the larger importance of preserving wildlands within a regional as well as a global context. First, Leopold revealed his regional bias when he chose only western reserves as candidates for wilderness area status: "Why should not the Gila area, and other similar areas, if possible one in each western state, be declared permanently roadless." In other words, he wanted to showcase the exceptional terrain in the West as region to justify preservation of wilderness as a symbol celebrating the closed frontier.

Second, Leopold also meant for the Gila Wilderness Area to be a beacon to the world as a "democratic" hunting ground. The use of wilderness to hunt was a poor man's right, he contended, and not just the privilege of big game hunters, or elitists, who could afford to pay for extravagant hunts to Africa, Siberia, Alaska, and British Columbia. From the Southwest's Mogollon Mountains to the northern Rockies' Jackson Hole, all Americans should have the right to track deer, bears, and other game that "do not always 'mix well' with settlement."[20]

Leopold's contribution to the history of wilderness was set in stone once the Forest Service established the Gila Wilderness Area in 1924. Perhaps his greatest achievement as a forester and philosopher, America's first wilderness area would come to symbolize to the rest of the world the nation's real desire to preserve its heritage as well as its unique wildlands. Likewise, his pleas for wilderness preservation marked an ideological shift from the Old West paradigm of frontier process to the New West paradigm of region. And even though the Gila would experience environmental hardships, it would remain intact for future generations to interpret and reinterpret as part of the evolving national psyche and the science of ecology and as a measure of humanity's role in the natural world.

The Formulation of Forest Service Wilderness Policy

The Forest Service's 1924 Gila Wilderness Area "Recreational Working Plan" initiated what would become a fairly consistent policy guideline

for wilderness areas until passage of the Wilderness Act forty years later. The plan recognized the special "primitive" characteristics of the Gila and its recreational values while also spelling out restrictions to protect these conditions: limitations on the construction of both roads and buildings.[21] The plan also recognized a new use—wilderness recreation—in the national forests.

Leopold understood, however, the weaknesses of the Forest Service's administrative designation of wilderness. The "Recreational Working Plan," for instance, was to expire in ten years, logging could be initiated at any moment, and permits could be awarded for construction of buildings and private roads. At the mercy of Forest Service policy, the Gila Wilderness Area could be reclassified to national forest status at the whim of a capricious district forester. And even though there was growing support in the Forest Service for wilderness areas, there was

This 1994 panoramic view of Watson Mountain on the southern border of the Gila Wilderness Area illustrates the piñon-juniper forest landscape in the area. The mythic Gila River is in the foreground. Courtesy Michael Berman

no guarantee of protection. Leopold made his feelings known in a 1926 letter to Forest Service officers, including Chief Forester William B. Greeley: "I find . . . among foresters outside the Service a strong belief that the mere embodiment of wilderness areas in Forest Service plans would not be a sufficient protection under changing personnel." Therefore, he continued, a "bill to authorize the president to permanently suspend the operation of the land laws in areas particularly valuable for recreational purposes" would be desirable.[22] "The one thing needful is for the Government to draw a line around each one and say: 'This is wilderness, and wilderness it shall remain.'"[23]

Forest Service officials responded favorably to the Gila idea. Despite the traditional commodity-based mandate of the agency, there was ample support among foresters for wilderness areas to protect certain "virgin" or "untouched" lands. District Forester Pooler, for example, wrote a prowilderness letter in February 1926 to Congressman Franklin W. Reed, chair of the Joint Committee on Recreational Survey of Federal Lands, in support of the bill authorizing the president to exempt recreational lands from federal land laws. "The whole argument for wilderness areas as such is based upon the theory that somewhere in the recreation plans in a Nation-wide sense, provision should be made for the small minority of recreationists who will be satisfied with nothing less than wilderness country which in the Southwest means rough, pack country with good hunting and fishing." The "ease-loving people," he continued, already had opportunities for motorized recreation on Forest Service and Park Service lands. Pooler then reiterated Leopold's reasons for closing the area to logging and development, stating that wilderness preservation "can not, in my opinion, be justified by mathematical [economic] calculations but on the recognition of the rights of minorities . . . . [W]ilderness areas should . . . be selected with a view to quality rather than quantity."[24]

Greeley, who was chief forester from 1920 to 1928, responded positively to the establishment of the Gila Wilderness Area, at the same time critiquing as well as modifying Pooler's ideas, in a letter to the district forester on December 30, 1926. "The facts and opinions available warrant the conclusion that there are conditions under which a body of land may yield the highest service if retained as nearly as practicable in a state of unmodified nature."

Like Leopold before him, Greeley was redefining Pinchot's commodity-based definition of "highest use," which had been the catchphrase for efficient exploitation of the nation's forests prior to the new plea for wilderness areas. As part of that redefinition, Greeley also argued that the new designation "may make important contributions to science, to education and to the spiritual stimulus and physical up building of the people, which surpass the benefits from complete utilization of material resources or from other forms of recreation."[25] Greeley was building on Leopold's list of nationalist and recreationist reasons for wilderness preservation by recognizing the "scientific" as well as the "subliminal" ("spiritual stimulus") contributions of wilderness to society.

Greeley did not enthusiastically share Pooler's identification with minority rights, however. "I have no sympathy for the viewpoint that *people* should be kept out of wilderness areas in any large number because the presence of human beings destroys the wilderness aspect.... [W]e should encourage public use of wilderness areas just as freely as in any other portion of the National Forests." Yet in the same letter, Greeley recommended regulations that would restrict access by "withholding of Forest roads not required for protection" and by "issuing no permits for recreation structures of any character within" areas "having wilderness values."[26] These recommendations suggest that Greeley believed that "minorities" (i.e., outdoor enthusiasts and hunters) would be the primary users of wilderness areas, but to state so openly would be a poor public relations move that would carelessly expose the Forest Service to criticism for protecting elitist rights and contradicting the democratic mandate of the federal agency.

Wildlife Management in the Gila

A close analysis of the management of the Gila Wilderness Area from the 1920s through the 1970s reveals that, if there were elitists who used the wilderness, they were ranchers and hunters. Backpackers and bird-watchers certainly entered the mix, but, in contrast to today, they played a secondary role in the 1920s. Protection of domestic cattle and deer, in fact, became a driving force that informed Forest Service policies in the Gila and in other wilderness areas in the United States. As Leopold's writings and the Forest Service's policies demonstrate, ranchers from the Diamond Bar, N-Bar, and other outfits believed they had a hereditary right to grazing permits in the Gila. The New Mexico Cattle Growers Association worked closely with the Forest Service to determine the number and sizes of grazing allotments.

At the same time, ranchers, predator hunters, and forest rangers cooperated to protect deer from fires, mountain lions, and starvation. In many ways, Leopold's pro–deer hunter stand in the 1920s illustrates a personal denial about the ecological health of the grass and woodlands and the symbiotic relationship of predators to prey. His early Boone-and-Crockett idealization of deer stood in the way of a more comprehensive and ecologically sound management strategy. And even though he would begin to realize in the 1930s that his earlier beliefs were wrongheaded, his ideas fueled an ecological crisis in the Gila Wilderness Area.[27]

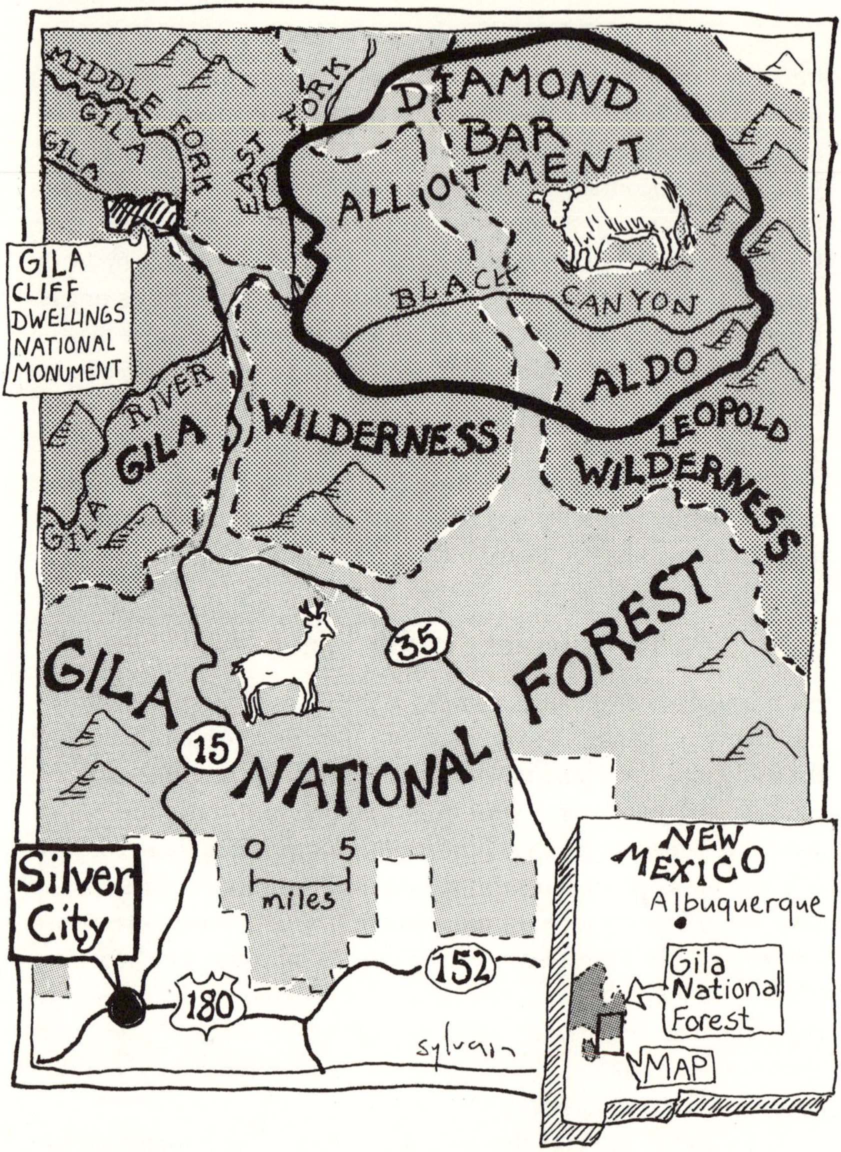

A *High Country News* map of the Gila and Aldo Leopold Wilderness Areas, Gila National Forest, New Mexico, drawn in 1994, slightly revised in 1999. Note the Diamond Bar allotment for grazing cattle, the Gila Cliff Dwellings National Monument (the site of mid-thirteenth-century Anasazi cliff house ruins), and Silver City, headquarters for the U.S. Forest Service's Gila National Forest. Courtesy Diane Sylvain and *High Country News*

From the beginning of his forestry career, Leopold showed an interest in "game management," a term he would redefine as "wildlife management" later in his life. During an eighteen-month illness that struck him down in 1913–14, he became deeply involved in game management issues. Soon thereafter he joined the New Mexico Game Protective Association, heading what many hunters believed to be the model game management program in the nation. As secretary of the association, chosen spokesperson, and political advocate, Leopold formulated strategies designed to optimize hunting conditions in the Gila and in other southwestern forests.[28]

The eradication of predators headlined Leopold's game protection plan. As early as the 1910s, he had worked with ranchers and other foresters to exterminate the Southwest's wolf population. By 1920, more than 270 had been destroyed, leaving fewer than thirty of the canine hunters in New Mexico. Likewise, Leopold labeled mountain lions a nuisance. In 1920 he made his views known at the National Game Conference in New York: "It is going to take patience and money to catch [for extermination] the last wolf or lion in New Mexico."[29] This pervasive attitude led to the destruction of millions of animals in the American West during the twentieth century.[30]

The ranch culture of southwestern New Mexico clearly supported Leopold's stand on predator control. A well-known figure in New Mexico by this time, Leopold won the confidence of Hispanic and Anglo ranchers in the southwestern part of the state. His mother-in-law had inherited the N-Bar Ranch, which was managed by Eduardo Otero, Leopold's wife's half-brother, further tying the forester to the ranching community.[31] Extermination of wolves, mountain lions, grizzlies, and coyotes was an accepted part of ranch culture and even became part of "the things that legends are made of."

Locals still tell fanciful stories of heroic tracking feats by legendary hunters such as Ben Lilly and "Bear" Moore. One of J. Frank Dobie's mythologized frontier characters, Lilly made his way to New Mexico via Louisiana, Texas, and Mexico in 1911 when in his midfifties. Already considered a member of a long line of great American hunters that included Daniel Boone and Teddy Roosevelt, he symbolized civilization's conquest of the wilderness despite spending most of his life in that same wilderness. A teetotaler who was said to have never hunted on Sundays so that he could read the Bible, Lilly lived in the Gila forests with his trusty bloodhounds. Perhaps the foremost authority on the habits and demeanor of bears, mountain lions, and other wildlife, Lilly served as a consultant to the U.S. Biological Survey, which had initiated a predator extermination program in the Southwest as early as 1915.

Ironically, one of Lilly's favorite hunting tales, "Ben Lilly's Longest Grizzly Chase," which he often told in his retiring years in the 1930s, was the time he had to break the Sabbath to track an unusually smart grizzly bear who had evaded him and his dogs for more than two days.[32] When Sunday morning rolled around, the old hunter initially decided to stay in camp. But after reading a portion of the Gospel, he changed his mind and continued stalking his biblical nemesis: "Ye shall keep my sabbaths, and reverence my sanctuary," he was said to have quoted. "I am the Lord. And I will give peace in the land, and ye shall lie down, and none shall make you afraid; and I will rid evil beasts out of the land."

Whether fate intervened or Lilly purposely read that passage was not questioned by commentators. The result was a bear fight that cost Lilly two of his beloved hounds and nearly his life. After finally wounding the bear, Lilly lost track of him, when suddenly the beast "sneaked out of the thicket and come back to hunt the hunter." When the grizzly lunged from above in a narrow canyon, Lilly was able to fire another round into the dying animal, who "knocked the hunter sprawling" some twenty feet with a "glancing blow from one massive paw." As one devotee remembered, "In listening to Ben's brief recital of the hunt, one got the feeling that the old hunter genuinely regretted killing the mighty grizzly."[33]

Aldo Leopold certainly would have understood Lilly's mixed feelings about killing the grizzly. His own famous story when he witnessed "a fierce green fire dying" in a mortally wounded she-wolf's eyes captures the same image of a man who cared for wildlife. Despite these spiritual pangs, however, both men agreed that the predators must be exterminated even though each of them realized the primordial connection between themselves and the animals. This latter innate instinct was suppressed in favor of the protection of domestic livestock and a hunter's paradise filled with mule and white-tailed deer.

The account of a mountain lion hunt in the Gila Wilderness illustrates the finality and less legendary results of killing natural predators. Clell and Dale Lee, hired guns with the Biological Survey, tracked and killed two mountains lions in March 1929. "Cold-trailed for three days," a female and male found themselves trapped between a pack of dogs and a precipice with a "sheer drop of about a thousand feet." The male temporarily escaped with dogs on his trail as the female was being attacked by another set of hounds. After a fierce fight with the dogs, the wounded female was able to climb a tree, but no more than ten feet above the ground. Hunter Clell Lee then attempted to rope the angry lioness. He was "intent on capturing the beast alive." But because of her size and tenacity, she was given "one shot from the hunter's automatic rifle and the fight was over."[34] The male met a similar fate later in the day.

The Lees justified the killings, as well as twenty-one other mountain lion slayings in 1928–29, because the predators were themselves hunting the "harmless" herbivores of the Gila Wilderness Area. Claiming that the lions killed two animals a week, Clell argued that most "frequently . . . the lion will kill young cattle" over pronghorn antelope and mule deer. Even though Leopold was beginning to rethink the logic of predator extermination by the time of these kills, it was too late to preserve the balance between predator and prey that ecologists were suggesting was needed in forest ecosystems.[35]

In the late 1920s wilderness managers' primary concern shifted from predator to deer population control. Eradication of predators and fire suppression had had the combined effect of causing both an incredible increase in deer as well as a slowing in the recovery of grasses and browse, which served as natural fodder. This cause-and-effect relationship had two devastating impacts: overgrazing of the grasses and overbrowsing of the trees, which, in turn, led to starvation of the deer as well as erosion problems. This Malthusian result reflected Leopold's ill-conceived strategy for protecting deer. For Leopold, fire also had to be "exterminated."

In "Wild Followers of the Forest" (1923), Leopold defined fire as "the enemy of the wild." Fire's immediate impact was to "surround and destroy grown animals and birds and kill them outright." Furthermore, he continued, "the greatest damage occurs through the destruction of eggs and young," who fall "easy prey to vermin and hunters." Perhaps more devastating, "fire also causes important disturbances among the food plants on which the game is dependent."

Underlying Leopold's philosophy in suppressing fires was an early commodity-based belief in "producing" deer through "game management" rather than working to create an ecosystem characterized by wildlife diversity and interdependency. Not realizing the potentially positive nutritional effects of fire on grasslands, Leopold incorrectly assumed that fires destroyed range and, thus, reduced the carrying capacity of the national forests. Because the goal of wilderness range management was to optimize conditions for optimal "game production," Leopold recommended that "deer [should] increase up to the carrying capacity of the yards." As a result, range for deer should be protected and, if possible, be allowed to expand to increase grazing areas within the forests. "When fire cut the capacity of the yards by half," as Leopold mistakenly assumed, "it likewise cut the number of deer." America's foremost game manager wholeheartedly believed in the mid-1920s that "Forest Fires Rob Game of Food."[36]

Indicative of his misreading of fire, Leopold argued that the Apaches had set fires not to create parklike conditions for grazers but to force deer into the smoke, where they could be easily killed. Leopold took E. W. Nelson, chief of the Biological Survey in the 1920s, at his word when the biologist claimed that deer congregated in smoke to seek "refuge . . . from the torment of flies."[37] If this was the Apaches' strategy, the theory did not take into account the broader reasons for setting fires: to improve game habitat, facilitate travel, reduce insects and pests, and deny potential enemies cover. Furthermore, the Indians realized that postfire landscapes produced luscious grasses, herbs, and berry shrubs for browse, a management strategy that created the parklike conditions that Euro-Americans had attributed to natural processes, not Apache burning.[38]

Not yet aware of the dynamics of ecosystem relationships, Leopold was devising a system of game management based on the same principles as timber management. "The fundamental thing in both game and forestry is that we, the public, must grasp the idea of *production*. If conservation does not mean production, then it means nothing at all." With this argument, Leopold was remaining faithful to his utilitarian training. Yet in the discussion that followed, he introduced principles that hint at a revolutionary wildlife management strategy. "I do not mean production merely in the limited industrial sense of boards, meat, dollars, or even tourists. I mean that attitude of mind which will not tolerate the existence of idle, lifeless land in America." Leopold believed that a "certain moral and esthetic competence" must inform wilderness management, with one of the primary goals being the production of diverse species "on the same forest lands."[39] Two of the chief components of Leopold's "land ethic" were surfacing: an ecological understanding of the natural world as a diverse community of biota and the formulation of an ethical code of human conduct toward the environment. "There must exist in the public mind that fundamental respect for living things and that fundamental aversion to unjustifiable killing and to unnecessary ugliness which in all lands and all times has been a necessary foundation for good morals and good taste." As wilderness scholar Max Oeschlaeger contends, Leopold came to the realization that "nature's economy" was intricately woven into the "human economy."[40]

No matter how forward-looking Leopold could be in his philosophical writings, the impact of his beliefs had a mixed result in the Gila. By 1927 the state game warden and Forest Service officials were deeply concerned about the skyrocketing deer population. If Thomas Malthus's ticking time bomb had not yet been detonated on humanity, it had exploded on the hoofed beasts of the Gila: deer, cattle, and other herbivores were eating themselves out of house and home.

Even though the New Mexico Game Commission held jurisdiction over game management in the state, the Forest Service, the state Game Commission, and the Biological Survey played significant roles in the evolving wildlife strategies. Leopold, J. Stokely Ligon of the Biological Survey, and Elliott Barker of the Game Commission all agreed that the Gila, like the Kaibab of northern Arizona and other national forests, was suffering from an overpopulation of deer. In the fall of 1927 deer counts showed thirty-eight to forty-three animals per square mile at a time when Leopold believed ten per square mile was excessive. Even more telling was the impact of the overpopulation: decimation of grasses, unhealthy fawns, and a depleted supply of mountain mahogany and live oak, important browse for deer and cattle. The desperate situation for the deer was apparent when foresters noticed that the ungulates were eating alligator juniper and piñon pine, a sure sign of extreme conditions.

When Ligon recommended opening the Gila Wilderness Area to five hundred hunters in the fall of 1927, Leopold agreed that thinning the herd was a possible solution. This marked a change in his thinking at a time when public opinion strongly favored leaving the deer alone. The massive starvation of deer that was occurring throughout the West, however, forced the Forest Service and state game commissions to act, or the forests themselves would be destroyed.[41]

As the deterioration of the condition of the range and forestlands in the Gila became obvious, game officials decided to award more than sixteen hundred deer-hunting licenses in 1927. Most of the hunters traversed the Black Canyon country, while others made their way to the upper reaches of the Gila River. The result was 629 deer bagged. This kill hardly made a dent in the population. Reports, in fact, suggested that numbers continued to escalate in 1928, reaching 27,045, an estimated increase of 31 percent over the previous year. In response, the Game Commission handed out more than 2,100 deer licenses in 1928, which resulted in 938 kills.[42]

Because of his idealized view of deer, Leopold himself had not yet grasped the dynamics of deer and cattle feeding. For example, he misread the proliferation of pine and juniper thickets as a positive sign of ideal grazing conditions when in fact their invading presence forced out more nutritious forage. He further erred by blaming only cattle overgrazing rather than deer browsing for this negative impact. Likewise, he was not aware of the importance of fires, especially lightly burning ones, in releasing surges of nitrogen into the soil, which in turn stimulated growth of grasses and shrubs for browse.[43] On the other hand, he did recognize that the absence of predators, which was the result of his own thinking, had caused an imbalance in species diversity.

Faced with deteriorating rangeland conditions, the Forest Service appointed a special investigative committee in 1929 to examine the deer problem in the Gila wilderness. Chaired by C. E. Rachford, chief of range management in the Forest Service, the committee recommended a management strategy for forest and forage, the assignment of a game biologist, the development of a scientific method of counting deer, and a study of the effects of predators and disease. A longer hunting season with increased bag limits (two deer per hunter) and the construction of more roads were also suggested. Leopold, who would publish his path-breaking textbook *Game Management* a few years later, concurred with these recommendations with the exception of the call to build roads.[44]

During the 1930s, the number of hunters in the Gila annually exceeded 3,200 despite pleas by local game associations to limit the number of does killed. Sportsmen did not realize that only twenty years earlier deer had been scarce. Now their numbers were growing exponentially. Increased hunting licenses had little impact despite 2,335 kills in 1931. The range felt the pinch. Indicative of the deteriorating situation was the location of grazing and browsing of some of the deer. During the winter, when most deer descended into the valleys, a large

Overgrazed Montoya Pasture in the Diamond Creek floodplain in the Gila Wilderness Area, 1994. Ranchers have traditionally grazed far too many cattle in the Gila, causing flooding, destruction of native grasses and trees, and other ecological nightmares for America's first wilderness area. Courtesy Michael Berman

portion of the Gila herd was still grazing in the high elevations of the Black Range, suggesting that little forage was left below the timber line.[45] Game managers did not acknowledge that wolves, lions, and grizzly bears thinned deer populations more effectively than human hunters.

The early game management strategies, however revolutionary for the time, were failing and would continue to plague the Forest Service for many years to come. In 1936 browse and forage conditions were worse than they had been in 1931. Despite this rangeland deterioration, the state stepped up predator control to exacerbate, rather than alleviate, an already chronic problem. The acute impact of overgrazing, which went unchallenged until the 1960s, would eventually highlight the late-twentieth-century battles between ranchers, the Forest Service, and environmentalists over cattle, deer, and wolves.[46]

Forest Service Regulations for the Wilderness

The increase in the number of hunters pointed to a growing controversy over road building in the wilderness. Hunters wanted easier access but were denied it because of Leopold's roadless criteria for the Gila Wilderness Area. In 1931, however, District Forester Pooler reversed an earlier decision to cease all road construction and improvements. The Forest Service opted to improve the North Star Road to open up the Gila to deer hunters. Leopold and Fred Winn, Gila National Forest supervisor, petitioned to keep the road closed except when protective measures were necessary, but they lost the appeal. The implications of the extension of the North Star Road would be momentous from the preservationists' standpoint. Leopold himself later lamented the fact in the original introduction to *A Sand County Almanac* (1948): "Here my sin against the wolves caught up with me. The Forest Service, in the name of range conservation, ordered the construction of a new road splitting my wilderness in two, so that hunters might have access to the top-heavy herd. I was helpless. . . . I was hoist on my own petard."[47] In retrospect, the "father" of wildlife management realized that extermination of predators had been a chief cause of the Malthusian effect on the Gila deer population, and the alleviation of this problem required roads.

Roads became a primary focal point for administrators of the Gila Wilderness Area from 1930 until passage of the Wilderness Act of 1964. Construction of the North Star Road, in fact, had compromised Leopold's roadless criteria for a wilderness area. As an important component of an emerging "purity doctrine," roadlessness became the measure of "real" wilderness for proponents of the Gila idea. In addition, the Forest Service viewed the roadless criteria as a way to curb Park Service claims to scenic wilderness spots. In response to this perceived threat, the Forest Service established the L-20 regulations in 1929, which redesignated most wilderness areas as "primitive areas."

The formulation of the new regulations was done internally. During the late 1920s, Leon F. Kneipp, assistant chief forester, had conducted an inventory of roadless areas in national forests. The report determined that there were seventy-four roadless areas with acreages of 230,000 or more for a total of 55 million acres, or about one-third of all national forest lands, that met the criteria for wilderness area status. Inherent in Kneipp's report, however, was a strong prowilderness bias, because in 1927 Assistant Chief Forester Morton Cheney had defined wilderness as an area of 500,000 acres or more. Kneipp reduced the wilderness designation to 230,000 acres. There was an agency-wide agreement that smaller designations would protect Forest Service lands from the outstretched arm of the Park Service, which itself had begun to preserve "primitive" areas, adding confusion to public lands management. In the end, wilderness values and the perception that the Park Service coveted Forest Service lands inspired the formulation of the L-20 regulations.[48]

Ideally, the new regulations would have protected wilderness and primitive areas from all encroachment. In the end, however, the purist criteria gave way to political and commercial pressures. Despite strong support for protection from Leopold, Kneipp, and Chief Forester Greeley, the greatest-good-for-the-greatest-number ideology of Forest Service utilitarians won the battle. Concerned with forging viable commercial communities in the national forests, most foresters supported the traditional mandates of the Forest Service. Primitive areas, in fact, were subject to virtually all the exploitative practices of the agency; the 1929 timber inventory of the Gila Wilderness Area suggests just such a threat.[49]

The limitations of the primitive area regulations became obvious during the process of reclassifying the Gila Wilderness Area as the Gila Primitive Area. In January 1930 Gila National Forest officers in Silver City completed the first draft of the "Report on Gila Primitive Area." The proposal, which served as the model for primitive area designations throughout the West, called for the Gila Primitive Area to cover an expanse of 695,296 acres. By this time, Forest Service officials had already reduced the original 755,000-acre Gila Wilderness Area, and more reductions were to come.[50]

Assistant Chief Forester Kneipp responded positively to the first draft of the proposal. He concurred with wilderness advocates that the primitive area designation was important for the nation, mainly because the Gila Wilderness Area had no long-lasting administrative protection. Many Americans feared the impact of the loss of "primitive conditions," which offered "educational, inspirational, and recreational values." To protect this national treasure, then, Kneipp recommended that the Forest Service limit development. "We are trying," Kneipp stated later in the letter, "to preserve a small cross section of the Old West."[51]

For Kneipp, protection of wilderness was a matter of clarification of cans and cannots in primitive areas: "the future administration of wilderness areas should be specific and, in large degree, prohibitive; that is, there preferably should be a series of unequivocal statements as to what will not be done, so that there can be no innocent infiltration of new conditions through which primitive values of the area eventually will be destroyed." For example, he proposed to Pooler to close logging at least until 1940. Since enough timber was available from non-wilderness-area lands in the Gila National Forest to meet anticipated logging industry needs for the next twenty-five years, he wanted greater protection while timber was still abundant.[52]

Kneipp, in essence, was formulating the purity doctrine that would serve as a Forest Service guide for countering both Park Service and environmentalist initiatives in the future. This evolving preservationist liturgy argued for strict limitations on human-made structures, especially roads, while also advocating the termination of logging in wilderness or, in this case, primitive areas. The Izaak Walton League, a national fishermen's association, agreed with this wilderness management strategy. "It would seem folly," the Silver City chapter argued in a news release, "to ignore a long list of automobile casualties on crooked and dangerous local roads, and spend taxpayers [*sic*] money on roads into a wild area that is much better kept primitive. . . . [Let's preserve] this last outpost of American Frontier Wilderness."[53]

Consistent with the Gila Wilderness Area recreational plan, the new regulations permitted grazing; 8,229 cattle and horses annually and 18,000 sheep during summer would be allowed to forage. Mining was also allowed under the provision that the operators had to prove their ventures to be economically viable; in essence, the Mining Law of 1872, which permitted mine development on nearly all of the public domain, could, in some instances, dictate primitive area policy.[54]

The utilitarians had won the day. The Gila Primitive Area, despite the new regulations, was still subject to the whims of the Forest Service's economic exigencies, which had driven national forest policy since 1905. Symptomatic of this philosophy was Assistant Chief Forester Morton Cheney's objection to a statement in the primitive area proposal that referred to range in the Gila as "fair to poor" and "par with the average of other Forests." Rather than initiate a discussion of this persistent problem, Cheney told James A. Scott, supervisor of the Gila National Forest, to mark out the phrase because "steps are being taken to remedy the situation" and "this would be an argument for the defense if the Gila were being accused of bad administration of grazing." He continued, "It does not appear necessary to take a 'crack' at the other Forests."[55] Such sensitivity to criticism suggests that at all levels Forest Service officers realized the devastating effects of grazing, but they were unwilling to accept responsibility for that endemic destruction.

The Gila was not immune to these threats either. The preservationists, who opposed cutting timber and constructing roads in wilderness areas, soon experienced disappointment over Forest Service management of America's first wilderness area. Gila National Forest officials themselves admitted to the ambiguous status of wilderness years later: "reports of the period [1930s] imply 'wilderness' to be only a temporary land use until timbering or something else became more economical."[56]

The realization that primitive areas were still subject to utilitarian exploitation inspired wilderness advocates to fight for still stricter preservationist regulations within the Forest Service. As a result, in 1939 the Forest Service published the famous U regulations from a draft written by Robert Marshall, founder of the Wilderness Society and head of the Forest Service's Division of Recreation and Land Uses in the mid-1930s. Designed to further secure national forest lands from Park Service reclassification, the new rules called for three wildland designations: U-1 wilderness, U-2 wild, and U-3 recreation. The new policy stipulated that U-1 wilderness had to be at least 100,000 acres and U-2 wildlands from 5,000 to 100,000 acres. The secretary of agriculture was authorized to create the units and make decisions concerning proposed boundary changes. For U-1 and U-2 classifications, the new restrictions prohibited road building, logging, motorized transportation, and special permit occupancy. Permits would continue to be awarded for grazing, water storage projects, and private road construction. The U-3 recreation classification, although rarely implemented, permitted logging.[57]

Under the U regulations, more than a million acres of wilderness and wild areas had been established in District 3 by 1940, with almost 700,000 more acres classified under L-20, making the Southwest the key wilderness region in the nation.[58] The new regulations also set the Forest Service apart from the Park Service in that the former recognized the multiple uses of the national forests (versus the "single use" of tourism in national parks), suggesting a more democratic administrative strategy.[59]

Local Protection of the Gila Wilderness Area

In the early 1950s, when wilderness advocates from all the national conservation organizations were mobilizing to block the Echo Park Dam in Colorado, the Forest Service proposed reducing the size of America's first wilderness area. The result was an anomaly. Local citizens overwhelmingly opposed the Forest Service's plan to reduce the Gila Wilderness Area from 563,000 to 375,000 acres. Initially, locals as well as Wilderness Society President Howard Zahniser approved of the reclassification. As early as 1951, Zahniser and others had wanted to eliminate newly tarnished boundary areas that were now rutted with four-wheel two-tracks. By early 1952, however, the conservationists had become concerned when they discovered that more than just roads threatened reductions. The Forest Service, in fact, proposed to open the Iron Creek Mesa area in the north to logging, a "proposal [that] may have been the whole undoing of the entire package."[60]

Local and national conservationists were incensed at the Forest Service's proposal to extract the Iron Creek Mesa forests from protected status. Herein lies an interesting twist to the history of the Gila Wilderness. Aldo Leopold's call for immediate action in support of wilderness preservation in the 1920s was finally coming to fruition in the early 1950s. Clearly, the impact of 1930s New Deal conservation programs and 1940s environmental destruction due to the war had galvanized a popular conservationism in the 1950s. At a time when affluence and wastefulness characterized postwar life, a regionwide movement to protect the natural resources of the Southwest was taking hold in a way never before seen in New Mexico.[61]

The manifestation of local support for wilderness protection was the so-called James Committee. Chaired by New Mexico Western College President H. W. James, the committee worked with conservation groups as well as with the regional forester to block logging on Iron Creek Mesa. Two subcommittees went into the field to study the forest in July 1952 and came to the conclusion that logging "would do too much harm." Subcommittee chair Marvin Fuller also lamented the destructive practices of four-wheeling hunters and suggested that the Forest Service build fences and post signs to deter further vehicular destruction. Most of all, Fuller contended, elimination of the area's protected status would set a "dangerous precedent": "In another 50 years, this wilderness could be a great thing in the southwest. If we set a precedent by allowing it to be cut now, there is a danger of constant chipping away at it in the future until we have nothing left."[62]

In the end, the Forest Service relented. Wilderness advocates did not allow the agency to use the purity doctrine as justification for the elimination of more than 100,000 acres. In accordance with Undersecretary of Agriculture Charles Brannan's 1953 order, the Gila was designated as a U-1 wilderness area to encompass 438,626 acres with an additional 137,388 acres on its eastern side still under L-20 administration as the Gila Primitive Area. Ironically, even though the Gila and Black Range Primitive Areas were still subject to utilitarian management principles, their total acreage plus that of the Gila Wilderness Area nearly equaled the acreage Leopold had first proposed to set aside in 1924 (758,000 to 755,000).[63]

Wilderness advocates were elated. Writing in the Wilderness Society's quarterly publication, the *Living Wilderness,* Weldon Heald called the Gila achievement a "smashing victory for wilderness preservation. . . . Like the embattled farmers of Lexington and Concord these fighting New Mexicans were inspired by a principle, and they fired heavy artillery heard round the conservation world from Bangor to San Diego."[64] National support for wilderness, Heald pointed out, was a given. What was surprising to him and others was the "almost unanimous local opposition that eventually reached the proportions of a crusade. 'Round Silver City for months before the hearing gas station attendants, school teachers, lawyers, judges, old timers, newcomers, and everybody else earnestly discussed the Gila Wilderness, and apparently not one had a good word to say for the Forest Service proposals."[65]

U.S. Senator Clinton P. Anderson of New Mexico concurred. "I rejoice," Anderson stated in the same issue of the *Living Wilderness,* "that this fine wilderness area was created by the hand of God and preserved by these pioneers, and if I had my way I would lay it down as the law that what God and these splendid men working jointly have brought about, let no man put asunder." For Anderson, a former secretary of agriculture, the Gila wilderness symbolized the new love affair the American public was having with natural areas, parks, and historic monuments. Joe Penfold of the Izaak Walton League topped off the celebration, calling the preservation of the Gila Wilderness Area "an important victory for the democratic principles of people getting together to solve mutual problems." Forest Service officials later wondered aloud whether the 1952 fight over the Gila was the jump start to congressional action: "Was this the beginning of the Wilderness Act?"[66]

At the public hearing on August 7, 1952, the James Committee, Forest Service officials, and Gila landowners agreed that the Copperas Canyon Road would be extended to the Gila Cliff Dwellings National Monument. Not until 1956, however, when President Eisenhower's highway program passed Congress and federal funds became available for such projects, did the road extension become an issue. Purists certainly were not pleased that more lands would be withdrawn from wilderness status to make room for the road. Further complicating matters was the Forest Service's initial refusal in 1956 to support construction of the road. Arguing from a purist's point of view, Regional Forester Fred Kennedy claimed that the Forest Service could not construct a road under the provisions of L-20. Under these regulations, the Forest Service could construct roads only in case of administrative needs such as fire control.[67]

Learning of this reversal, locals mobilized against the apparent betrayal of their 1952 agreement. Gila landowner Clyde Ely fired off letters to Senator Anderson, Howard Zahniser, and Forest Service officials while simultaneously excoriating the agency for "reneging" in editorials in his own newspaper, the *Silver City Daily Press*. Ely claimed that private landowners had a right to have reasonable access to their homes while at the same time providing a service to visitors of the Gila Cliff Dwellings National Monument. Anderson, Zahniser, Penfold, and others, including C. R. Gutermuth of the Wildlife Management Institute, Kenneth B. Pomeroy of the American Forestry Association, and Richard H. Stroud of the Sport Fishing Institute, each concurred that the road should be constructed as soon as possible.[68]

An Early Call for Ecosystem Management of Wilderness

Despite the general public's growing appreciation of the Mogollon peoples who had built the Gila cliff dwellings in the late thirteenth century, wilderness purists still lamented the elimination of more lands (7,800 acres) to build an access road to the ruins. The chipping away continued in the face of widespread opposition in 1952. In fact, a closer look at the supporters of the Gila reveals an interesting commercial element to the story. Landowners hoped to increase the value of their holdings, while others, such as the Chamber of Commerce, service clubs, the American Legion, and women's clubs, hoped the new highway would attract large numbers of paying tourists. Hunters and ranchers hoped to protect the production of their respective commodities, deer and cattle. Preservation of wilderness was closely linked to commercial exploitation of nature whether indirectly through New West tourism or directly through Old West harvesting of deer and grazing of cattle.

This realization was not lost on biologist Richard E. Warner, who wrote "The Dismembered Wilderness" for *Pacific Discovery* in 1962. In this essay, Warner articulated the ecologist-preservationist viewpoint through a historical evaluation of the Gila. The biologist not only attacked the "dismemberment" of America's first wilderness area, but he also criticized the Forest Service for its questionable wildlife and fire suppression policies. Uppermost in Warner's mind was the need for congressional protection of wilderness to put an end to eliminations so that foresters could focus specifically on ecological management strategies not subject to the whims of the market economy.

Warner's criticisms of Forest Service wildlife and fire management strategies in the Gila symbolize a dramatic shift in the conservation-environmentalism paradigm. Grounded in ecology, the young biologist questioned the overstocking of the Gila rangelands with livestock and the suppression of fire in the forests. More grass eaters and more efficient fire suppression would not fix a damaged ecosystem. By applying accepted ecological principles, Warner logically concluded that it was necessary to reverse Forest Service policies formulated during Leopold's generation. In essence, Warner was first among the generation of wilderness managers—wildlife biologists, soil scientists, hydrologists, and forest engineers—who countered the mature Leopold's land ethic principles, to "think like a mountain," against the utilitarian postulates he formulated while a young forester.

Warner believed that the best way to heal the ailing Gila was to restore and maintain an ecological equilibrium between the grazers and the carrying capacity of the range. Consequently, when the New Mexico Department of Game and Fish reintroduced 254 elk from 1954 to 1957, he contended that their growing numbers (500–1,000 by 1962) would contribute to an already problematic Malthusian effect on deer, especially with the continued "removal of large predators." Meant to replace the extinct Merriam elk, the Rocky Mountain elk symbolized nothing more than the "production of game for the hunter," who frequently broke the roadless criteria to encroach deeper into the wilderness. In short, Warner argued that the mistreated Gila rangelands could not afford to have yet another commercial herbivore eating their diminishing sources of forage.[69] The elk proved far more threatening to cattle because of the diseases they carried, and, like timber wolves, ranchers called for their eradication in the 1980s and 1990s.

What Warner may not have realized in 1962 was that he was not taking into account that deer and cattle were deeply woven into the idea of wilderness, the former symbolizing the first contact of "frontiersmen" with nature, the latter representing civilization's conquest of nature. As historian Paul Hirt suggests, "a conspiracy of optimism" had crept into postwar Forest Service policy that not only allowed continued overgrazing to take place in wilderness areas but prompted "intensive management," which included massive cutting of timber and widespread development of recreational facilities in the national forests. The Forest Service's removal of junipers, reseeding of native and nonnative grasses, and limits on numbers of cattle and sheep on the Gila rangelands also reveals the agency's faith in its ability to "manage" overgrazed grasslands, just as replanting ponderosa pine on forestlands shows a strong Forest Service belief that, despite calls to protect old growth forests, tree "farms" could be grown to support the frenzied postwar construction industry.[70] As Warner astutely noted, faith in technology and science masked the long-term destruction of the vegetative landscape of the Gila country.

Warner also took the Forest Service to task for its fire control policies. The previous fifty years of fire suppression, he contended, had restricted the natural burning process. Since records had been kept, fully 95 percent of Gila fires had been sparked by lightning. The visible result of this "continual firing" could be seen in portions of the Gila least affected by fire suppression: "trees spaced some distance apart and ground cover" in grass and scattered scrub. This naturally determined outcome "kept accumulations of fuels to a minimum" and, therefore, produced cooler fires which "inflicted only minor damage on the trees."[71]

This ideal ecological balance, however, did not exist in the rest of the Gila. During the previous fifty or more years, fire suppression had allowed vast areas of the forest floors to be overgrown with fuels, and "grassy meadows" had been replaced by "stunted pine trees," "increased scrub and brush," and "growth of thick, tangled stands of stunted 'doghair' pine in areas formerly of large ponderosa pine in open, park-like stands." The heat from this buildup of fuel, once torched, could be devastating, as was the case between 1951 and 1956 when 80,000 acres of prime Douglas fir and ponderosa pine burned. Warner's assessment was correct. Since the 1940s, the Southwest region had been the most successful at fire suppression in the Forest Service because of the implementation of new techniques such as bulldozed fire breaks and a public relations campaign headed by Smokey the Bear.[72]

Warner concluded his discussion by warning the Forest Service not to allow its longtime fire suppression policy to compromise wilderness values. A case in point, the biologist claimed, was the agency's recent construction of an airstrip to fly in firefighters, or "Hot Shots." Proposed in 1955 as a 2,700-foot runway at McKenna Park, the airfield met with fierce local and national opposition but was finally constructed in 1959 with the excuse that "the airstrip was absolutely essential for fire control and administrative purposes."

The Gila and the Wilderness Act of 1964

For Warner, legislative protection was needed to eliminate these kinds of encroachments and to allow for ecological management, or "we can resign ourselves to the gradual but inevitable melting away of our remaining wilderness resources."[73] Once a clearly defined legal definition could be established, the science of understanding the biotic relationships in wilderness areas could be pursued with less interference. In essence, a biocentric voice was needed to fully appreciate and interpret the natural world. That voice could best be interpreted through the principles of foundational ecology, which called for perpetual examination of the nature-culture relationship, instead of the tenets of imperial ecology, which allowed for utilitarian exploitation of nature.[74] As fate would have it, wilderness managers were able to take a major step in that direction only two years later, when Congress passed the Wilderness Act of 1964.

The fight for passage of the Wilderness Act began in 1955, when Howard Zahniser introduced the first draft of the bill in Washington, D.C. Zahniser, the son of an evangelical Free Methodist minister, was a conservation veteran who had zealously fought to protect wilderness since the early 1930s. By the early 1950s, he had gained national prominence during the battle to protect Dinosaur National Monument from the proposed development of the Echo Park Dam. In the bill he called for a biotic communalism. Even though humans lived outside the wilderness, in reality they were part of a symbiotic relationship born of "profound humility, to recognize one's littleness, to sense dependence and interdependence, indebtedness and responsibility . . . as members of a great community of life."[75]

If Zahniser's thinking galvanized the national wilderness movement in 1955, it also served as a rallying point for opponents. In New Mexico in particular, miners, loggers, ranchers, and water power advocates began to formulate the antiwilderness ideology of "sagebrushers" and "wise users" who for the remainder of the century called upon western states to take over federal lands. They would have their say as well, because Senator Clinton P. Anderson, who in the 1950s served on the Committee on Interior and Insular Affairs before chairing that body in the 1960s, called for hearings in his home state to initiate a public discussion of the issues.

Despite blocking early passage of the bill, Anderson supported congressional protection of the Gila and other wilderness areas in the West. Yet he realized that support in the Southwest was mixed. Ranchers and sheepherders, for example, feared loss of traditional grazing rights, arguing that wilderness "locked up" rangelands. Ranchers were also willing to misrepresent the bill to bolster opposition. At a meeting of the Catron County Farm and Livestock Bureau in 1959, for example, speakers claimed that the bill would "extend New Mexico's Gila Wilderness without limit." These future sagebrushers also misled their fellow ranchers when they claimed that the bill put an end to grazing and predator control programs in the Gila.[76]

Even stronger opposition to the wilderness bill surfaced in the New Mexico State Engineer's Office. In November 1958 State Engineer Steve Reynolds headed a long list of wilderness opponents at an Albuquerque field hearing chaired by Senator Anderson. Anderson opened the meeting with a call for testimony on the potential advantages and disadvantages of wilderness designation even though he made it clear that he supported establishment of wilderness areas on "some of our public lands." After hearing various arguments for and against wilderness areas, Anderson called on Reynolds. The state water specialist asserted that wilderness designations would "lock up" New Mexico's natural resources and reduce tourism to satisfy a few elites. He also claimed that congressional protection of the Gila wilderness would doom the proposed Hooker Dam. As part of the Colorado River Storage Project, the reservoir would provide water for farmers, ranchers, and the mining industry, sacrificing only one small valley in the southern part of the wilderness area. And even though the dam would never be built because of the presence of rare birds such as the elf owl, Wied's crested flycatcher, and the Mexican hawk, water developers were still fighting for a dam on the Gila River in the late 1990s.

Despite regional opposition to the bill, the nation as a whole strongly supported congressional protection of wilderness areas. A growing number of American urbanites began joining environmental groups in unprecedented numbers; they viewed wilderness in primarily recuperative terms as a place to hike, view beautiful scenery, and relieve the stresses of modern life.[77] Congress also recognized this growing use by passing the Multiple Use Sustained Yield Act of 1960. In addition to the traditional national forest uses (logging, grazing, and mining), the new law gave "outdoor recreation" and "wildlife and fish purposes" equal billing.[78]

Congressional recognition of these two newly mandated uses clearly demonstrates the shifting emphasis from utilitarian conservationism to preservationist environmentalism, while at the same time fueling the conspiracy of optimism that gave wilderness managers a false sense of security about their ability to balance these sometimes conflicting uses. The Forest Service was now caught between an emerging environmental movement and an entrenched resource-exploiting group. Indicative of the growing influence of environmentalists were the dramatic membership increases in national organizations from tens of thousands in the 1950s to millions by the late 1960s.[79] At the same time, the nation's first citizen validated the movement by becoming an advocate himself.

When President John F. Kennedy took office in early 1961, he made it clear to his advisors, Anderson among them, that he planned to make environmental protection one of the principal thrusts of his administration. One of his chief goals for the New Frontier environmentalism was a wilderness protection act.[80]

Over the next three years, the wilderness bill was discussed in the national arena at nineteen different public hearings. Prowilderness testimony, much of which was coordinated and molded by Anderson, strongly supported congressional review of wilderness. As Anderson so adeptly made the public aware, the wilderness bill represented something quite unique in conservation history. He even suggested that the protection of wilderness was a counterbalance to the recent prodevelopment stand of the Park Service. Rather than protect its unique natural areas from encroachment, Chief Conrad Wirth's Mission 66 program called for the commercialization of some of America's choice scenic spots, and, according to Chief Forester Richard McArdle, the wilderness law would "tie his [Wirth's] hands in building roads and other developments needed in the national parks."[81] Purists in the Forest Service, in fact, were lamenting the recent development of the Gila Cliff Dwellings National Monument, which had attracted more than 25,000 visitors by 1965, thus threatening the integrity of the Gila Wilderness Area.[82]

In the meantime, virtually all the national environmental organizations were busily lobbying Congress in support of the bill. Howard Zahniser, David Brower, Rachel Carson, Joe Penfold, Stewart Udall, and the whole range of wilderness advocates mobilized on an unprecedented scale. Regional chapters of the Wilderness Society, the Sierra Club, the Izaak Walton League, the League of Women Voters, and grassroots supporters sent an avalanche of mail to their congressional delegations. The result was spectacular: the so-called Conservation Congress passed four pivotal environmental acts in September 1964, including the Wilderness Act, the Land and Water Conservation Act, the Canyonlands National Park Act, and the Public Land Law Review Act.[83] With the first of these acts, wilderness became the greatest symbol of a new biocentric voice, an emerging ecological realization of the language of nature that was destined to be interpreted and reinterpreted for generations to come, a place, the act proclaimed, "where the earth and its community of life are untrammeled by man, where man himself is a visitor who does not remain."

The new law provided for the first time congressional protection of wilderness areas in the public domain under the Wilderness Preservation System. Those areas already designated under the U regulations were automatically reclassified as wilderness areas. Uses were not as limited as purists would have liked, since grazing would continue indefinitely, prospecting was allowed until December 31, 1983, and the president could approve hydroelectric projects. On the other hand, logging was no longer permitted, and the commodity versus preservation dichotomy that had entangled the Forest Service with users was further complicated because the new law allowed commercial backpacking, white-water rafting, skiing, and other recreational uses.

One of the surprise positive effects of the wilderness law centered on the process of designating new entries. Environmentalists had hoped to incorporate a provision that would give the president the authority to singlehandedly designate new wilderness areas. This measure, however, was blocked by Representative Wayne Aspinall of Colorado, who had chaired the House's Interior Committee during the wilderness debate, because he feared an "imperial presidency." In the long run, this lost battle proved to be only a temporary defeat, because Congress, rather than the president, had to confirm new wilderness areas, which requires a long, very public process that usually captures the media's and the nation's attention. In hindsight, environmentalists have come to appreciate this highly deliberative and publicly visible democratic process.

Leopold would have been proud. Americans had appealed to their highest representative body, Congress, to achieve protection for the very wilderness that so many believed had been instrumental in the development of the nation's democratic heritage. In an ironic twist of history, however, no longer did the human-wilderness relationship starkly culminate in democracy, as Turner had argued. Rather, environmental advocacy and action formulated and redefined the democratic process, and wilderness protection was the ultimate result. In the years to come, the maturation of the democratization of America's national forests would also culminate in legal recognition of Hispanic and Native American claims to the national forests of the Southwest.

Like many paradoxes in U.S. history, wilderness still symbolized American exceptionalism both because of the path that led to its protection and because those preserved lands now became a beacon of democratic environmentalism to the rest of the world. At the same time, wilderness areas became just another type of anthropomorphized landscape, like cityscapes, farmlands, national parks, Indian reservations, air force bases, and other western landscapes, that society had platted on its checkerboard grid of the earth's surface. Yet, contrary to historian William Cronon's contention that wilderness is merely a human invention, these protected wildlands were a manifestation of the contextual "idea of wilderness" (a human construction) with the natural "reality of wilderness" (a natural construction), special places that Leopold himself claimed could never be created but only preserved.[84]

The Aldo Leopold Wilderness Area

Aldo Leopold had become the embodiment of the wilderness achievement in the twentieth century. He was not only responsible for the preservation of America's first wilderness area, but he also authored the land ethic. Consequently, when a provision in the wilderness law provided for review of all primitive areas before 1974 to determine whether they were eligible for wilderness status, it was fitting in 1970 that the Forest Service propose redesignating the Black Range Primitive Area as the Aldo Leopold Wilderness Area. A flood of support poured into the District 3 and Gila National Forest offices.

Earth First! founder Dave Foreman headed the environmental delegation, which included Wilderness Society and Sierra Club representatives, that supported establishment of the Leopold wilderness. He circulated memoranda to numerous chapters of national environmental groups and among grassroots organizers. The democratic winning of wilderness, he believed, was a methodical, well-planned process that, even though it required years of sustained mobilization, testimony, and lobbying, would result in a victory for wilderness and for American democracy. In a "confidential" memorandum of February 17, 1975, for instance, Foreman and his colleague Debbie Sease devised a four-point strategy to convince Congress of the necessity for the Leopold wilderness. The organizers needed to "discuss strategy" in order to remain on the same page. They must treat the Gila wilderness, even though it would have two designations (i.e., Gila and Aldo Leopold), as "one wilderness." Probably most important was maintaining an awareness of events in the wilderness areas themselves through field trips. And lastly, they needed to provide direct input into the formulation of the Gila Wilderness Area management plan currently being drafted by the Forest Service.[85]

The growing influence of the environmentalists was obvious. No longer was Forest Service policy the only window through which wilderness could be viewed. Rather, dozens of environmental groups, with millions of members mobilizing in every western state to influence their elected officials, formulated wilderness policy. Letter-writing and telephone campaigns, use of scientific expertise at public hearings, mobilization of millions of private citizens, and careful use of the legal system assured that a biocentric voice would be heard. And by the late 1970s, environmentalists had become both the principal watchdogs of and most vocal public participants in Forest Service policy and actions.

Ultimately, the hard work and commitment paid off for wilderness advocates; by 1983 New Mexico alone had seventeen wilderness areas comprising more than 1.4 million acres, or about 10 percent of the nation's 14.6 million acres (a figure that would exceed 33 million acres by 1990).[86] The Aldo Leopold Wilderness Area, which was first put before Congress in 1975, included more than 200,000 acres when President Jimmy Carter finally signed it into law in 1980. Wilderness had become synonymous with the Southwest.

In some ways, the Gila Wilderness Area, with its attached Aldo Leopold Wilderness Area, had come full circle. The acreage of the two legislatively protected areas totaled 759,889. This was 4,889 acres larger than Leopold's original 1924 designation. Former Chief Forester McArdle claimed that the Forest Service deserved credit for this "big wilderness" because of a "strong continuity of policy and purpose" since the 1920s. There were other continuities as well. Despite controlled reductions in numbers of herbivores, ranchers and sheepherders still grazed too many cattle and sheep, and deer and elk were still subject to Malthusian pressures. Hunters ranged far and wide slaying predators despite pleas to reintroduce wolves, mountain lions, and grizzly bears. Grazing and hunting remained Old West practices in an emerging New West milieu; the image of the southwestern cowboy and frontier deerslayer endured to blur the idea and reality of wilderness, while also reflecting the Forest Service's increasing willingness to satisfy its multiple users—resource extractors as well as resource protectors.

Forest Service managers clearly accepted the difficult task that faced them, realizing that new pressures would emerge, especially as visitation continued to rise from the few thousand who came annually in the 1930s to the more than 30,000 who came in the 1970s; national forest visitation on the whole, in fact, had increased from 18 million visitors in 1946 to 100 million in 1965 to 250 million in 1985.[87] Likewise, the perpetuation of the original mission of the Forest Service—to implement wise-use exploitation of natural resources—could not be completely reversed, as the example of cattle as "commodities" and deer as "products" suggests. The Forest Service was caught between two competing wilderness philosophies: the utilitarian and the preservationist. Because of the agency's commitment to the ideals of democratic use of forest resources, its officers would find themselves being vilified by ranchers and other extractors for restricting commodity uses and by

environmentalists for allowing just such uses. Justifying management plans and actions in an increasingly hostile environment to the multitude of wilderness users was going to be the challenge of the future.

In the final analysis, the Gila Wilderness Area has come to represent many things to many different people. Former Chief Forester McArdle claimed in 1974 that "wilderness has been an elusive term. It means whatever the user wants it to mean."[88] District 3 Regional Forester M. J. Hassell explained his definition of wilderness in 1979 to a backpacker who complained about the ruggedness of the Gila's trails: "Part of the purpose of wilderness is to allow natural processes to operate. In all of our literature on the subject of wilderness, we stress that visitors should be prepared to accept wilderness on its own terms."[89]

Cattle grazing on the Diamond Bar allotment in the Aldo Leopold Wilderness Area, Gila National Forest, New Mexico, 1994. The Wilderness Act of 1964 established wilderness areas throughout the United States, especially in the American Southwest. The federal law, however, did not eliminate destructive grazing practices. Cattle, deer, elk, and horses, most of which have been placed in the wilderness by humans, have ravaged the range and riparian zones (rivers and their floodplains). Note the narrowleaf cottonwood, ponderosa pine, and other species of southwestern trees. Courtesy Michael Berman

Despite this ambiguity, Leopold's first call for wilderness to preserve national identity, to provide recreational services, and to serve as an ecological laboratory has remained intact. In his book *The Gila Wilderness* (1988), for example, English professor and wilderness lover John A. Murray argues that these same uses make the Gila an enduring symbol

of the reality of American nature. “The wilderness experience, both in its egalitarian spirit and democratic opportunity, is so much a part of our history as to almost be synonymous with it. We are, quite literally, a frontier society, and our history is largely a history of the frontier.” But more than nationalist character building, Murray postulates, the wilderness provides a multitude of opportunities for recreation. Furthermore, “science and technology have not, as some originally feared, made wilderness obsolete but have on the contrary given new significance to our relationship with nature and have informed that partnership with an understanding unprecedented in the history of the human race.” This primordial connection to nature combined with a faith in science places wilderness on a pedestal. “The wilderness,” Murray concludes in ideal-

Most recreationists and tourists come to the Gila and other wilderness areas in the Southwest to visit idyllic scenes like the one here of the Middle Fork of the Gila River, New Mexico, 1994. Courtesy Michael Berman

istic Thoreau-like terms, "holds the key to understanding our past, as well as ensuring that we as a species continue to have a future."[90]

Murray's celebrationist argument for wilderness harks back to Jefferson's ideals of democracy, Turner's frontier thesis, and Aldo Leopold's call to wilderness. This ideological continuity reflects a deep American belief in the nation's exceptional character and manifest role in the world because of its western wilderness heritage. It is debatable whether this way of thinking is accurate or even just for all the nation's peoples.[91] What is perhaps more telling is the individuation of wilderness. As one reporter-backpacker wrote during the fiftieth-anniversary celebration of the Gila in 1974, "the real magnificence of the Gila rests in its unspoiled natural beauty—a hawk lazily circling over a jagged cliff, a trout scurrying past a horse's hoof in an idyllic streambed, a pillar of stone standing guard over the valley below, a gentle waterfall dropping into a clear, green pool."[92]

Some say that a kind of natural virtuosity emanates from a wilderness. There individuals can commune with an ecogod, a creator deity, or the living pantheons of nature. Others see dollar signs in rangeland, deer, and white water. Scientists prefer to examine the dynamics of an ecological laboratory. Earth First!er Dave Foreman used the Gila for profit, recreation, and science: "my guide service has merged with the University of Wilderness and I will be operating the Southwest Campus of this national non-profit educational institution" in the Gila.[93]

In the end, the result is the layering of multiple meanings on top of the already complex multiple uses of wilderness, creating a web of ideas and realities that constitutes the story of the Gila wilderness. Whether America's first wilderness area will be able to hold up under the increasing pressure of these multiple demands will be the responsibility of our and future generations. And even though wilderness areas are in large measure a human construct, their naturalness, ecosystems, and biota connect humanity to the larger community of life. Humans must learn to better understand this culture-nature relationship to develop what ecologist-historian Nancy Langston calls "an ethic of attentive humility rather than an ethic of control."[94]

Postscript

During the 1990s, the Gila wilderness came under increasing environmental pressure from its users. The result was a pitched battle between environmentalists and ranchers with the Forest Service again caught in the middle. The Gila Watch and Gila Biodiversity groups, self-appointed watchdog organizations that came on the scene in the late 1980s, have delivered an ultimatum to wilderness exploiters: if you mistreat the Gila, your abuse will be described in local newspapers and in the *High Country News* and even on national television. The long-standing practice of allowing too many cattle to graze in the river valleys, in particular, has fueled the war. Out of this conflict there emerged above the crowd a new feminine voice as well—Gila Watch's undaunted spokesperson, Susan Schock. Schock has shocked ranchers, Forest Service officers, and even other environmentalists with her candid and vociferous criticism of Gila Wilderness Area mismanagement of the rangelands. In response, ranchers have formed their own advocacy group, the People for the West! In tackling the Diamond Bar Ranch, which Kit and Sherry Laney, fourth-generation Catron County ranchers, are operating as a small corporation, Schock chose a tough couple with longstanding roots in the ranching community to pick on. Feeling cornered, the Laneys and other ranchers have armed themselves to the teeth, hoping to avoid being "put out to pasture" themselves. Some predict an Old West (ranchers) versus New West (environmentalists) gunfight at the Diamond Bar corral.

Despite this continuing threat, Schock claims that ranchers do not deserve the right to perpetually abuse the Gila's grasslands, especially in floodplains where a multitude of bird species thrives. The Forest Service has been able to get the ranchers to reduce cattle numbers but not to fence threatened riparian zones. It is too late anyway, Schock laments. The damage to the Gila wilderness Area has been done, and "people don't know what we've lost. We memorialize this shit as the West."[95]

Like many Americans (and non-Americans), Schock views wilderness as synonymous with the West. However, many have come to see ranching as a threat to the protection of wilderness area ecosystems and resent the frontier heritage that dictates wilderness management. Because of the uncompromising philosophy of deep ecology, the traditional commodity-based use of the Gila wilderness as a ranch is coming under searing scrutiny. Whether the environmentalists will be able to thwart the ranchers and their frontier mentality remains to be seen, especially in light of the Forest Service's entrenched multiple-use dictum. Under this democratic mandate, ranchers, environmentalists, and growing numbers of recreationists and tourists each have rights to the Gila wilderness. In the end, the persistent question remains: do human rights take precedence over nature's rights?

NOTES

1. For the best biography of Leopold, see Curt Meine, *Aldo Leopold, His Life and Work* (Madison: University of Wisconsin Press, 1988).

2. For a discussion of Apache lifeways in the Gila country, see Dan L. Thrapp, *Victorio and the Mimbres Apaches* (Norman: University of Oklahoma Press, 1974), and Donald E.

Worster, *The Apaches, Eagles of the Southwest* (Norman: University of Oklahoma Press, 1979).

3. Christopher J. Huggard, "The Impact of Mining on the Environment of Grant County, New Mexico to 1910," *1994 Annual of the Mining History Association* (1994): 5–7.

4. See Robert D. Baker, Robert S. Maxwell, Victor H. Treat, and Henry C. Dethloff, *Timeless Heritage: A History of the Forest Service in the Southwest* (College Station, Tex.: Intaglio Press, 1988), 77.

5. Aldo Leopold, "The Wilderness and Its Place in Forest Recreational Policy," *Journal of Forestry* 19, no. 7 (November 1921): 719.

6. Ibid., 719, 721.

7. Gifford Pinchot, *The Fight for Conservation* (New York: Doubleday, 1910), 48–49.

8. Leopold, "The Wilderness," 719.

9. Aldo Leopold, "A Plea for Wilderness Hunting Grounds," *Outdoor Life* 56, no. 5 (November 1925): 348.

10. Quoted from Meine, *Aldo Leopold,* 201.

11. Aldo Leopold, "The Last Stand of Wilderness," *American Forests and Forest Life* 31, no. 382 (October 1925): 599–604; idem, "Wilderness as a Form of Land Use," *Journal of Land and Public Utility Economics* 1, no. 1 (October 1925): 398–404; idem, "A Plea for Wilderness Hunting Grounds," 348–50; idem, "Conserving the Covered Wagon," *Sunset Magazine* 54, no. 3 (March 1925): 21, 56.

12. Leopold, "Wilderness as a Form of Land Use," 398.

13. Ibid., 400–401.

14. Ibid., 401.

15. Paul Sutter, "Driven Wild: The Intellectual and Cultural Origins of Wilderness Advocacy during the Interwar Years" (Ph.D. diss., University of Kansas, 1997), 95.

16. Leopold, "The Last Stand of Wilderness," 603.

17. Ibid., 602.

18. Leopold, "A Plea for Wilderness Hunting Grounds," 348–49.

19. Leopold was criticized in the mid-1920s by Howard R. Flint in "Wasted Wilderness," *American Forests and Forest Life* 32 (1926): 407–11. Flint believed the new designation limited access to national forest lands to "himself [Leopold] and the elect few." In response, Leopold argued that he believed in "facing nature alone" without the amenities of civilization—"to get away"; see Sutter, "Driven Wild," 113–15.

20. Leopold, "A Plea for Wilderness Hunting Grounds," 350.

21. See "Recreational Working Plan," March 28, 1924, in Gila Wilderness Area files, USFS Offices, Silver City, N.M. (hereafter cited as GWA).

22. Aldo Leopold to District Forester Frank C. W. Pooler, March l, 1926, 4, GWA.

23. Quoted from Meine, *Aldo Leopold,* 206.

24. See Baker et al., *Timeless Heritage,* 49; quote from District Forester Frank C. W. Pooler to Congressman Franklin W. Reed, Joint Committee on Recreational Survey of Federal Lands, February 25, 1926, l, 4, GWA.

25. Chief Forester W. B. Greeley to District Forester Frank C. W. Pooler, December 30, 1926, l, GWA.

26. Ibid., 2.

27. Susan Flader, *Thinking Like a Mountain: Aldo Leopold and the Evolution of an Ecological Attitude toward Deer, Wolves, and Forests* (Columbia: University of Missouri Press, 1974), 101.

28. Ibid., 1–35.

29. Quoted from ibid., 3.

30. See Donald Worster, *Under Western Skies: Nature and History in the American West* (New York: Oxford University Press, 1992), 34–52.

31. Flader, *Thinking Like a Mountain,* 77.

32. Norman G. Woolsey, "Ben Lilly's Longest Grizzly Chase," *True West* (March–April 1958): 18–19, 39.

33. Ibid., 39; see also J. Frank Dobie, "The Story of the Man Who Died Twice," *Silver City Enterprise,* September 20, 1940.

34. "Government Hunters Take Two Big Lions in Late Expedition," *Silver City Enterprise,* March 29, 1929.

35. Flader, *Thinking Like a Mountain,* 154.

36. Aldo Leopold, "Wild Followers of the Forest," *American Forestry* 29, no. 357 (September 1923): 519.

37. Ibid., 518.

38. See Douglas W. MacCleery, *American Forests: A History of Resiliency and Recovery* (Durham, N.C.: Forest History Society, 1994), 6; see also Nancy Langston, *Forest Dreams, Forest Nightmares: The Paradox of Old-Growth in the Inland West* (Seattle: University of Washington Press, 1995), 41–47.

39. Leopold, "Wild Followers of the Forest," 568.

40. Max Oeschlaeger, *The Idea of Wilderness: From Prehistory to the Age of Ecology* (New Haven: Yale University Press, 1991), 218; see also Roderick Nash, *Wilderness and the American Mind* (New Haven: Yale University Press, 1967).

41. Flader, *Thinking Like a Mountain,* 81–89.

42. "Forestry Department Says Hunters Killed Most Big Game in 1928," *Silver City Enterprise,* January 11, 1929; see also Flader, *Thinking Like a Mountain,* 116.

43. Langston, *Forest Dreams, Forest Nightmares,* 29–31; Flader, *Thinking Like a Mountain,* 88–89.

44. Memorandum for Files, Assistant District Forester Hugh G. Calkins, March 30, 1927, 1, GWA; see also MacCleery, *American Forests,* 36; Susan Flader, "Aldo Leopold and the Evolution of Ecosystem Management," in W. Wallace, *Sustainable Ecological Systems: Implementing an Ecological Approach to Land Management* (Flagstaff, Ariz.: USDA, Forest Service, 1994), 15–19, and "'Let the Fire Devil Have His Due': Aldo Leopold and the Conundrum of Wilderness Management," in David W. Lime, ed., *Managing America's Enduring Wilderness Resource* (Minneapolis: Minnesota Extension Service, 1990), 88–95.

45. Flader, *Thinking Like a Mountain,* 110; "Roads and the Wilderness" (ca. 1935), a brochure of the Izaak Walton League, Silver City Chapter, GWA; "Estimate That Over 2000 Hunters Are in District This Year," *Silver City Enterprise,* October 23, 1931; "State Game Warden and Game Specialist Make Report on Area," *Silver City Enterprise,* April 8, 1932.

46. Worster, *Under Western Skies,* 47–48.

47. Quoted from Flader, *Thinking Like a Mountain,* 102.

48. J. Douglas Wellman, *Wildland Recreation Policy* (New York: Wiley and Sons, 1987), 133–35; see also Dennis M. Roth, *The Wilderness Movement and the National Forests* (College Station, Tex.: Intaglio Press, 1988), 2–3.

49. Wellman, *Wildland Recreation Policy,* 135.

50. "Report on Gila Primitive Area" (1930), 2–3, GWA.

51. Assistant Chief Forester L. F. Kneipp to District Forester Frank C. W. Pooler, May 20, 1930, 1, GWA.

52. Ibid., 2.

53. "Roads and the Wilderness."

54. "Revised Report on Gila Primitive Area," March 9, 1933, GWA; see also Richard E. Warner, "The Dismembered Wilderness," *Pacific Discovery* 15, no. 4 (July–August 1962): 4.

55. Memorandum, Assistant Chief Forester Morton M. Cheney to Gila National Forest Supervisor James A. Scott, March 17, 1933, GWA.

56. Memorandum, B. Williamson, Gila National Forest, to Director, Recreation Staff Unit, "Gila Wilderness Boundary History," June 30, 1977, l, GWA.

57. Wellman, *Wildland Recreation Policy,* 152–54; Roth, *The Wilderness Movement,* 4–5.

58. See memorandum, Regional Forester Fred H. Kennedy to Chief Forester Richard E. McArdle, May 28, 1959, attached "Wilderness Type Areas," GWA.

59. See Paul W. Hirt, *A Conspiracy of Optimism: Management of the National Forests since World War Two* (Lincoln: University of Nebraska Press, 1994), 36–38.

60. "Gila Wilderness Boundary History," 2.

61. See "Forester Society, Southwestern Region to Meet Here Tomorrow and Saturday," "History of Land Management in County," "'N.M. Cattleman of '52' Says Good Management Means Good Conservation," "Hollimon Farm Gets Results from Work Started in 1942," "Scouts Join Conservation Movement with Improvements at Camp Lindauer," "Conservation Workshop Featured Class in New Mexico Western Summer Session," all in *Silver City Enterprise,* November l, 1951, September 17, 1953, September 24, 1953.

62. "Subcommittee Sees No Need for Cut of Gila Wilderness," *Silver City Enterprise,* July 24, 1952.

63. See "Gila Wilderness Area," order of the Secretary of Agriculture, January 15, 1953; J. W. Penfold, Izaak Walton League, to District Forester C. Otto Lindh, February 12, 1953, 2–3, GWA.

64. From the brochure "Save the Gila" (Fall 1972), Collection MSS488, Box 2, Center for Southwest Research, University of New Mexico.

65. Ibid.

66. Anderson quoted from ibid.; J. W. Penfold, Izaak Walton League, to Regional Forester C. Otto Lindh, l; memorandum, Richard H. Spray, Assistant Director Outdoor Recreation, to Director, Recreation Staff Unit, June 30, 1977, 4, GWA.

67. Regional Forester Fred H. Kennedy to Claude Wood, Administrative Assistant to U.S. Senator Clinton P. Anderson, February 6, 1956, 2, GWA; see also "Forest Service to Get Fleet of Helicopters," *Silver City Daily Press,* August 8, 1945.

68. Memorandum, Regional Forester Fred H. Kennedy to Chief, USFS, August l, 1956; C. R. Gutermuth, Vice President of Wildlife Management Institute, to John Sieker, Chief of Division of Recreation and Land Uses, USFS, November 5, 1956; Kenneth B. Pomeroy, Chief Forester, American Forestry Association, November 5, 1956; Richard H. Stroud, Executive Vice President of Sport Fishing Institute, to John Sieker, November 9, 1956, GWA.

69. Warner, "The Dismembered Wilderness," 5–6.

70. See Hirt, *A Conspiracy of Optimism*; Ronald Lee Stewart, "Wilderness Heritage of the Arid Southwest" (Ph.D. diss., University of New Mexico, August 1970), 101.

71. Warner, "The Dismembered Wilderness," 6.

72. See Baker et al., *Timeless Heritage,* 56–57.

73. Warner, "The Dismembered Wilderness," 8.

74. See Oeschlaeger, *The Idea of Wilderness,* 214, 230.

75. Quoted from Richard Allen Baker, *Conservation Politics: The Senate Career of Clinton P. Anderson* (Albuquerque: University of New Mexico Press, 1985), 102.

76. Ibid., 117.

77. See Mark Harvey, *A Symbol of Wilderness: Echo Park and the American Conservation Movement* (Albuquerque: University of New Mexico Press, 1994), 53.

78. Baker et al., *Timeless Heritage*, 65.

79. See Kirkpatrick Sale, *The Green Revolution: The American Environmental Movement, 1962–1992* (New York: Hill and Wang, 1993).

80. Baker, *Conservation Politics*, 127–28. Kennedy revealed his concerns for the environment when he wrote the introduction to Secretary of the Interior Stewart L. Udall's book *The Quiet Crisis* (New York: Holt, Rinehart and Winston, 1963), an influential history of environmental thought.

81. Richard E. McArdle with Elwood R. Maunder, "Wilderness Politics: Legislation and Forest Service Policy," *Journal of Forest History* 19, no. 4 (October 1975): 173.

82. Baker, *Conservation Politics*, 132–33.

83. Ibid., 219–21.

84. See William Cronon, "The Trouble with Wilderness: Or, Getting Back to the Wrong Nature," *Environmental History* 1, no. 1 (January 1996): 16; see also Leopold, "Wilderness as a Form of Land Use," 399.

85. Memorandum, Dave Foreman and Debbie Sease (Gila Wilderness Committee), February 17, 1975, 1, in Collection MSS488, Box 2, Center for Southwest Research, University of New Mexico.

86. Baker et al., *Timeless Heritage*, 29.

87. MacCleery, *American Forests*, 44, 48.

88. McArdle with Maunder, "Wilderness Politics," 168.

89. Regional Forester M. J. Hassell to U.S. Senator Harrison Schmitt, July 2, 1979, 1, GWA.

90. John A. Murray, *The Gila Wilderness: A Hiking Guide* (Albuquerque: University of New Mexico Press, 1988), xii–xv.

91. See Cronon, "The Trouble with Wilderness," 7–28.

92. Phil Niklaus, "Gila Wilderness 50 Years Old on June 3," *Albuquerque Journal*, May 11, 1974.

93. Dave Foreman to Senator Pete V. Domenici, June 17, 1975, l, in Collection MSS488, Box 2, Center for Southwest Research, University of New Mexico.

94. Langston, *Forest Dreams, Forest Nightmares*, 280.

95. Quoted from Tony Davis, "A Struggle for the Last Grass," *High Country News*, May 2, 1994, 14.

6

"Where there's smoke"

Wildfire Policy and Suppression in the American Southwest

JOHN HERRON

In 1990 drought conditions in the American Southwest worsened as a light winter gave way to an even milder spring and, finally, to a moisture-free summer. For the residents of southern Arizona, the heat of the first few days of summer proved especially stifling. As the dry Arizona wind blew through the Valley of the Sun, the mercury at Sky Harbor International Airport already hovered near 115 degrees. On June 26 the temperature climbed to a new record high—122 degrees. While air conditioners worked overtime and backyard pools were filled to capacity in the Desert City, in Payson, a small town forty miles to the northeast, the heat wave produced far more dramatic results. Situated in the scenic Tonto National Forest and framed by the Mazatzal and Sierra Ancha Mountains, Payson has long served as a cool retreat for tourists and Phoenix residents seeking refuge from the oppressive summer heat. But on this day, while Phoenix baked, Payson burned.

At 12:30 P.M. on June 25, 1990, a lightning storm ignited a small blaze just under the Mogollon Rim, a prominent geographic ridge that flows through the region. By 1:15 P.M., the abundance of nearby forest litter and the availability of heavy fuels pushed the fire to five acres. Less than an hour later, flames had spread to cover nearly fifty acres. At 4:15 P.M., Forest Service observers estimated that the size of the young burn, now named the Dude Fire, approached one hundred acres as it spread rapidly through large stands of ponderosa pine, manzanita, and scrub oak. In response to the growing fire, at 6:00 P.M. regional officials ordered eighteen crews—more than 360 firefighters—and a highly experienced professional fire management team to assist in the expanding suppression efforts. Cooler temperatures led to a small anticipated decline in fire activity in the late evening hours, but brisk canyon winds pushed the now strengthening fire to more than 1,900 acres in the early morning hours of June 26. Throughout the night, fire-fighting crews arrived at a hastily constructed base camp several miles from the head of the blaze. At 2:30 A.M., one of those crews, the Perryville Firefighting Crew from the Arizona Department of Corrections, arrived and received an assignment to clear brush and debris along an access road that snaked up nearby Walk Moore Canyon. Working all morning, the crew used the access road as an anchor point and began constructing a fire line (a trail-like break in the forest fuels) along the fire's eastern edge.

Pausing briefly to rest and eat a meal at 10:00 A.M., the crew soon returned to the fire line, working back down to the access road to make needed improvements and to widen this important firebreak. By 1:00 P.M., nearby lookouts had reported a marked increase in fire activity as the temperature rose and the humidity dropped from the effects of the midday sun. At 1:45 P.M., a fire observer noted a small but growing spot fire *below* the Walk Moore Canyon access road. Fire managers hoped to use this road to contain any further fire spread, but ten minutes later, before alarmed Forest Service officials could react, the spot fire, driven by winds from an approaching thunderstorm, blew up. Now about halfway down the fire line, members of the Perryville Crew immediately began running toward the access road, hoping to find safety. Unable to reach the road as an advancing wall of fire cut off their retreat, eleven members of the crew turned back uphill in an effort to outrun the wind-driven flames. Trapped on all sides, at 2:23 P.M. the panicked members of the Perryville Crew deployed their safety fire shelters (a fiberglass tent designed to protect firefighters from heat and flames). High winds and excessive temperatures battered the huddled crew for nearly twenty minutes. After the flames finally subsided, sometime around 3:05 P.M., five seriously burned crew members stumbled out of the smoke to the access road. Moments later, Forest Service officials walked up the cooling canyon to find the bodies of the six remaining members of the Perryville Crew.[1]

Beyond this dramatic loss of life, the Dude Fire remains significant for the changes in attitude and public policy that followed it. Previously, few regional foresters questioned the environmental legacy of American fire policies, but a century of aggressive fire suppression has contributed to heavy fuel loads and modern fires that burn with an intensity previously unseen in the American Southwest. Wildfires usually burn in a pattern of mixed mosaics and consume only a fraction of the total vegetation within the fire perimeter. But postfire studies of the Dude incident illustrate that it burned more than two-thirds of the surrounding vegetation, and in many areas the intense heat totally killed all plant life.[2] Further complicating this picture, the Dude Fire also burned through a summer home subdivision, evaporating more than fifty houses in the process. This mix of the human and the nonhuman, now a common element in southwestern forests, has led many southwesterners to question where nature stops and civilization begins. Like the massive Yellowstone conflagration of 1988 or the equally devastating year of 1994, when thirty-three people died fighting fires across the United States, the "fire shock" of the Dude incident prompted government agencies, environmentalists, and even the general public to reevaluate their relationship with nature. The reign of flame in the West throughout the 1990s was extraordinary. This fiery epoch challenged many of our assumptions about forest structure, environmental balance, and human interaction with natural forces. A critical element of this restructuring process is an examination of the role that fire and, importantly, fire suppression should play in this newly established order.

Throughout its history, the U.S. Forest Service, the federal agency most involved in wildfire policy, has struggled to develop a comprehensive national plan for fire suppression.[3] In an effort to balance human expediencies and natural preservation, the Forest Service worked to establish fire suppression specifications that would best meet the needs of an expanding population and still maintain a healthy forest environment. Yet as the Dude Fire illustrates, the Southwest presents an unusual problem for forest managers trying to establish national management guidelines. The region's interesting blend of diverse environmental and historical landscapes makes the Southwest America's most unique and complex fire environment.

Bordered on the east by the Llano Estacado and on the west by the Colorado River, the Southwest is characterized by a varied mix of mountains, mesas, canyons, and desert basins. This series of valleys and plateaus, defined by the Chihuahuan and Sonoran Deserts, the Mogollon Rim, the Colorado Plateau, and the western slopes of the Basin Range, exhibits a range of climatic variation. Moisture-rich peaks, known as sky islands, reach elevations of 12,000 feet and higher and support a wide spectrum of vegetation. Valley floors and desert bottoms offer a far more arid but no less diverse landscape. Even within similar geographic ranges, an abundance of sunshine translates into vastly divergent moisture levels, vegetation, and wildlife support on neighboring exposed slopes or sheltered faces.[4]

Not surprisingly, given this topographic variety, the Southwest also possesses a unique climatic environment. Strong seasonal weather patterns fluctuate between extreme aridity and powerful summer monsoonal rains. Large-scale meteorological patterns and topography provide the ideal conditions for annual summer storms. Moisture-starved basins and plateaus can receive nearly half their annual precipitation in a single storm cycle. Subtropical high pressure cells forming in the central Rocky Mountain region allow moist easterly flows to enter the Southwest from the Gulf of Mexico. These fast-moving systems interact with the mountain landscape and the heated desert to produce generous summer rains.[5] The foothills of Albuquerque, New Mexico, for example, receive only fourteen inches of rain a year, but more than six of these inches fall during the monsoon season.[6] In Arizona the contrast is even greater. For most of Arizona, annual rainfall fluctuates between eight and fourteen inches, with nearly half the yearly totals coming in July, August, and September.[7] High winds, while not always matching the velocity of the Great Plains or Pacific Coast, are still a constant feature of this seasonal spasm of violent weather. These monsoons are carried into the region by highly charged thunderstorm clouds. In the Southwest, the mix of high- and low-pressure storm systems pushes cloud bases so high that precipitation often evaporates before it reaches the ground. Although moisture does not reach the earth's surface, lightning does.

If the magnitude of the Dude Fire makes it extraordinary among southwestern fires, its source of ignition was quite ordinary. The highest concentration of lightning fires in the United States is in the American Southwest. The Mogollon Rim of central Arizona, in fact, receives the most lightning strikes on the North American continent.[8] One obvious result of this incredible lightning load is an increase in wildfires. As environmental historian Stephen Pyne has noted, "the Southwest's national forests average more fires per year than any other region; they have the second-highest rate of burned acreage, and critical fire weather occurs here with greater frequency and persistence than anywhere else in the nation."[9] On one summer day in 1960, for example, lightning sparked more than 140 fires. On July 18, 1970, lightning kindled 100 fires; and on the next day, 100 more.[10] Lightning caused more than 2,000 southwestern fires in 1974, 1976, and 1977. More recently, even in the slow fire year of 1992, lightning started 7 of the 10 largest fires that burned across the Southwest. In 1993, 42 of 61 fires that consumed more than 1,000 acres were lightning based. One year later, of the 53 fires 2,000 acres or more in size, 47 were ignited by lightning. And finally, in 1995, of the fires that raged across the Southwest destroying more than 10,000 acres, lightning was the chief cause of ignition.[11]

For Forest Service planners and fire managers, however, a diverse climate and large lightning load are only part of the puzzle of the southwestern fire scene. Fires burn in the Southwest as much from human intervention as from natural causes. Humans have shaped and reshaped their world with fire for millennia, and recognizing this history is as important to understanding what fire means to the Southwest as the influences of climate and terrain. The human presence in the Southwest dates back more than 12,000 years. A procession of indigenous hunter and then hunter-gatherer civilizations occupied the region for thousands of years, but while these ancient peoples introduced their own fire culture into the Southwest, the arrival of Euro-Americans resulted in the most significant changes for the fire regime and the Native Americans who lived there.[12]

Before Hispanic and Anglo entrance into the region, native inhabitants' use of fire was, in the most understated of terms, extensive. As an integral part of native culture, fire served various uses in hunting, domestic life, agriculture, communication, horticulture, and warfare.[13]

The significance of these fires is illustrated in the study of fire scars in the tree ring record. Tree analysis proves that low-intensity fires burned in much of the region an average of every two years. Even more protected sites burned every three to five years, and rarely did a subregion of the Southwest go longer than a decade without fire.[14] This frequent burning did much to ensure the health of the forest ecosystem. Regular burns replenished grasslands, cleared forest clutter, and opened wide meadows that supported diverse flora and fauna. Fire also checked invading stands of pine and helped to maintain a mix of softwood and hardwood trees. The movement of Euro-Americans into the Southwest altered the fire cycles of the region and restructured this landscape. These newest residents of the region brought a starkly different understanding of fire and its role to the Southwest, but a more significant change involved the introduction of livestock. Domesticated livestock dramatically altered this indigenous fire culture and set in motion a rapid transformation of the Southwest's landscape.

Brought by the Spanish in the late sixteenth century, sheep and cattle became a major component of the ecosystem of the new Southwest. By the early 1600s, Pueblos had acquired Spanish stock, and by the 1660s, the Navajos and Apaches had integrated the horse and other domesticated animals into their pastoral societies.[15] Among the southwestern native groups none adopted Spanish ranching practices as well as the Navajos, and they successfully built large herds throughout the eighteenth and nineteenth centuries.[16] Trader Charles Bent stated that during the Mexican War, the Navajos possessed 30,000 cattle, 500,000 sheep, and 10,000 horses.[17] While the veracity of Bent's estimates are suspect, the fact that the Navajos maintained extremely large herds is not. Once these animals were introduced into the Southwest, their numbers rapidly increased, especially after the widespread arrival of Anglo Americans into the region. As late as 1870, botanist Charles Cooper estimated that only three small Anglo-owned cattle herds existed in Arizona. By 1880, however, ranchers were grazing more than 140,000 cattle in the same area. The numbers for New Mexico are even more dramatic. New Mexico rangelands, for example, supported 150,000 cattle in 1870, more than 500,000 in 1880, and upwards of 1 million by 1886.[18] While this increase in southwestern cattle is significant, the growth of the sheep industry in the region is even more impressive. In the decade of the 1850s, New Mexico ranchers sent nearly 500,000 head of sheep just to the mining towns of California. By the mid-1880s, more than 5 million sheep grazed on New Mexico rangelands, prompting a Bureau of Animal Industry report to assert that "New Mexico is the mother of the sheep industry of the Rocky Mountain region and the Great Plains."[19]

This massive influx of livestock into the Southwest radically altered patterns of regional culture and economic subsistence. Equally significant, however, were the changes made to the natural environment. Where grasses once provided fuel for annual fires, overgrazing transformed the landscape into devastated terrain. In 1826 James Pattie surveyed the Gila River drainage and stated that after just one day of exploration, "we were fatigued by the difficulty of getting through the high grass which covered the heavily timbered bottom." Joseph Rothrock, chief botanist with the Wheeler Survey of Arizona and New Mexico, described the thick grasses of the Southwest as "luxuriant bunch-grasses covering the ground as thickly as it could stand." While visiting the Blue River Valley in southwestern New Mexico, Aldo Leopold, who began his Forest Service career in the Southwest, wrote, "All the old settlers agree that at the time of settlement, stirrup-high grama grasses" covered the region.[20]

During the last two decades of the nineteenth century, however, the pattern of the southwestern grasslands and forests changed significantly. In 1893 Governor L. C. Hughes of Arizona wrote, "in nearly all districts, owing to overstocking, many weeds have taken the place of the best grasses." As herds of animals trampled the forest into dust, the forest floor became "as bare and compact as a road bed."[21] In many cases, recovery seemed virtually impossible as erosion moved in where overgrazing left off. Grass-covered stream banks and tree-anchored ravines gave way to deep-cut arroyos and water-washed gullies. Floods from this destruction ravaged the Southwest in the late nineteenth century. One of the most dramatic examples of this ecological crisis was in Silver City, New Mexico, where raging floods turned Main Street into a fifty-foot chasm known as the Big Ditch. As one historian describes the transformed scene, "desert succulents, mesquite, juniper, and chaparral replaced grasses in desert basins and across high plateaus; brush congealed into jungles; open forests, once dappled with glades of sun and shadow, snarled with downed logs, dense tangles of under story, and young groves of pine and fir reproduction 'thick as the hair on a dog's back.'"[22]

Adding another layer to this destructive mix was the growth of the southwestern timber industry. Railroad construction, mining, and new community building introduced a need for forest products, and by 1879 at least thirteen large-scale sawmills operated across the Southwest.[23] The focused harvest of old growth forest further upset the forest/fire balance as tree density exploded into gnarled thickets of growth that clogged southwestern forests. The diversity of vegetation, so much a part of the southwestern landscape, slowly vanished as balance surrendered to imbalance.

The environmental changes caused by fire exclusion, however, were not a primary concern for most southwesterners. Regardless of the introduction of livestock grazing, the increase of the industrial timber harvest, the expansion of the reservation system, and the resulting decline of native burning, fires still burned in the Southwest, and most southwesterners believed that these fires led to unnecessary forest destruction. Southwestern forest managers reacted to the continual burning of regional forests by implementing a clear policy of aggressive fire suppression. Timber was a valued commodity, and protecting forest resources from fire became the primary goal for area officials. As the twentieth century approached, this southwestern amalgam of geographic variety and cultural competition provided a significant challenge for regional fire managers.

The effort to manage fire and natural resources in the United States has, obviously, not been limited to the Southwest. Driven by the zeal of Progressive Era convictions and supported by a strong faith in science and technology, forest managers across the country tried to implement a systematic program of fire control that would best support the prevailing Forest Service dogma of utilitarianism. Imported from Europe, largely as a result of the experiences of German-trained foresters, this new conservation philosophy gave birth to professional forestry in America and a growing belief in federal responsibility for resource preservation.[24] In the first decades of the twentieth century, achieving this wise-use goal was based on scientific management of the forests and directly translated into fire suppression through aggressive initial attack.

The early years of American forestry, however, were marked more by policy experimentation and inefficiency than by ecological understanding. The first plans to reduce the fire threat, in fact, consisted of nothing more than minimally equipped forest rangers and district fire guards crisscrossing their ranges and extinguishing blazes. Recognizing these manpower shortages, early-twentieth-century southwestern ranger guidelines advised that in the event of large wildfires or "in exceptional cases of great emergency," rangers could enlist local residents in suppression but "weren't to guarantee the pay of such persons."[25] Removing the fire threat remained a primary directive for federal wildfire policy, but inexperience and limited funding made fire-fighting rangers easy targets for criticism. In 1907, for example, the head of the Southwest Wool Growers Association, E. S. Gosney, stated that the southwestern "forests can never be protected by forest rangers. I have known stockmen to fight fires for days and extinguish them without ever seeing a ranger!"[26] Rancher criticism of the Forest Service would eventually grow into a perennial ritual, but it appears Gosney's criticism did have merit. In that same year, a New Mexico resource survey of regional forests revealed that labor shortages forced many solitary rangers to patrol expansive districts of more than 200,000 acres.

After several years of organizational confusion, an improved infrastructure did finally come to the region with the arrival of the first southwestern regional forester, Arthur C. Ringland. One of Gifford Pinchot's original students recruited to study forestry, Ringland graduated from the Yale School of Forestry in 1905. Joining most of his classmates in the West, he began his professional career as a ranger in the Lincoln National Forest in southeastern New Mexico and became regional forester in 1908. Writing to his forest supervisors, Ringland suggested that they perform a "careful study of the conditions of the Forest" and urged "the adoption and use of a definite fire plan." Ringland pushed for the construction of lookout stations, telephone lines, and fire patrol roads and the organization of fire-fighting volunteers. Revealing an overwhelming desire to eliminate the fire threat from southwestern forests, most foresters responded positively to Ringland's requests, resulting in the immediate implementation of fire suppression in the 1910s.[27]

One of the often overlooked factors that contributed to the early success of federal fire suppression policies in the Southwest was its territorial status. While northern neighbors struggled to balance federal policy objectives against local agricultural, ranching, and timber interests, the majority of lands in the Southwest were federally owned. Arizona and New Mexico did not enter the Union until 1912, a fact that helped establish the supremacy of federal objectives, such as Forest Service fire suppression policies, over private desires. The end result was that by the end of Ringland's term in office in 1916, total fire suppression had become the dominant policy in the region. Few southwesterners disagreed with Chief Forester Henry Graves's comment that "the necessity of preventing losses from forest fires requires no discussion. It is the fundamental obligation of the Forest Service and takes precedence over all other duties and activities."[28]

Ironically, this plan of increased fire suppression succeeded in the Southwest, not necessarily due to any inherent policy brilliance but more to a combination of ecological factors that were already at work in the regional forests. First, the overgrazing that so altered the southwestern landscape also functioned to fortify the artificially created ban on wildfire. With little left to burn, little burned. Second, while the rest of the West experienced a number of horrific fire years, most notably in 1910, when more than 5 million acres burned in America's national forests, the Southwest remained relatively immune to this fiery fate. A wet cycle, characterized by above average winter and spring precipitation and reduced summer drought, dampened fire fuels over most of the second decade of the century. Eventually, high fire years did occur, but most often with limited devastation. The dry and windy spring of 1917, for example, produced an active fire season, but only enough to prompt forest officials to issue a reminder to all rangers to be on the lookout for any unattended campfires.[29]

Confident in the effectiveness of fire exclusion, by the opening of the 1920s southwestern foresters had made fire prevention a top priority. Supported by the national administration of new Chief Forester William Greeley, professional foresters reformulated fire management. Gone were the inefficiency and policy experimentation that had characterized the first two decades of the century and in their place came a standardization of practices, terminology, and methods for an evolving national fire prevention program.[30]

In this restructuring process the Southwest became a leading innovator in new suppression tactics. As part of the older fire policy, large suppression efforts were considered unnecessary risks and fiscally unwise, especially on "low-value" or back country lands. This "let burn" stance operated well in the Southwest, but in 1926 the region changed its approach to wildfire. Southwestern fire managers conducted a five-year experiment with a policy of all-out suppression of wildfire. Rather than allow some fires to burn unchecked, foresters now attacked every fire they could reach. Even with limited resources, the experiment was judged a complete success, and the results of the aggressive fire policy proved to regional as well as national officials that immediate and decisive suppression translated into fewer acres burned and reduced firefighting costs.[31] The decision to move to all-out suppression did not please everyone, and the complete exclusion of fire in the Southwest did have its detractors, notably, Aldo Leopold.

Leopold aired his criticisms of fire exclusion in a 1924 article in the *Journal of Forestry.* In "Grass, Brush, Timber, and Fire in Southern Arizona," Leopold argued that the exclusion and suppression of fire disrupted the ecology of the Southwest. He outlined the devastating environmental changes that had marked the Southwest since the arrival of white settlement and rhetorically questioned whether "our administration applied these facts" when formulating fire policy. Leopold concluded that "15 years of Forest administration were based on an incorrect interpretation of ecological facts." Leopold also assured his readers that while mistakes were made, "this error of interpretation has now been recognized and administrative policy corrected accordingly." Believing that the Forest Service now understood the importance of fire in natural balance, the famous conservationist argued that in the next decade of fire policy, damage to the Southwest would be kept to a minimum.[32]

But Leopold remained a faithful employee of the Forest Service, and to argue that he now favored a more flexible policy of fire suppression would be a mistake. Just four years earlier, for instance, Leopold had authored a scathing attack on the idea of "light burning," the belief that frequent low-intensity burns maintained the health of the forest ecosystem. In the early 1920s, private timber companies in the Southwest criticized the Forest Service for allowing public forests to become overgrown with brush and reproduction, making them vulnerable to disease and devastating conflagrations. Responding to this criticism in an editorial that appeared in *Southwestern Magazine,* Leopold supported the Forest Service doctrine that fire was an unmitigated forest evil, and he attacked the "light burning propaganda" as misguided, uninformed, and dangerous. Wildfire, Leopold believed, destroyed the important regrowth needed for future forest regeneration and should be contained at every opportunity.[33] As the evolution of Leopold's perception of fire illustrates, slow changes were under way in the Southwest. Leopold's faith in a more enlightened fire policy never materialized, however, and southwestern foresters stepped up their policy of fire exclusion in the 1930s.

Setting the stage for the fire policy of the 1930s was the administration of President Franklin D. Roosevelt. Deeply committed to the conservation of America's lands and wildlife, FDR took a personal interest in the management of forested lands. Roosevelt's appreciation for trees dated back to the earliest days of his youth growing up in New York State. The family estate in Hyde Park included several hundred acres of forests, and, as a boy, Roosevelt spent countless hours exploring these woods and their wildlife. This boyhood interest blossomed into a grown-up passion, and, as a result, in 1910 Roosevelt added a thousand acres to the family holdings. Over the next thirty-five years, Roosevelt supplemented the family forest with more than 500,000 additional trees. Roosevelt did not plant these trees for their scenic beauty, however, but for their potential industrial use. As early as his first State of the Union address, Roosevelt introduced his forest conservation philosophy to the American public. "Forest protection is not an end of itself," Roosevelt stated, "it is a means to increase and sustain the resources of our country and the industries that depend on them. . . . Whatever destroys the forests threatens our well-being."[34] To Roosevelt and his expanding circle of like-minded conservationists, nothing destroyed the forest like wildfire.

Given this background, it is not surprising that one of his first conservation measures was the appointment of Ferdinand Silcox as head of the Forest Service. Silcox, a fervent New Dealer, witnessed firsthand the devastating western fires of 1910 and thereafter remained deeply committed to fire exclusion. In the 1920s timber companies made significant inroads in their effort to influence forest policy, but with Roosevelt's support, Silcox successfully reestablished federal authority and increased control of public forestlands.[35] This move met with a warm reception in the Southwest, because regional fire managers had tried to create an efficient fire policy to regulate forest growth and still produce a substantial timber harvest. Key to the success of this vision was a massive fire suppression plan. As regional foresters were well aware, however, wanting to reorganize fire management plans and having the ability and resources to do so were two very different things. Now, with an increase in federal presence and the coming of the New Deal, the once remote possibility of creating a comprehensive fire policy seemed assured. In the span of only a few years in the mid-1930s, everything southwestern foresters had hoped to accomplish since the turn of the century became part of a maturing modern fire program.[36] New

Deal conservation programs, especially the crown jewel, the Civilian Conservation Corps (CCC), transformed the fire regime of the Southwest once again.

In the 1930s America struggled to recover from a series of financial disasters. The stock market crash of 1929, the decline in industrial production, and high unemployment left the economy in ruins. But no less significant was the environmental devastation of the era. The Dust Bowl of the Great Plains and the resulting destruction of millions of acres of farmland displaced thousands of small farmers, prompting a great exodus from midwestern America. Given this combination of factors, FDR believed that America's successful recovery had to include a plan to restore both the social *and* environmental well-being of the nation. As a result, many of Roosevelt's alphabet soup relief programs, such as the Tennessee Valley Authority (TVA), were founded on the premise that environmental and social improvement would go hand in hand. The massive water projects of the TVA were an attempt to assert human control over nature. In return, the production of power would change and improve the social conditions and living standards of the residents of the region. The chief goal of the TVA program was to refashion the natural landscape for human uses.[37] This same foundational belief in the need to restructure the human and nonhuman landscapes spread to other fields as well, most notably, forestry. Continuing the pattern established more than three decades earlier and supported by a new faith in the ability of social forces to direct natural forces, foresters in the Southwest devised a plan for the strengthening of regional forests, a plan that centered around protection from fire.

In 1933 Forest Service officials published *A National Plan for American Forestry*. An extensive study of forest conditions across the country, this report focused on the questions of watershed protection, recreation, timber harvests, and, of course, fire. Following the theme of the era, the authors concluded that "one of the major problems in American forestry is to rebuild depreciated lands that have already suffered severely from over cutting and burning, and success in recapturing such forest values must be predicated on keeping fires entirely out or within reasonable check."[38] In the Southwest the job of "keeping fires entirely out" fell to the Civilian Conservation Corps as the unprecedented influx of CCC manpower and money altered fire policy and control techniques. Prior to the CCC's emergence on the scene, southwestern foresters had relied on local settlers, loggers, and ranchers to assist with fire suppression. Reflecting this condition, the regional fire suppression experiments of the 1920s were designed to curb fire-fighting costs. In the 1930s federal monies and the willingness to spend them meant that economic restrictions melted before the expansion of southwestern fire control.

Given the region's small population, most southwestern CCC enrollees arrived from the Midwest and East and were then assigned to camps established throughout the area. The number and location of camps varied annually due to the changing nature of project work, but on average New Mexico supported fifteen camps populated by more than three thousand men, while Arizona had twenty camps filled with nearly four thousand enrollees. Representative are the figures from June 1935. A federal survey in that summer listed more than two thousand camps across the United States, with twenty-two in Arizona, seventeen in New Mexico, and fourteen in Nevada.[39] From these camps, CCC enrollees participated in a host of conservation projects across the region, including road improvements, dam construction, and trail maintenance. Among the important functions of the CCC was fire suppression.

The original guidelines for CCC management called for these crews to fight fires only in emergency situations, but it quickly became clear that fire fighting would dominate the activities of many CCC troops. This movement into fire suppression is easily understood, because roughly 75 percent of all camps fell under the direction of the Department of Agriculture, and of those, more than half worked for the Forest Service.[40] Utilizing the labor pools at its disposal, the Forest Service in the Southwest soon made fire suppression the chief task for many CCC troops. Reflecting this trend, in 1935 CCC troops spent more than 5,330 days in fire suppression in the Southwest, and this figure continued to rise into the early 1940s. As the program escalated, southwestern enrollees kept fire numbers to an all-time low. In 1938 the region reported more than fifteen hundred fires, with only forty-two exceeding ten acres. In 1939 more than two thousand fires occurred in the Southwest, yet only fifty-four covered more than ten acres.[41] Reaffirming the results of the regional fire policy, Chief Forester Silcox noted with satisfaction that the "largest and most important contribution of the Civilian Conservation Corps . . . has been in protecting the forests from fire."[42]

Still, not everyone shared the enthusiasm of Silcox and most southwestern foresters regarding the success of the CCC. While CCC enrollees built roads, fire towers, and telephone lines that aided in fire suppression efforts, those same activities increased access to many of the region's more isolated natural sites. Bob Marshall, the leading supporter of American wilderness in the 1930s, expressed his reservations about the activities of the CCC in a letter addressed to Chief Forester Silcox. "The bulldozers are already rumbling up into the mountains," Marshall wrote. "Unless you act very soon, eager CCC boys will have demolished the greatest wildernesses which remain in the United States."[43] Yet as long as those "eager CCC boys" curtailed fire activity in the Southwest, criticism like that from Marshall would go largely unheeded. Reluctant to place any limits on CCC activities, Forest Service officials continued to expand fire suppression programs throughout the region.

Along with the massive resources and man-days that the CCC mobilized against southwestern fires, the CCC molded the fire regime of the region in other important ways as well. Foremost, the CCC dramatically altered the fire-fighting infrastructure in the Southwest. When thousands of men poured into CCC camps across the region, hundreds of supervisors were imported to manage and direct them. In early 1933 Southwestern Regional Chief of Operations Hugh Calkins contacted the leading forestry schools in the nation in an urgent search for professionally trained foresters. As a result of this plea for help, by the summer of that year more than eighty supervisors, the largest increase in forest management personnel in the region's history, came to the Southwest.[44] As the CCC expanded, so did the need for supervisory administration, but when the program disbanded and the CCC enrollees went home, many of these supervisors stayed. The lessons they learned about the success of fire exclusion stayed as well.

The second and probably most significant legacy of the CCC's work in fire suppression was the introduction of the organized crew. Replacing volunteer crews, the CCC crews were better trained, better equipped, and, as a result, more effective. Quickly, Forest Service officials realized that organized crews were the best tools to reduce regional fire dangers.[45] Like their predecessors of a previous generation, however, foresters in the Southwest of the 1940s built their fire policies around circumstances that could not last. After a number of favorable weather years gave credibility to a short-sighted fire policy in the early century, now an overabundance of resources supported regional fire managers who remained confident in their ability to eliminate the fire threat in their forests. Unfortunately, both policy plans failed accurately to reflect the needed balance between wildfire and the landscape of the American Southwest. The ecological disruption would continue for another four decades.

The changes in fire policy in the 1930s and 1940s led to other costs as well. Like most relief agencies, one of the primary goals of the CCC program was to spend federal funds to stimulate the economy. The Forest Service heartily welcomed an increase in its budget, but as a result, the agency expanded faster than its capacity to manage that growth. While not always ecologically sound, the fire policy of the early twentieth century had at least evolved slowly in response to regional conditions and resources. New Deal largesse encouraged the Forest Service to expand fire protection in size and scope. Forestlands that had always been subject to wildfires were now the site of fire lookouts, access roads, and fuel breaks. Once these lands fell under the protective jurisdiction of the fire exclusion policy, the Forest Service felt obliged to maintain them. As many of these labor-intensive projects fell into disrepair and the number of burned acres crept steadily upward in the 1940s, however, fire managers discovered that matching the fire protection levels of the 1930s would be extremely difficult. Largely as a result of New Deal conservation, southwestern foresters made their beds on aggressive fire suppression; unfortunately, sleep proved elusive.

Adding to the growing pressure to reduce the regional fire threat was America's entrance into World War II. As the nation moved closer to war, Americans reexamined their allegiances and began the difficult process of sorting friends from foes. In this effort, enemies at home were deemed just as threatening as enemies abroad. Illustrating how environmental policies reflect the diplomatic climate, added to the list of American adversaries was wildfire. In July 1942 Secretary of Agriculture Claude Wickard stated that "in wartime, the control and prevention of forest fires is a first line defense job on the home front . . . every fire in our fields and forests . . . is an enemy fire."[46] Statements like these crystallized the perception that fire, a clear threat to American stability, should be removed from the landscape. Shortly after Wickard's comments on fire danger, the Japanese Army floated incendiary balloons over the West Coast to set forest fires and divert attention from the war. Few landed and even fewer burned the forests, but if the American public needed any further convincing about the evils of forest fires, they had it.[47] The belief that human effort could control natural forces was a message the Forest Service had been supporting for years, and during the war it was reworked into a positive public relations strategy.

In the twenty years preceding the war, the Forest Service, in coordination with a host of other federal agencies, tried to convince the American public of the need to prevent wildfires. A mix of propaganda posters, radio commercials, newsreels, even official legislative acts was used to get this message across, but the effort met with limited success. In 1942 Forest Service officials reorganized this effort by helping to create the Cooperative Forest Fire Prevention (CFFP) campaign. The specter of world war gave the program the impetus it needed to spread the word about forest fire. Just as wood was receiving a classification as a "critical war material," Clint Davis of the CFFP outlined the threat of wildfires in great detail to the American people:

> The danger of uncontrolled forest fires hampering an all-out rearmament program was well recognized. Lumber—billions of feet of it—was needed for cantonment construction, crating, aircraft carrier decks and numerous other defense needs. . . . Important watersheds had to be protected from fire to prevent any interruption of water flow for domestic use and hydroelectric power so essential in keeping the wheels of industry humming around the clock. Forest fires were also recognized as a threat to defense and security measures within the nation's boundaries. Columns of smoke would nullify the effectiveness of air raid detection systems. Smoke screens might well protect submarines and other means of enemy harassment. Forest fires had already proven a problem in the vicinity of airfields and artillery training grounds. Smoke interference often made it necessary to shut down essential training programs for pilots and gunners for periods of days. Another threat, that of the enemy firing our forest, was always prevalent.[48]

Now, supporting the war effort meant preventing forest fires. In a liaison with the War Advertising Council, the Forest Service linked fire prevention programs with "Meatless Mondays," scrap rubber drives, and the purchase of war bonds. Much to the delight of the Forest Service in the Southwest and elsewhere, fire suppression policies had finally reached a national audience. And the message was clear—to battle wildfires was to fight patriotically against Imperial Japan and Nazi Germany.

Taking advantage of this public relations success, southwestern foresters went on a two-decade-long search to upgrade their suppression program and stamp out fire altogether in the American Southwest.

As part of this effort, regional officials returned to the concept of the organized crew that had worked so well during the CCC years. Forest Service officials worked to design crews with multifaceted fire suppression capabilities. The most skilled and visible of these crews were known as the smoke jumpers. Originating in Washington State in 1939, the smoke-jumper program introduced fire suppression to the American public on a grand scale. Smoke jumpers were more than just a valuable suppression tool, however. They were rugged firefighters attacking fiery conflagrations from the air, parachuting down into remote lands to battle wildfires. Evoking images of wartime heroism, the smoke jumpers quickly captured the imagination of the American public. In 1949 smoke jumpers staged an exhibition on the White House lawn, and in 1951 the program reached the pinnacle of cultural stardom when Twentieth Century–Fox released a major motion picture based on the experiences of a smoke-jumper crew.[49]

Buoyed by field success and popular approval, the smoke-jumper program expanded across the West in the late 1940s after first being implemented in the Southwest in 1947. The smoke-jumper program was a perfect fit to the fire regime of the Southwest with its abundance of lightning fires in remote lands. As part of an "experimental trial system," eight smoke jumpers from Missoula, Montana, arrived at a small aerial base in Silver City, New Mexico. The experiment proved successful, especially in the initial attack of fires in the Gila and Apache National Forests, and resulted in an expansion of the program in the 1950s. In 1954 smoke jumpers in the Southwest made 105 jumps on more than forty fires; by 1956 they had made more than 225 fire jumps. Two years later, in 1958, the number of southwestern jumpers had tripled, and aerial firefighters made nearly 275 jumps on more than a hundred fires throughout the region.[50] Despite their perceived ubiquity, the smoke jumpers limited their activities to the mountainous region surrounding the central Arizona–New Mexico border. Smoke jumpers excelled in battling small, remote blazes but lacked the resources to attack large, regional fires. The smoke-jumper program filled an important niche in the regional fire plan, but it was unable to satisfy the suppression demands of regional foresters.

Southwestern foresters found part of the solution to the broader fire crises and regional labor shortages in local Native American tribes. Just a year after the arrival of the smoke jumpers, federal officials created a new fire-fighting program, the Southwest Forest Fire Fighters (SWFFF). In 1948 the BIA organized and trained a fire-fighting crew of Mescalero Apaches in central New Mexico. The next year the Mescaleros were assigned to a major fire on the Lincoln National Forest, adjacent to their reservation. The effectiveness of the crew dispelled any early skepticism surrounding the program, and the BIA, in conjunction with the Forest Service, began to build other Native American fire-fighting programs throughout the region. Crews were fielded from the Hopi and Navajo reservations in Arizona and from the Zuni, Jemez, Zia, Santo Domingo, and Taos Pueblos of New Mexico. Specially trained and properly equipped, the twenty-five-man SWFFF crews were in high demand in the Southwest. In the first years of the program, the SWFFF crews faced, and passed, many difficult fire challenges. During a severe fire bust in 1950, SWFFF crews were sent to southern California and performed extremely well. The program grew quickly, and by 1951 more than two thousand southwestern Indians were fighting fires across the greater Southwest. As their reputation spread, so did the range of their travels. In 1952 more than six hundred SWFFF crew members traveled outside the region to fight blazes in Montana, Idaho, Colorado, and Washington.[51]

Restructured in 1953, the SWFFF program expanded to include Hispanic crews from the northern New Mexico communities of Pecos, El Rito, Questa, and Peñasco. A few years later, the Papago Indians organized a crew on the Coronado National Forest, as did the Jicarilla Apaches and the Utes of the Four Corners region. Beyond this southwestern development, the SWFFF program became a model for the formation of other Native American crews across the West and Alaska.[52]

The growth of the smoke-jumper program and the organization of the SWFFF crews matched a renewed emphasis on suppression in official Forest Service policy guidelines. In the 1950s fire plans from forests across the region incorporated fire exclusion into the daily management of southwestern forests. Echoing the sentiments of foresters from more than forty years earlier, officials on the Kaibab National Forest in Arizona called for "prompt and aggressive suppression of all fires" and stated that "fire suppression takes priority over all other work."[53] As the growth of New Deal conservation programs demonstrated, the commitment to fire exclusion in the Southwest often had little to do with the principles of fire management. Similarly, in the 1950s and 1960s the commitment to the "suppression of all fires" was shaped, in part, by another unique southwestern contribution to American culture—Smokey the Bear.

As stated earlier, as part of the campaign to alert the American public to the growing dangers of unregulated fire, the Forest Service launched an advertising campaign in the early 1940s in an attempt to mobilize the country behind the dual themes of fire prevention and suppression. After a slow start, the program gained extraordinary national exposure when, in 1950, a small black bear cub was recovered after a major fire on the Lincoln National Forest. In what turned out to be an incredibly popular human-interest story and a major public relations coup for the Forest Service, a New Mexico game officer, Ray Bell, and his young daughter Judy nursed the badly burned bear back to health. Regional fire officials sent the now famous bear to the National Zoo in Washington, D.C. The story of this young bear, who became the official "Smokey the Bear," became a national sensation, and Smokey very quickly became *the* symbol for fire suppression in America.

Forest Service promotion of Smokey the Bear has grown into big business. His likeness appears on sweatshirts, litter bags, games, records, sheet music, comics, balloons, lamps, and, interestingly, hickory-scented lighter fluid.[54] Illustrative of the significance of Smokey the Bear in American culture were the results of a national public opinion survey conducted in 1968. The survey results showed that 98 percent of all children, 95 percent of all teenagers, and 89 percent of all adults recognized Smokey and understood his message of fire exclusion.[55]

Unfortunately, the message Smokey the Bear taught to most Americans was to fear fire, not to understand it. Smokey's own life experience, in fact, suggested to the average citizen that fire was a destructive force that had no place in American forests. The Forest Service established Smokey the Bear as the authority on fire; when he told us, "Only you can prevent forest fires," we could hardly disagree. As a consequence, the Smokey the Bear campaign greatly influenced fire management strategy in the Southwest. Since the Southwest was the home of the original Smokey (and his mate as well as his successor), southwestern officials felt even more compelled to rid the region of the fire threat. If Smokey became the ultimate symbol for fire suppression, the Southwest became the ultimate proving ground for fire exclusion practices. Victims of their own successful advertising campaign, the Forest Service in the Southwest was compelled to follow the now well-worn path of increased fire exclusion.

Since the earliest days of solitary rangers patrolling their ranges on horseback in search of fire, the key to fire exclusion in the Southwest had rested in adequate levels of manpower. The CCC, smoke jumpers, and SWFFF crews had performed well, yet each of these crews also had serious limitations. The CCC crews, for example, were temporary, with a limited range of capabilities and training. The smoke jumpers were an improvement, but their utility was limited largely to small fires in isolated areas. The arrival of the SWFFF program translated into an expansion in size and training of southwestern fire crews, but as the program moved into the 1960s, the availability and quality of SWFFF crews varied widely. In the continuing search for a labor source to meet the demands of fire exclusion, the Forest Service in the Southwest turned to yet another type of fire-fighting crew—the "Hot Shots."

During the early years of southwestern forestry, the idea that the Forest Service should create a special department devoted solely to fire fighting would have been as unacceptable as it was unnecessary. Sixty years of aggressive fire attack, however, had created an environment where restricting the fire threat had indeed become a full-time job. In response to these growing suppression needs, the Forest Service, under the direction of Chief Forester Edward Cliff, created the Interregional (later Interagency) Fire Suppression Crew, or Hot Shots, in 1961. The original concept centered on the creation of a force of professional firefighters who, unlike the majority of Forest Service employees, would work exclusively in fire suppression. As these twenty-person crews materialized in the 1960s, their primary mission was "to provide mobile and highly trained crews" who were "highly motivated and reliable" and stood "capable of providing a well disciplined, adaptable work force to deal with a wide variety of fire situations."[56] If this description sounds like a military unit, it should. The concept of a "rapid deployment force" of firefighters was modeled on the example of specialized military units, and as their name implies, the Hot Shots welcomed this comparison since they viewed themselves as the elite corps of the Forest Service.

The Hot Shot program, like its predecessors, grew quickly. The Forest Service expanded the Hot Shot program from only nine crews in the nation in 1963 to more than thirty crews by 1977. The Southwest became especially fertile ground for the program, and, by the mid-1980s, the region maintained eighteen professional Hot Shot crews. Highly skilled and extensively trained in the latest techniques of fire suppression, Hot Shot crews battled all types of fires in a variety of environments. A 1978 federal survey of the program revealed that Hot Shot crews constructed more fire lines per shift with fewer recorded accidents than even the most optimistic Forest Service estimates. That same survey also noted that Hot Shot crews in the Southwest were a 50 percent improvement over all other suppression crews previously used in the region.[57] The Hot Shot program represented the pinnacle of fire suppression in the Southwest. Always stationed within thirty minutes of a major airport, Hot Shot crews could respond to fires across the region (or even across the country) to eliminate the fire threat with professional precision.

From the 1950s on, southwestern fire-fighting crews built an impressive record of suppression. In the end, however, for all their efforts and skill, these firefighters were simply a reflection of the current fire suppression standards. As the rise of the Hot Shot program illustrates, regional fire managers tried to control the fire environment of the Southwest by using more and more advanced means of suppression. In the effort to eliminate fire, southwestern foresters developed extensive "pre-attack strategies," designed high-tech detection systems, and organized "interdepartmental planning infrastructures." By 1970, some of the southwestern foresters who had formulated these sophisticated policies began to question just what connection efficient fire suppression had to southwestern forest stability. One of the reasons for this reflective pause in forest policy was the growing number of southwestern foresters who watched modern fires destroy greater and greater acreages despite increased investment in suppression efforts.

Despite agency insistence on fire exclusion, voices of caution had existed in the Forest Service for nearly three decades. In 1943 Harold Weaver, one of the individuals responsible for organizing the SWFFF program, wrote in the *Journal of Forestry* that the "complete prevention of forest fires has certain undesirable ecological and silvicultural effects. Conditions are already deplorable and are becoming increasingly serious. Fires, when they occur, are exceedingly hot and destructive and are turning extensive areas of forests into brush fields."[58] More than half a century later, Weaver's insights seem prophetic. Weaver, and others like him, pushed for a more accommodating forest management policy that considered the role of natural forces like wind, floods, and fire in attempts to maintain forest health. Few would support such a position in the 1940s or even the 1950s, but fortunately, foresters were slowly beginning to realize that changes to the structure of the southwestern forest, in this case fire exclusion, came with an incredibly high environmental price tag. The great achievements in fire suppression that had occurred since the 1930s had turned regional forests into a powder keg. Open, parklike forests had once dominated the Southwest, but now heavy fuel loads, dense understory growth, and a near solid canopy made these forests incredibly susceptible to wildfire. By the late 1960s and early 1970s, the once radical view of beneficial fire began to work its way into Forest Service policy.

A key part of this process of change appeared in the 1967 southwestern region *Multiple Use Management Guide*. As a traditional piece of Forest Service literature, management guides provided information on a wide range of forestry activities such as proper logging and timber practices, watershed protection, wildlife management, and, of course, fire policy. Following the example of the guides that preceded it, the 1967 manual renewed the call to more aggressively seek complete fire suppression. Reflecting the shift to a new fire strategy, however, regional officials also suggested that foresters look to fire management rather than total suppression as the best way to balance forest needs with the presence of wildfire. By working with fire and not against it, a growing percentage of foresters believed that the elusive goal of ecological stability could be restored to the forests of the Southwest. Supporting this position, Southwest Forest Inspector C. K. Collins noted in 1967 that "forest plans . . . dating back to 1911, show some of the trends toward complete fire protection, which has us in trouble today. . . . Fire plays a major role in the silviculture used by nature."[59] In an attempt to understand this role, foresters attempted to increase the role of fire in the Southwest.

The commitment to these new fire management principles became more official in practice with the publication of a regional fire plan in 1972.[60] While regional Forest Service officials still supported large-scale suppression, they wanted to limit these efforts by reducing the regional fire danger. For example, half a century of extensive logging in the Southwest had resulted in large collections of scattered slash and debris. These massive fuel concentrations had powered many of the largest fires of the 1950s and 1960s. The new fire management plan called for the elimination of these debris stockpiles through the use of "controlled burning," the same style of controlled fire that Aldo Leopold had vehemently argued against in the 1920s. Rather than continue the struggle to totally suppress fire, after more than sixty years of exclusion the Forest Service had come to realize the importance of reintroducing fire to the southwestern landscape.

As part of the growing awareness of and concern for the environment in America in the 1970s and 1980s, Forest Service officials began to reevaluate the natural role of fire. Fire exclusion, so long the dominant paradigm in regional forest management, now seemed to pose as serious a threat to forest health as did overcutting, road building, and overgrazing. With fire now reinstated as a basic element on the periodic table of nature, many foresters in the Southwest believed that aggressive fire fighting should be applied more wisely to match forest conditions. Building on the concept of controlled burning, the Forest Service also began to implement other strategies such as "prescribed fire" and "natural burns." Like a medicine cabinet antidote for forest ills, prescribed fire became a common Forest Service practice in the 1970s and 1980s. This policy was seen as the best way to keep fire in the forested Southwest while still maintaining human control over fire behavior. Unfortunately, the growing use of prescribed fire in the region led to the creation of a false fire dichotomy for many foresters. Now, controlled or prescribed fire became "friendly fire," while uncontrolled wildfire remained an "enemy" to the forest. This stance against natural fire hampered efforts to increase the presence of fire in the Southwest and set the stage for the final policy shift of the 1990s.

In October 1978 the Department of Agriculture prepared a management guide for American wilderness areas. Designed to reflect "emerging state of the art approaches" to land management, this publication was written to give land managers a blueprint for administering the nation's wildlands. With regard to wildfire the guide stated that "fire exclusion is really a large-scale ecological experiment because . . . we have no way of foretelling the ultimate consequences. If the policy succeeds, there will probably be major changes in both vegetation and wildlife populations and perhaps also major breakdowns in the productivity, diversity, and stability of the ecosystem."[61] Probably changes? Perhaps breakdowns? With the benefit of more than two decades of hindsight, it is safe to dispense with such caution. By the late 1980s and early 1990s, even the staunchest detractors of fire had realized the need to bring fire back into the Southwest. Today, the same types of federal management guides are far more succinct: "fire exclusion . . . is a clear recipe for disaster."[62]

The challenge facing the Southwest is how to devise more enlightened management plans that consider fire as central to forest health. Unfortunately, the same complex mix of culture and geography that makes the fire regime of the Southwest so unique now makes the use of fire as part of the larger program of forest stabilization extremely difficult. Like the controversy surrounding the reintroduction of wolves to our national parks, the problem for foresters in the Southwest is that many researchers believe that fire exclusion has caused such extensive ecological damage that many southwestern woodlands may not return to "normal" for centuries. Simply bringing fire back will not instantly revitalize regional forests. In fact, some southwestern forest ecologists believe that "naturally occurring fires will likely never play a role in most ecosystems again because of the changes that have occurred with Euro-American settlement."[63]

Most ecosystem analysts are not as pessimistic about the future of southwestern forests, but even so Forest Service reports—prompted by events like the Dude Fire—reveal that foresters are facing a "fairly narrow 15- to 30-year window" to remedy the problems of the region's forests before large conflagrations, insects, and disease make forest management a moot issue.[64] As with the success of any federal program, questions of economic feasibility will most likely determine the extent to which the Southwest will incorporate fire into regional forests. Other factors, from smoke management to the ever present danger of escaping burns, continue to make forest officials reluctant to embrace an aggressive fire policy.

As we enter a new century, however, there are encouraging signs that local, state, and federal officials are beginning to significantly reevaluate their understanding of fire in the Southwest. Prompting these changes was a number of fires that charred large sections of the region. In May 2000, for example, a controlled burn at the Bandelier National Monument exploded into the Cerro Grande fire, the largest in New Mexico history. It burned more than 50,000 acres, destroyed more than two hundred homes in nearby Los Alamos, and left damages in the hundreds of millions of dollars. After the fire, angry (and now homeless) residents demanded to know how a small, controlled burn could become so devastating. Claiming federal responsibility, Interior Secretary Bruce Babbitt blamed the fire on a "seriously flawed" burn plan and "multiple failures" in regional fire policy, and he promised to revise forest management goals in the Southwest. New Mexico's congressional delegation did not want to wait for Babbitt's reappraisal, and in the weeks after the Cerro Grande wildfire, they drafted federal legislation that would, in the words of Senator Jeff Bingaman, "avoid these types of catastrophes in the future." A proposal by Senator Pete Domenici included the reintroduction of small-scale logging by family companies and "historical practices" like the collection of dead wood for firewood to reduce forest clutter. Perhaps, in the search for improved forest management policies, New Mexico politicians will avoid knee-jerk reactions and factor complex historical and ecological contingencies into future management plans.

For fires will burn in the Southwest. The issue before the Forest Service and the residents of the region is how to relate to fire within an evolving culture that is concerned with nature. Fire suppression has not (and will not) eliminate forests from the American Southwest. The question before regional foresters is not whether forests will survive but, rather, whether the forested landscape of the region will be one that we want and value and that includes a high degree of diversity and stability. Instead, will it be degraded, unstable, and, in the end, susceptible to catastrophic fires and disease? Putting fire back into the twenty-first-century landscape is going to be much more difficult than was taking it out in the nineteenth century. Fire must play a larger role in regional ecosystems if southwestern forests are to thrive in the future.

NOTES

1. For more information on the Dude Fire, see Eldon W. Ross et al., "Accident Investigation Report: Dude Fire Incident, Multiple Fire Fatality" (Tonto National Forest, 1990). Copies available from the Regional Headquarters in Albuquerque, N.M., on a Freedom of Information Act request. Additional information on the sequence of events comes from the experiences of the author and interviews of eyewitnesses, July 1990.

2. John N. Rinne and Daniel G. Neary, "Fire Effects on Aquatic Habitats and Biota in Madrean-type Ecosystems: Southwestern United States," in *Effects of Fire on Madrean Province Ecosystems*, U.S. Department of Agriculture General Technical Report RM-GTR-289 (Fort Collins, Colo.: Rocky Mountain Forest and Range Experiment Station, 1996), 138–40.

3. Other agencies, including the National Park Service, the Bureau of Land Management, the Bureau of Indian Affairs, State Forestry Departments, and the Fish and Wildlife Service, have also formulated fire management plans, but for this essay I use the U.S. Department of Agriculture, Forest Service as the primary agency involved in fire management.

4. The literature on the topography of the Southwest is obviously immense, but for a quick overview of the major regional features, see Henry P. Walker and Donald Bufkin, *The Historical Atlas of Arizona* (Norman: University of Oklahoma Press, 1986), and

Warren A. Beck and Ynez D. Haase, *The Historical Atlas of New Mexico* (Norman: University of Oklahoma Press, 1969).

5. Again, the literature on weather patterns is legion, but for more information, see H. R. Beyers and R. R. Braham, *General Meteorology* (New York: McGraw-Hill, 1959); Sverre Petterssen, *Weather Analysis and Forecasting* (New York: McGraw-Hill, 1956); and Donald Fuquay, A. R. Taylor, R. A. Howe, and C. W. Schmidt, "Lightning Discharges That Caused Forest Fires," *Journal of Geophysical Research,* April 20, 1972, 2156–58. For additional information, see Stephen J. Pyne, *Fire in America: A Cultural History of Wildland and Rural Fire* (Princeton, N.J.: Princeton University Press, 1982), 514–16.

6. National Climatic Center, *Local Climatological Data,* vol. 99, nos. 1–12 (Asheville, N.C.: National Oceanic and Atmospheric Administration, 1996).

7. National Climatic Center, *Annual Climate Information,* vol. 83 (Asheville, N.C.: National Oceanic and Atmospheric Administration, 1995).

8. Steve Yurich, *The Natural Role of Fire,* U.S. Department of Agriculture, Northern Region, Missoula, Mont. (Washington, D.C.: Government Printing Office, 1973).

9. Stephen Pyne, *World Fire: The Culture of Fire on Earth* (New York: Henry Holt, 1995), 283.

10. Ibid.

11. U.S. Forest Service, *Facts about the National Forest in the Southwest* (Albuquerque, N.M.: USDA, Forest Service, September 1983); *Southwest Area Annual Fire Report for 1992* (Washington, D.C.: U.S. Department of Agriculture, 1993), 11–13; *Southwest Area Annual Fire Report for 1993* (Washington, D.C.: U.S. Department of Agriculture, 1994), 14–20; *Southwest Area Annual Fire Report for 1994* (Washington, D.C.: U.S. Department of Agriculture, 1995), 13–18; and *Southwest Area Annual Fire Report for 1995* (Washington, D.C.: U.S. Department of Agriculture, 1996), 14–18.

12. The literature on the prehistory of the Southwest is especially rich. For more information, see Nancy Fox, Helen Bumenschein, and Alden Hayes, *The Prehistory and History in the Southwest* (Santa Fe, N.M.: Ancient City Press, 1985).

13. For information on Native Americans and the environment, see Richard White, "American Indians and the Environment," *Environmental Review* 9 (Summer 1985): 101–3; Richard White, "Native Americans and the Environment," in William Swagerty, ed., *Scholars and the Indian Experience: Critical Reviews of Recent Writing in the Social Sciences* (Bloomington: Indiana University Press, 1984), 179–204; Richard White and William Cronon, "Ecological Change and Indian-White Relations," in *Handbook of North American Indians,* gen. ed. William C. Sturtevant, 20 vols. (Washington, D.C.: Smithsonian Institution Press, 1978–98), 4:417–29. For more specific information on Indian burning, see Richard White, "Indian Land Use and the National Forests," in Harold K. Steen, ed., *Origins of the National Forests: A Centennial Symposium* (Durham, N.C.: Forest History Society, 1992), 173–79, and John Herron, "We Were Born to It: Native Americans, Fire, and Environmental Cultural Continuity," in John Herron and Andrew Kirk, eds., *Human/Nature: Biology, Culture, and Environmental History* (Albuquerque: University of New Mexico Press, 1999).

14. Thomas W. Swetnam, "Fire History and Climate in the Southwestern United States," in *Effects of Fire Management on Southwestern Natural Resources,* U.S. Department of Agriculture General Technical Report RM-191 (Fort Collins, Colo.: Rocky Mountain Forest and Range Experiment Station, 1990), 6–17; Thomas W. Swetnam and Christopher H. Baisan, "Fire Histories of Montane Forests in the Madrean Borderlands," in *Effects of Fire on Madrean Province Ecosystems,* 15–36; and Conrad Bahre, *A Legacy of Change: Historic Human Impact on Vegetation in the Arizona Borderlands* (Tucson: Uni-

versity of Arizona Press, 1991), 124–42.

15. See Donald Meinig, *The Southwest: Three Peoples in Geographical Change, 1600–1970* (New York: Oxford University Press, 1971); and Jack D. Forbes, *Apache, Navajo, and Spaniard* (Norman: University of Oklahoma Press, 1960).

16. The more nomadic native groups of the Southwest took longer to adopt Euro-American livestock, and in fact, up until the 1870s, many of these groups, like the Apaches, continued to use the seasonal burning of the region to create landscape mosaics more to their liking.

17. Charles Bent, "Letter," in *Report in Relation to California and New Mexico*, 31st Cong., 1st sess., 1851, S. Exec. Doc. 18.

18. Charles F. Cooper, "Changes in Vegetation, Structure, and Growth of Southwestern Pine Forests since White Settlement," *Ecological Monographs* 30 (1960): 129–64.

19. William D. Rowley, *U.S. Forest Service Grazing and Rangelands* (College Station: Texas A&M University Press, 1985), 17–18.

20. Cooper, "Changes," 132.

21. Ibid.

22. See Christopher J. Huggard, "The Impact of Mining on the Environment of Grant County, New Mexico to 1910," *1994 Annual of the Mining History Association* (1994): 3–8; quote from Pyne, *World Fire*, 288.

23. Robert D. Baker, Robert S. Maxwell, Victor H. Treat, and Henry C. Dethloff, *Timeless Heritage: A History of the Forest Service in the Southwest* (Albuquerque, N.M.: USDA, Forest Service, August 1988), 19.

24. For more information, see Harold K. Steen, *The U.S. Forest Service: A History* (Seattle: University of Washington Press, 1976).

25. Baker et al., *Timeless Heritage*, 109.

26. *Albuquerque Journal*, October 9, 1907, 2.

27. Edwin A. Tucker and George Fitzpatrick, *Men Who Matched the Mountains: The Forest Service in the Southwest* (Albuquerque, N.M.: USDA, Forest Service, 1972), 59.

28. Quoted in Pyne, *Fire in America*, 260.

29. Baker et al., *Timeless Heritage*, 110.

30. For more information on the administration of William Greeley, see Pyne, *Fire in America*, 262–72.

31. Earl W. Loveridge, "Fire Suppression Policy of the U.S. Forest Service," *Journal of Forestry* 42, no. 7 (July 1944): 551.

32. Aldo Leopold, "Grass, Brush, Timber, and Fire in Southern Arizona," *Journal of Forestry* 22, no. 6 (October 1924): 6–7.

33. For a transcript of the *Southwestern Magazine* article and editorial commentary, see David E. Brown and Neil B. Carmony, eds., *Aldo Leopold's Southwest* (Albuquerque: University of New Mexico Press, 1995), 139–45.

34. Quoted in Paul W. Hirt, *A Conspiracy of Optimism: Management of the National Forests since World War Two* (Lincoln: University of Nebraska Press, 1994), 32.

35. Stephen Fox, *The American Conservation Movement: John Muir and His Legacy* (Madison: University of Wisconsin Press, 1981), 183–90.

36. Pyne, *Fire in America*, 275.

37. Fox, *The American Conservation Movement*, 189.

38. Quoted in Pyne, *Fire in America*, 324. For the complete report, see E. I. Kotok et al., *A National Plan for American Forestry*, 2 vols., 73rd Cong., 1st sess., 1933, S. Doc. 12.

39. See Alison T. Otis, William D. Honey, Thomas C. Hogg, and Kimberley K. Lakin,

The Forest Service and the Civilian Conservation Corps: 1933–1942 (Washington, D.C.: U.S. Department of Agriculture, August 1986); John C. Paige, *The Civilian Conservation Corps and the National Park Service, 1933–1942: An Administrative History* (Washington, D.C.: U.S. Department of the Interior, 1985); Clyde P. Moose, Teri A. Cleeland, Patrick J. Putt, and W. H. Mall, *People and Places of the Old Kaibab* (Albuquerque, N.M.: USDA, Forest Service, September 1990); and Edwin A. Tucker, *Sourcebook of Southwestern Regional History*, books 1–3, Cultural Resource Management Report No. 12 (Albuquerque, N.M.: USDA, Forest Service, July 1992).

40. John A. Salmond, *The Civilian Conservation Corps, 1933–1942: A New Deal Case Study* (Durham, N.C.: Duke University Press, 1967), 174.

41. *Statistics: National Forests, Southwest Region, Arizona and New Mexico* (Albuquerque, N.M.: USDA, Forest Service, 1940), 7–9.

42. Quoted in Pyne, *Fire in America*, 365.

43. Quoted in Fox, *The American Conservation Movement*, 210.

44. Tucker and Fitzpatrick, *Men Who Matched the Mountains*, 162–63.

45. Pyne, *Fire in America*, 360–70.

46. Quoted in ibid., 176.

47. Jack Shepard, *Forest Killers: The Destruction of the American Wilderness* (New York: Weybright and Talley, 1975), 293; Pyne, *Fire in America*, 395–97.

48. Clint Davis, "Only You Can Prevent Forest Fires," *American Forests* 57, no. 4 (1951): 6–10, 40–41.

49. Pyne, *Fire in America*, 373–75.

50. *A History of Smokejumping* (Missoula, Mont.: USDA, Forest Service, 1968), 11–16.

51. Pyne, *Fire in America*, 378.

52. Unfortunately, information on the SWFFF program is both limited and dated, but for more information, see K. O. Wilson, "Operation Redskin," *Fire Control Notes* 13 (July 1952): 1–3; Clarence K. Collins, "Indian Firefighters of the Southwest," *Journal of Forestry* 60, no. 2 (January 1962): 87–91; and Pyne, *Fire in America*, 378–80.

53. Baker et al., *Timeless Heritage*, 115.

54. Shepard, *Forest Killers*, 288.

55. Arthur A. Brown and Kenneth P. Davis, *Forest Fire: Control and Use*, 2d ed. (New York: McGraw-Hill, 1973), 292.

56. *Interagency Hotshot Crew Operations Guide* (Boise: USDA, Forest Service, 1989), 4.

57. Lynn Biddison, "The Changing Role of Fire Management," *Fire Management Notes* (Winter 1977–78): 20.

58. Quoted in Tom Knudson, "How the West's Asbestos Forests Were Turned into Tinderboxes," *High Country News*, March 6, 1995, 9.

59. Quoted in Baker et al., *Timeless Heritage*, 116.

60. See Biddison, "The Changing Role of Fire Management," 19–21.

61. John C. Hendee, George H. Stanley, and Robert C. Lucas, *Wilderness Management*, U.S. Forest Service Miscellaneous Publication No. 1365 (Washington, D.C.: U.S. Department of Agriculture, 1978), 265.

62. Christopher H. Baisan and Thomas W. Swetnam, *Interactions of Fire Regimes and Land Use in the Central Rio Grande Valley*, U.S. Department of Agriculture General Technical Report RM-RP-330 (Fort Collins, Colo.: Rocky Mountain Forest and Range Experiment Station, 1997), 19.

63. Stephen S. Sackett, Sally M. Haase, and Michael G. Harrington, "Prescribed Burning in Southwestern Ponderosa Pine," in *Effects of Fire on Madrean Province Ecosystems*, 178.

64. W. Wallace Covington and Margaret M. Moore, "Southwestern Ponderosa Forest Structure: Changes since Euro-American Settlement," *Journal of Forestry* 92, no. 1 (January 1994): 45.

7

Struggle in an Endangered Empire

The Search for Total Ecosystem Management in the Forests of Southern Utah, 1976–1999

THOMAS G. ALEXANDER

Before the late 1970s most Forest Service employees had tended to view their jobs as managing the traditional functions codified in the Multiple Use Sustained Yield Act of 1960. Some foresters called this approach "functionalism."[1] Although foresters had understood since Gifford Pinchot served as chief forester after 1905 that they must measure their success by the condition of the land, especially of watersheds, and they spoke of multiple use, they tended to classify functions such as logging, grazing, recreation, wilderness, and wildlife separately, though they recognized their interrelationship.

As foresters prepared the forest plans under the National Forest Management Act between 1976 and 1990, they began to recognize even more than before that these functions bore an inseparable relationship to each other. If, for instance, grazing changed the condition of the land, it also affected other components of the ecosystem such as the watersheds, wildlife habitat, recreational opportunities, and the health of trees. With this recognition, many foresters began to speak of managing ecosystems rather than functions. In assessing this change, Supervisor Hugh C. Thompson of the Dixie National Forest in southern Utah's red rock country called ecosystem management "the next plateau above multiple use management."[2]

Recognizing the change in perception that had placed ecosystem management in a central position, Dale Robertson, appointed as chief of the Forest Service in 1987, called on national forest personnel to promote holistic ecosystem management. By this he meant that employees must learn to judge the effects of management decisions by the health of entire ecosystems rather than solely by the health of individual features such as forests, ranges, or even watersheds.

Jack Ward Thomas, whom President Bill Clinton appointed as chief in 1993, echoed the same theme. Clinton selected Thomas from the ranks of wildlife biologists; the previous occupants of the chief's chair had been foresters, with the exception of one engineer. Moreover, in a tradition dating back to Theodore Roosevelt but often neglected by recent presidents, Clinton and especially his vice president, Al Gore, took part in national forest decisions. For instance, Clinton helped negotiate a solution to the northern spotted owl controversy, and he approved salvage sales of fire- and insect-damaged trees.[3]

In 1997 the new chief forester, Michael Dombeck, expected to advance the cause of ecosystem management even more. A fisheries biologist and former acting director of the Bureau of Land Management, Dombeck was the second chief without previous national forest line experience. He said that he would hold each forest supervisor "accountable for key areas such as streamside condition and health, water quality, watershed health, noxious weed management, and endangered species habitat management and protection."[4]

Reflecting on the emphasis on ecosystem management, some foresters like R. M. "Jim" Nelson, supervisor of the Toiyabe National Forest in western and southern Nevada, argued that the Forest Service had only two reasons to exist. First, he said, the service had to manage the ecosystem. Second, since the agency received its budget and directions from Washington, employees also had to manage their relations with the current administration and with Congress.[5] In view of the diverse groups that pressed their views on foresters and on Congress and the administration, the two duties seemed frequently to conflict.

Nevertheless, in an attempt to implement holistic ecosystem management, some foresters began to reenvision their roles in quite untraditional ways. In 1993 Nelson argued that officers should perceive national forest users as tools for meeting "ecological objectives" rather than as customers or as shareholders. For instance, he said, ranchers herded their livestock onto the forest. Rather than viewing cows as producers of red meat, he thought the Forest Service should consider them as one of the components of the ecosystem just like deer, elk, birds, plants, recreationists, and loggers. In his view, each of these was one of the tools foresters had at hand to use in achieving management objectives. If any of these components did not contribute to the objective of sound ecosystem management, foresters had to find a way to fit them in in order to develop and maintain "healthy, sustainable, and diverse ecosystems for public benefit and use."[6]

The conflict between holistic ecosystem management and the drive to cut more trees surfaced frequently within the Forest Service, perhaps nowhere more forcefully than at the Sunbird Conference held in Tucson in November 1989. At the conference, a number of forest supervisors from the interior West (the Northern, Rocky Mountain, Southwestern, and Intermountain Regions) emphasized the conflicts over timber management.[7] Their message in the words of Kent Taylor, then supervisor of the Fishlake National Forest in southern Utah and a participant in the conference, amounted to this: "We shouldn't be cutting trees, or we're cutting too much timber." In Taylor's view, some in the Washington office (he specifically mentioned Assistant Chief Forester George Leonard) wanted to "get the cut out no matter what." In order to emphasize the need for ecologically beneficial forest management, a number of the supervisors drafted a letter to Robertson in which they charged that the Washington office did not involve them sufficiently in decision making. The signers pointed out that values both in the public and among foresters had changed and that anger over failure to reduce the pressure from Congress and the Washington office to cut more timber had reached a flash point. Allowable sale quantities, a term that had replaced previous designations such as allowable cut, they said, had remained far too large for the health of the ecosystems.[8]

Though most viewed the conference as a revolt of the supervisors against the pressure to damage the ecosystem through excessive cutting, a number of supervisors, including Hugh Thompson, viewed it as a dialogue with Chief Robertson. In Thompson's view, "The chief listened to our concerns and to our complaints. He responded to them. It was not a revolt. I think it was an honest dialogue."[9] Nevertheless, in the year following the Sunbird Conference, pressure to increase the volume of logging brought about the resignation of Northern Regional Forester John Mumma. Mumma, who had served as director of fish and wildlife in the Intermountain Region and had worked in the Washington office and as the Northern Region's deputy for resources before assuming the job of northern regional forester, had developed a reputation as an extremely conscientious and popular officer.

In the Northern Region, Mumma had found what appeared to him an insoluble problem. In part because of the large bodies of inventoried roadless areas, which were closed to logging, and in part because of a perceived need to protect fragile ecosystems, Mumma found it impossible to meet the harvesting goals that Congress and the Washington office had set. Under the circumstances, Mumma told his forest supervisors to protect the ecosystem rather than cut too much timber. Threatened with a disciplinary transfer, Mumma resigned.[10]

As events surrounding the Sunbird Conference and Mumma's resignation show, the struggle for holistic ecosystem management took place in the context of a conflict between partisans of a wide range of views on proper forest management, including some environmentalists bent on eliminating all logging and grazing from the national forests, on the one hand, to certain logging companies and ranchers who demanded maximum commodity production, on the other. Supervisors were left with the unenviable task of balancing irreconcilable demands while trying to develop and maintain healthy forest ecosystems.

In some cases, the Washington office did not make their task easier. In addition to punishing ecologically friendly foresters like Mumma, Washington siphoned more money to support its bureaucracy. Despite Congress's stingy appropriations during the late 1980s, the Washington office staff actually increased by about 60 percent from fewer than eight hundred people to nearly thirteen hundred during Robertson's tenure. The larger Washington office staff together with the staffs of the regional and supervisor's offices ate up about 65 percent of the Forest Service's budget before any money got to the ranger districts for management and mitigation.[11]

As the various national forests prepared their plans during the 1980s, the officers quickly learned that the majority of the public perceived the purposes of the forests much differently in the late 1980s than in the 1950s. An increasingly urban population had become more concerned with access to recreation of various sorts than with supplying more timber and grazing opportunities. Many of these urbanites tended to view the protection of wildlife and what they perceived as pristine forests and meadows as more important than any commodity functions. Their views were not uniform, however, since some demanded increased access for the expansion of ski resorts or off-highway vehicles.

In this context, George Olsen, former Intermountain Region director of lands and recreation, said, "The public told us through that whole forest planning cycle [from 1976 to 1990] that they felt that the national forests were not being managed in adequate balance." Olsen said that the public wanted a greater emphasis on outdoor recreation and environmental values, which they associated with recreational experiences, not with commodities.[12] By the early 1990s, Congress had designated one of every six acres in the national forest system as wilderness, and the average American spent twelve hours each year in the national forests. Public patronage of such noted attractions as the national parks and of theme parks like Disneyland and Magic Mountain paled by comparison.[13]

In a shift already under way by the 1960s as grazing and timber diminished in importance in the Mountain West, by 1983 recreation had become the "largest single program component in [the Intermountain] region's budget."[14] Responding to these changes, the composition of the national forest work force changed as well. Although many of the Forest Service personnel still came from the ranks of traditional timber and range management specialists, an increasingly large number included soil scientists, hydrologists, wildlife managers, landscape architects, and ecologists.[15]

Although recreation assumed an unprecedented importance in the Intermountain Region, in line with their multiple-use mandate, the officials of southern Utah's forests along with others in the national forest system believed they should manage amenity values along with commodity values within the total ecosystem. Often this proved extremely difficult since many users did not understand the difference between national forests and national parks, which had no mandate to manage for multiple use. Moreover, the public often confused the BLM, which tended to manage areas with few forest values, with the Forest Service. Thus, in order to implement total ecosystem management, forest officers had to consider the relationship between recreation and other amenity uses such as wildlife, wilderness, scenic vistas, and water quality as they related to such commodity uses as timber, mining, oil and gas extraction, and grazing.[16]

Unlike Nelson, who viewed animals, plants, and people as tools, most foresters preferred to think of the public as customers or owners who could contribute to ecosystem management. By the 1980s, however, getting forest users to help with management proved more difficult than in previous years in part because of their changing backgrounds and in part because of the larger number of people who demanded goods and services. This was an extremely serious problem, especially in high-demand recreation areas like southern Utah. The Dixie, Manti-LaSal, and Fishlake National Forests, units in the Intermountain Region, lie adjacent to Utah's major recreational marvels. These include national parks like Zion, Bryce, Canyonlands, Capitol Reef, and Arches; national monuments like Cedar Breaks, Rainbow Bridge, Natural Bridges, Escalante–Grand Staircase, and Hovenweep; recreation areas like Glen Canyon; and the mountain biking and jeep trails of southeastern Utah. They also serve as home to or lie near statutory wilderness like Pine Valley, Box-Death Hollow, and Dark Canyon; de facto wilderness like Fishlake Hightop and Grand Gulch Primitive Area; and numerous BLM Wilderness Study Areas. Much of the habitat for wildlife and the sites for recreation in southern Utah rest within the national forests.[17]

Moreover, the management of recreation proved extremely difficult since recreationists have tended to believe that, in contrast to logging, grazing, and mining, their activities have caused little or no damage to forest ecosystems.[18] Recently, however, environmental scientists have raised serious objections to this view. Questioning this perception, in a letter to the *High Country News* Richard L. Knight, a professor of wildlife conservation at Colorado State University, argued that "outdoor recreation was the second largest cause (after dams) for the decline of endangered and threatened species." Knight listed skiers, snow-boarders, kayakers, rafters, climbers, bikers, and hunters as examples. Recreationists were, he suggested, "more numerous than chain saws . . . [and] more intelligent than cows."[19]

An article by David S. Wilcove and his associates published in *BioScience* in August 1998 attempted to quantify the threats by various activities to the plants and animals listed under the Endangered Spe-

A party of deer hunters with off-road vehicles, Fishlake National Forest, Utah, 1990. Much controversy has surrounded motorized vehicle recreational use in the national forests. Courtesy U.S. Forest Service, Intermountain Regional Office, Ogden, Utah; photo by Bert Lowry

cies Act of 1973. In what may seem a counterintuitive conclusion, the study found recreation a greater threat than grazing, logging, or mining and nonvehicle recreation a greater threat than off-road vehicles. Relying on available data, the study quantified the major threats to the percentage of species listed under the ESA as outdoor recreation, including off-road vehicles (27 percent of listed species), grazing (22 percent), nonvehicle outdoor recreation (14 percent), logging (12 percent), and mining (11 percent). According to this analysis, outdoor recreation threatened a significantly higher proportion of plants (33 percent) than of animals (17 percent).[20]

Under the circumstances, the diverse composition of the recreation public generated both ecosystem damage and management difficulties. In contrast to the rural background of early forest users, by the 1980s most recreationists came from urban backgrounds. Since most urbanites knew little from personal experience about actual forest conditions, national forest officers recognized that if they intended to get cooperation from the public, they had to provide education in aspects of forest management.[21] In attempting to accomplish this goal, the Forest Service inaugurated programs such as Tread Lightly to minimize the damage caused by motor vehicles and Leave No Trace to educate forest users in "outdoor ethics." In promoting the Tread Lightly program, a term that originated with Cliff Blake of the Intermountain Region, foresters cajoled the automotive industry into acting as an ally. Regional officials convinced car manufacturers that they might suffer financially if land abuse forced the closure of large blocks of national forest land to motor vehicles. Nevertheless, in an attempt to control the damage caused to watersheds by off-highway vehicles, forest planners prepared travel management plans, often as they drafted the forest plans. National forest employees also offered travel maps and Tread Lightly orientations to schools and other organizations. They also released information to newspapers to provide details to the public on the various programs.[22]

In some cases, the adoption of travel management plans stoked the fires of controversy. For instance, people who lived near the Fishlake had regularly driven off-highway vehicles virtually everywhere in the national forest. Supervisor Kent Taylor estimated in 1993 that "every year we were getting about one hundred miles of new roads or access travel ways, . . . mostly during hunting season."[23] Significantly, the only appeal on the Fishlake's 1986 forest plan came not over the relatively inconsequential volume of timber forest officers planned to harvest but over vehicle access to a roadless area. As part of the plan, Taylor and his staff decided to close all of Fishlake Hightop west of Fish Lake to motorized vehicles. Objecting to this proposal, Richfield Mayor Sue Marie Young and a number of her constituents condemned the closure of a jeep trail that led to the top of the plateau. Taylor, on the other hand, believed that if the jeep trail remained open, forest officers could never successfully control off-highway vehicles on the plateau.[24]

After the Richfield residents objected, the Republican members of Utah's congressional delegation (Senators Orrin Hatch and Jake Garn and Representatives Howard Nielson and Jim Hansen) appealed the forest plan. Dick Carter of the Utah Wilderness Association intervened in the appeal on behalf of the Forest Service. The appeal went to Regional Forester J. S. "Stan" Tixier. Negotiating a compromise, the parties agreed

A road closed under the road closure program, Fishlake National Forest, ca. 1990. Threats to the environment have forced the Forest Service to protect roads, trails, and mountain terrain formerly subjected to the impact of motorized vehicles on forest ecosystems. Courtesy U.S. Forest Service, Intermountain Regional Office, Ogden, Utah; photo by Bert Lowry

to leave the road open but to prohibit motorized vehicles on the plateau.[25] Although the parties to the negotiations agreed to the compromise, it is unclear just how forest officers would control off-highway vehicle movement once drivers reached the summit. In view of the excessive damage inflicted by motorized vehicles, a number of the national forests have closed areas to various types of travel.[26] For instance, the Fishlake closed large blocks of the Salina Creek drainage to motorized vehicles.[27]

Conflicts over various recreational uses, like the conflict between commodity and amenity users, constituted part of the continuing battle for the soul of the national forests. Backpackers often dislike all-terrain vehicles. Some campers who enjoy wild parties, loud music, and blaring televisions taint the outdoor experience of those who value solitude and sublime vistas. These and similar conflicts became so serious that the Forest Service adopted regulations to manage controversial practices. Astoundingly, the adoption of regulations to control noise generated opposition among such local rights advocates as Chuck Cushman of the American Land Rights Association and Met Johnson, at the time an outspoken state legislator from southern Utah and leader of the legislature's Cowboy Caucus. These objections apparently derived from the laissez-faire perspective of those who opposed any interference with the use of public lands.[28]

Although the forests experienced a rapidly increasing flood of users, by the early 1990s Congress generally expected the Forest Service to support its recreation program from reduced appropriations. Moreover, with the exception of relatively modest campground, summer home, marina, special attraction, and ski area fees, the Forest Service had failed "to address the issue of pay for play."[29] By contrast, congressional appropriations had subsidized logging at relatively high levels, and logging roads or roads constructed for other purposes generally served a dual purpose by providing access for the recreational public as well.[30] Congress subsidized grazing as well by charging permit fees far below the prevailing market rate for comparable private land, and an attempt to raise them to the market level during the first term of Clinton's administration withered under adverse political pressure.

Some observers recognized that Congress might have made funds available for the purchase of recreational facilities from the Land and Water Conservation Fund. After all, payments for offshore oil drilling brought about $900 million into the fund each year. Instead of appropriating most of the money to meet needs on the public lands, however, Congress sidetracked 78 percent for deficit reduction and other programs, leaving only about $200 million a year for programs on the public lands.[31]

In an attempt to meet budgetary shortfalls while the number of recreationists increased, the Forest Service began to change the way it managed campgrounds. Before 1985 the Forest Service had established campgrounds in places that seemed to fit public recreational patterns. Like virtually all national forest activities, including logging and grazing, campgrounds failed to pay their own way. In a perennial problem that had become increasingly severe as pressure for the amenity uses of the public lands increased, the cuts in federal budgets coupled with the perceived need to allocate larger sums to planning, management, and conflict resolution left insufficient funds to keep recreational facilities in a condition the public expected.

Because of increasing demand in an age of tight budgets, the Forest

Service began to rely on volunteers and eventually on private enterprise. As a first step, it brought in volunteers to work as "campground hosts." Then in 1985 it began to invite proposals from private businesses to schedule the camping spaces, collect the fees, and provide the various campground services. In an ultimately successful experiment, in 1985 L & L, based in Orem, Utah, contracted to manage two campgrounds on the Uinta National Forest. The experiment proved so successful that by 1993 the company had assumed the management of nearly six hundred national forest campgrounds in seventeen states.[32] In 1989, in an attempt to provide better service to recreationists, the Intermountain Region negotiated an agreement with the state of Utah to provide a campground reservation system for national forests and state parks. By calling an 800 number, people could make reservations for sites in various campgrounds.[33]

When L & L began managing the campgrounds, it found them in disorder and disrepair. A shortage of funds had sidetracked plans for repairs and improvements. National forest employees assigned to other functions generally left work at 5:00 P.M., about the time that most people sought a camping site for the evening. The Forest Service had the authority to collect campground fees, but Congress directed the fees into the federal treasury, and it often failed to benefit the campgrounds where the money was collected. Moreover, in a number of cases, fees collected for special recreational uses ranged below the market value of the goods and services furnished. A study released early in 1997 demonstrated that rental fees for marinas and cabin sites were generally lower than the market rate.[34] Instead of charging a market-based rental on lands used for ski areas, the service charged a percentage of the receipts.

One example from the Ashley National Forest demonstrates the attempts to solve some of these problems. Though both the Ashley National Forest and the Flaming Gorge National Recreation Area, a unit in the forest, collected campground fees, neither had the funds to keep camping facilities in repair. As a result, a typical campground at Flaming Gorge offered the public weathered picnic tables, scattered haphazardly throughout a weed-choked area, and a gravel-covered and decaying boat ramp. The Ashley National Forest addressed the problem by turning to L & L. After contracting to manage several campgrounds at Flaming Gorge, L & L repaired the facilities and earned a good reputation for itself and for the Forest Service in the view of many who used the campgrounds. The company solved part of the monetary shortfall by hiring retired couples to work as campground hosts. The hosts collected fees and monitored campground use. The retirees seemed satisfied with a small income that supplemented their retirement and Social Security checks in return for a national forest experience and a place to park their recreational vehicles that included hookups for their utilities.[35]

Many observers recognized that the Forest Service and other land management agencies could have offered a much more satisfactory experience if Congress had authorized them to collect additional user fees and to apply the proceeds to improvements at the point of collection. In 1993, for instance, a modest recreation user fee of $3 per day on central Utah's Uinta National Forest would have generated $9.3 million. This amount was $8.9 million more than the $389,000 per year the Uinta received in receipts for its meager timber sales. Randal O'Toole, executive director of Cascade Holistic Economic Consultants in Portland, Oregon, and a frequent critic of the Forest Service, advocated the adoption of such a fee primarily to restructure the incentives for forest use and to promote true multiple use.[36]

Utah Congressman Jim Hansen, who had often criticized both the Clinton administration's management of the public lands and the lobbying of environmentalists for more wilderness, suggested a similar approach. In an address to the Society of American Foresters in April 1996, Hansen said that he was "considering a bill that would require people to buy an annual pass before they could use public lands for recreation." Hansen's proposal generally received qualified support from representatives of the Izaak Walton League of America, the National Audubon Society, and the Forest Service.[37] Opposition to the fees came from a number of sources. Some liberals who generally support environmental causes opposed fees because they feared such charges might deny poor people access to national forest amenities. Others opposed them because they believed that access to the public lands should be free and unfettered.[38] Some disputed fees since they believed that dispersed recreation is not a consumable use and that the government ought to support nonconsumptive and not-for-profit uses through taxes paid for national forest administration.[39]

A major problem with each of these views is that, as the studies cited indicate, recreation damages an ecosystem, and funds are needed to repair or mitigate that damage. In December 1996, in part through

Hansen's efforts, Congress approved a three-year Demonstration Areas experiment that allowed the BLM and the national forest and national park systems to collect fees for recreational services on two hundred sites. The law permitted the agencies to use the fees to repair facilities at the units where they were collected but not to construct new ones. This seemed clearly necessary, since by mid-1997 the Forest Service faced a one-billion-dollar maintenance backlog.[40]

Under the Demonstration Areas program the service began to charge for access to certain areas. Late in 1996 the Forest Service in cooperation with the state of Utah and the county governments proposed to charge access fees for the Flaming Gorge National Recreation Area, Mirror Lake Highway in the Uinta Mountains, American Fork Canyon east of Utah Valley on the Uinta National Forest, Fishlake Campground on the Fishlake National Forest, and Joe's Valley Campground west of Orangeville on the Manti-LaSal National Forest.[41]

In some cases, instead of paying for services through user fees, the service has rationed access.[42] In 1996, following a six-year study of overcrowding and over the opposition of outfitters, the Ashley National Forest and the BLM limited the number of hand-powered rafts and boats and prohibited motorized craft on the Green River between Flaming Gorge Dam and the Colorado state line.[43]

As recreation assumed a larger role in the national forests, critics continued to complain that the Forest Service bent over backward to accommodate commodity users while ignoring those who favored amenities. Some forest officers have agreed that in the past the Forest Service had "put more emphasis on commodities than non-commodity values on the forest" largely because "Congress established" priorities that favored them. By the early 1990s, however, foresters like Gary Schaffran argued that the change in emphasis to "ecosystem . . . management with commodities as byproducts of that management" had eliminated this condition. The new emphasis on the ecosystem, he said, was "a much better philosophy for the country as a whole and for the forest."[44]

Schaffran's views mirrored those of the entire national forest system expressed in the drafting and implementing of forest plans during the late 1980s. In a press release in May 1993, the Forest Service's associate chief, George Leonard, pointed out that timber harvests on national forests had fallen steadily since FY 1989. Previously, the service had sold an average of 10 to 12 billion board feet annually.[45] Leonard estimated that the allowable sale quantities for FY 1994 would range between 4.1 and 4.6 billion board feet. Leonard attributed the decline to ecological considerations, citing specifically protection for threatened and endangered species such as spotted owls and salmon. At the time, the Forest Service anticipated further severe cutbacks in the volume of logging throughout the Intermountain Region.[46]

In spite of the decline in allowable sale quantities, the conflict between environmentalists and commodity interests escalated. In September 1994 an anonymous caller warned officials of the Fishlake National Forest that someone had driven spikes into Engelmann spruce trees scheduled for harvest on Thousand Lake Mountain. Conducting an inspection, the officials found six-inch spikes embedded in trees throughout a sixty-six-acre sale area.[47] In offering a ten-thousand-dollar reward for information leading to the conviction of the saboteurs (he used the term "terrorists"), Ralph Goddard, Fishlake law enforcement officer, cited the danger of spiking. He pointed out that spikes can shear off the teeth of sawmill saws, at times causing the disintegration of the blade and throwing shrapnel throughout the mill.[48]

In planning this sale, the Fishlake had used ecologically friendly procedures. A minor timber-producing forest, the Fishlake's allowable sale quantities were only about 3 million board feet per year. The Fishlake had conducted a four-year study of the sale area that included public hearings and the preparation of an Environmental Impact Statement (EIS). Opponents of the sale filed half a dozen protests, but forest officers concluded that logging offered the best way of controlling a spruce bark beetle infestation in the drought-stressed and dense timber stand. The timber purchaser, Thousand Lake Lumber Company, intended to shape the timber into cabin logs. Lloyd Chappell, owner of the company, said that in spite of the sabotage tactics, his firm planned to remove the spikes and proceed with the sale.[49]

This ecological sabotage came on the heels of the solution to some serious difficulties on the Fishlake. Supervisor Kent Taylor had joined the staff of the Fishlake in 1979 as forest officers embarked on preparing its forest plan. After hiring geologist Andy Godfrey as forest planner, Taylor put the planning process on a fast track. Since everybody "found out very early that public meetings were somewhat frustrating," Taylor tried to deal with the polarization through one-on-one contacts. In spite of the stress of addressing the demands of a far from unified public, no one, he said, actually suffered a mental breakdown, but "it

A lumber contractor uses heavy equipment to remove logs in Fishlake National Forest, ca. 1990. Courtesy U.S. Forest Service, Intermountain Regional Office, Ogden, Utah; photo by Bert Lowry

caused divorces," separations, and resignations. Forest officers completed their plan on June 13, 1986.[50]

Just as controversial as the Thousand Lake Mountain sale on the Fishlake were a number of sales on other forests of southern Utah, often undertaken for ecosystem restoration and to recover some economic value from beetle-infested trees.[51] By the late 1980s land management agencies had abandoned the expensive and controversial methods, including massive spraying with DDT, malathion, and other insecticides or piling and burning, that had been used during the 1940s, 1950s, and 1960s in an attempt to control insect infestations. All these methods were generally considered dangerous as well as ineffective, though foresters continued to spray trees in unusual and critical areas in and near campgrounds.[52] Spraying and treatment with biological controls and natural enemies are very expensive, and, according to forest entomologists, these methods do not always remedy the conditions that have caused such infestations. Logging does not always correct the problem either, since the beetles live in virtually every forest and can move to uninfected trees.

Conditions on the Dixie National Forest exemplify the problem. Like a number of forests in the Intermountain Region and throughout the interior West, by the early 1990s many stands on the Dixie consisted of old growth or old second growth that had lived beyond maturity.[53] Because of the post-1900 suppression of wildfires, the stands had grown, as retired Forest Service entomologist William H. Klein put it, "beyond maturity, becoming overly dense and begin[ning] to stagnate, since each tree must compete with its neighbor for light, moisture and nutrients."[54] Supervisor Thompson estimated that the stand densities averaged 250 to 350 square feet per acre. Under these conditions, the trees became targets for "a myriad of damaging natural agents," including bark beetles and fire.[55]

Beetles live in virtually every forest. In a forest subject to wildfire, wind throw, and other natural forces, the Engelmann spruce bark beetles generally attack "trees damaged by wind, snow, lightning, and . . . man," although outbreaks may occur at unpredictable times. Even before the Smokey the Bear campaigns, foresters on the Dixie as on most other national forests had suppressed wildfires. Weakened by decades of competition with other trees and five years of insufficient moisture, the spruce trees could not employ their natural defense mechanisms like secreting pitch to ward off the beetle attacks, and the beetles established thriving colonies in live trees.[56]

Contrary to the often-expressed opinions of critics about timber management on the Dixie, Thompson had reduced the timber cut during his tenure as supervisor. Reductions to compensate for environmental damage had placed the Dixie's allowable sale quantities at 26 million board feet in the 1986 plan. After the approval of the plan, Thompson argued that the forest's ecological system could only stand a cut of 17 million board feet. The regional office approved those reduced allowable sale quantities.[57]

In part because of the reduction in allowable sale quantities to 26 million board feet, the Dixie experienced very little opposition to the 1986 forest plan. The appeals on the plan, lodged by the Utah Wilderness Association and the Wilderness Society and led by Dick Carter and Mike Medbury, proved quite simple to resolve. None reached a level higher than the regional director of planning and budget and the deputy regional forester.[58]

In line with the reduced allowable sale quantities, in 1991 Thompson approved a harvest on 1,607 acres of timber in Tippets Valley near

Cedar Breaks National Monument. As he offered the sale, Thompson learned quickly that a new set of critics were less willing to negotiate and were much more sophisticated in their opposition strategy than Carter and Medbury. Since the approval of the forest plan, a number of new people had moved into Iron and nearby counties. These people sought amenity values, particularly forest beauty and wildlife protection, and they had little sympathy with timber sales or other commodity uses. Thompson characterized them as people who had no historic or cultural connection with southern Utah or the forest. At the same time, some of Thompson's critics considered him to be extremely difficult to work with.[59] Recognizing the need for consensus with these new critics, Thompson sought to negotiate agreements with the various forest users during the new planning cycle between 1986 and 1993.[60] In an attempt to address the concerns of amenity users, the Dixie staff continued to conduct various types of public discussions.[61]

Then disaster struck. During the winter of 1992–93 abnormally high snowfall drifted around the trees, sheltering the nesting beetles from freezing weather. By the summer of 1993, beetle larvae had girdled the trees in a massive attack on the forest.[62] In what they perceived as a spirit of total ecosystem management, Thompson and his staff concluded that they must conduct a salvage logging operation to replace the decadent and closely packed forest with a healthy stand of wider spaced, young, and vigorous trees.[63]

Not surprisingly, as forest officers began to plan the salvage sales, Thompson encountered a polarized public. Most people in long-established towns with sawmills, like Panguitch and Kanab, wanted forest officers to offer more trees for sale. By contrast, partisans of amenity uses organized the Friends of the Dixie National Forest to try to scrap the existing plan and emphasize recreation, scenic and wildlife protection, and plant species variety. Responding to opposition from these environmental groups, who were fearful of the loss of scenic values and of damage to habitat for the northern goshawk, Mexican spotted owl, and other wildlife, Thompson rescinded the proposed 1991 Tippetts Valley sale.[64] Then, after two more years of study, in 1993 Thompson approved the sale of 5.3 million board feet in the Tippetts Valley. The management prescription included selective cutting, tree thinning, and replanting.[65]

In view of the beetle epidemic, in early July 1993 Mark Van Every, a spokesman for the Dixie, announced that spruce bark beetles and mountain pine beetles had attacked 80 to 90 percent of the trees sixteen inches or larger in a region of the Markagunt Plateau stretching from Cedar Breaks National Monument and Brian Head ski resort on the west to Panguitch Lake on the east. He announced that forest officers would propose salvage sales to try to remove the infested and immediately threatened trees. With the projected salvage sales, the Dixie planned to increase the allowable sale quantities from 17 to 25 million board feet, a level just slightly lower than that approved in the 1986 forest plan.[66] Van Every said, however, that in planning the sales, forest officers would protect the watersheds by mandating helicopter yarding on steep slopes.[67]

Opposition arose immediately, and by the fall of 1993 an embattled Thompson had become convinced that the Dixie administration would "not . . . get consensus." On November 22, 1993, opposition groups served Thompson with papers on a suit filed to block the Tippetts Valley sale. The appellants challenged the Forest Service's EIS. The statement, they said, had failed adequately to protect the northern goshawk and the yellow-breasted chat, to analyze the impact of the sale on riparian areas and wildlife habitat corridors, and to determine the cumulative effects of past, present, and future timber sales on the Mammoth Creek watershed and on visual, recreation, and water quality as well as aquatic ecosystems.[68] Dismayed by this opposition, Thompson contrasted these new opponents with those who had challenged the 1986 forest plan. The new opponents, he said, were "very intelligent and dedicated, if misguided, people." They were less willing to negotiate than Carter or Medbury had been. Thompson said that if "you took [it] at face value, you'd say that we're the dumbest bunch of bastards that ever existed."[69]

Recognizing the visual and ecological changes that massive salvage sales would generate, a number of amenity groups, including the Southern Utah Wilderness Alliance (SUWA) and the Friends of the Dixie National Forest, questioned the need for timber harvesting. Ken Rait, SUWA issues coordinator, said, "This is just the latest excuse to cut timber down there." "This is not a timber salvage program, this is a timber-industry salvage program." "Bugs," he said, "are part of the forest ecosystem. But if their levels have increased, does it mean the bug's predators are lacking?" The infestation's severity, he said, is a reflection of Forest Service mismanagement, since it could have introduced natural predators to control beetle populations.[70]

Thompson considered Rait's charges irrational and "unprofessional." He invited Rait and others to inspect the damage and view for themselves the need for the salvage sales.[71] Thompson and his staff took various parties, including newspaper and television reporters and Friends of the Dixie, on inspections of the forests. He and other foresters took axes and chipped the bark to reveal bark beetle larvae busily girdling the trees.[72]

The proposal to cut the infested trees and the environmentalist opposition elicited a response from Klein. The retired entomologist said that both the Forest Service and Rait were wrong but that the service offered the only feasible remedy for the beetle infestation. Klein pointed out that neither the quick importation of natural predators nor the salvage sales would address the underlying ecological damage on the Dixie. Rather, he believed that the beetle infestation had resulted from long-established management prescriptions—especially fire suppression—that had failed to generate a healthy ecosystem. In a healthy forest, he argued, beetles subsisted on dead trees and snags.[73] This is, of course, only partly true, since infestations can break out because of optimum conditions for the breeding and nesting of the beetles, as, for instance, in a heavy snow pack that protects the larvae from winter freezing. It is, however, true that healthy trees have defense mechanisms like secreting pitch to rid themselves of beetles.

Moreover, in addition to removing the infested trees, the salvage sales, designed to realize some economic value from the dead and dying trees, served a silvicultural and economic objective. Clearly, the Forest Service could have achieved the objective of eliminating the infested trees by destroying the trees in some way such as burning. Leaving the trees in place could have preserved habitat for certain types of wildlife, but it is difficult to see just how such an action would have helped reverse the ecological damage caused by previous national forest policy and practice.[74]

Before the Forest Service had begun to manage the forests, in part because of Native American land management practice and in part because of natural conditions, fire had helped to keep the stands healthy by thinning the trees and removing much of the heavy undergrowth and snags that otherwise harbored excessive beetle populations. From the early twentieth century through the 1980s, when the Forest Service began to authorize prescribed burning, foresters had suppressed virtually all fires. By doing so on the Dixie, foresters had fostered conditions in which extremely dense and overage trees competed with each other for moisture, sunlight, and nutrients. Five years of drought in southern Utah had weakened these already unhealthy trees to such a degree that they could not ward off the attacks of beetles.[75]

Given the drought and thick growth–induced stress, in Klein's view, attacks by beetles or uncontrollable wildfires such as the 1988 Yellowstone fires had become almost inevitable. Klein believed that attempts to control beetle infestations in overage and weakened forests through natural and artificial means had generally failed. Klein said that Rait was "irresponsible" in his charges and in his belief that natural predators could control the infestation. The technology to solve the problem through natural means, he said, "does not" exist "and probably never will."

Klein acknowledged that Rait was right in his belief that the Forest Service could simply allow the beetle infestation to run its course. He pointed out, however, that by "doing nothing, bark beetle populations will not decline until they decimate their food supply." On balance, Klein concluded, a "far more responsible alternative is proposed by the Dixie National Forest officials: Salvage the dead, infested and immediately threatened trees so as to reduce stand density, thereby encouraging the smaller, younger trees with the potential for growth."[76] Thompson and his staff had chosen this alternative not to save the lumber companies but because they thought that through it they could induce the development of a healthy ecosystem. After all, Thompson had originally decided to reduce the allowable sale quantity below that approved in the 1986 plan largely because of foresters' concern for the ecosystem.

Following these tactics, the Dixie placed the sale of 4 million board feet of timber near Rainbow Meadows on the Markagunt Plateau on the fast track. Forest officers offered the timber for sale after a quick Environmental Assessment (EA) rather than after preparing an EIS with full public comment. Of the sale, 2.8 million board feet consisted of diseased Engelmann spruce and 1.2 million board feet of healthy trees. Trying to block the sale, the SUWA, the Friends of the Dixie National Forest, and American Wildlands filed a suit alleging that forest officers had failed to follow the regulations established by the Appeals Reform Act of 1992, which they said required the Dixie National Forest to consider appeals. They also charged that the refusal to allow appeals on such salvage sales struck "at the very heart of our system of representative, democratic government."[77] In practice, the environmentalists did not question the fact that previous fire suppression had caused much of the infestation. They did, however, question the Forest Service's insistence that logging was the best way to generate a healthy ecosystem. Questioning the thoroughness of the EA, SUWA's Ken Rait said, "It doesn't make sense to cut down the forest to save the trees." SUWA staff attorney Heidi McIntosh said, "The Forest Service is hiding behind an insect as the latest excuse to get the cut out."[78]

Predictably, spokespersons for commodity interests, who seem to have understood the forest as a total ecosystem as little as the environmentalists did, responded quite differently. Allen Henrie, city manager of Panguitch, where Kaibab Forest Products operated a mill, said, "It's a good management tool for the Forest, especially with the bug problem they have." He, too, seems to have been unaware of the problems caused by fire suppression.

The environmental groups lost the appeal, and the Dixie National Forest began to plan nine salvage sales on the Markagunt Plateau, not to save the trees, as Rait suggested, but to try in the short run to recover some economic value from the infested trees and in the long run to induce recovery in an ecosystem in which beetles would pose less of a threat.[79] In practice, however, the Dixie had a difficult time achieving its objective, because companies often failed to cooperate by purchasing the timber offered for sale. For instance, one proposed sale in the Sidney Valley east of Brian Head consisted of 22 million board feet. Of this volume approximately 4 million board feet consisted of healthy trees removed to check the progress of the infestation. Lumber companies were so reluctant to bid on the sale that the Dixie managers had to offer it several times, in each case reducing the sale price.[80] Following the decision to proceed with the Sidney Valley sale, the SUWA, backed by Friends of the Dixie National Forest president Brandon Fowler, filed an appeal with Intermountain Regional Forester Dale Bosworth. In the appeal, they insisted that the Dixie managers needed to prepare a full EIS rather than the less detailed EA. Bosworth rejected their appeal.[81]

In view of the seriousness of the beetle epidemic throughout the West, the Forest Service sought general relief in an attempt to generate healthy ecosystems. By 1995 the Forest Service's Washington office had concluded that the damage induced by fire suppression, the dry years, and the subsequent beetle attacks and fire outbreaks had become unacceptably serious. As a result, the Forest Service asked Congress for authority to expedite salvage logging on the national forests. Over the opposition of environmental groups, some scientists, and a group of smoke jumpers, Congress passed and President Clinton signed salvage logging legislation in the summer of 1995. This led the Forest Service to harvest 3 billion board feet of insect-infested or burned timber throughout the national forest system from late 1995 through January 1997, when the act expired. Under the act, opponents had to prove the sales were "arbitrary and capricious" before they could secure an injunction to halt the logging.[82]

In March 1994 the SUWA, the Friends of the Dixie National Forest, and the Native Ecosystems Council secured a preliminary injunction from the U.S. District Court for Utah to halt the sales. In August 1995, perhaps because of the recently passed legislation, Judge Thomas Greene lifted the injunction. In a postmortem on Greene's decision, SUWA attorney Heidi McIntosh said that she still believed that the sale was wrong and that a harsh winter in 1995–96 might solve the problem by killing the beetles. In contrast, Forest Service and lumber company officials blamed the environmental groups for the death of additional trees caused by the delay in the timber harvests.[83]

Similar salvage sales of beetle-infested, fire-killed, or wind-thrown trees took place on other forests in the region. In virtually every case, environmentalists opposed the sales, generally citing roadless area, scenic, or wildlife habitat protection as their grounds.[84] In December 1993, faced with the death of more than 90 percent of saw timber and about 80 percent of all the Engelmann spruce along Skyline Drive on the Wasatch Plateau, southern Utah's Manti-LaSal National Forest managers contemplated the sale of 40 to 60 million board feet of timber.[85]

As with the Dixie, the Manti-LaSal faced considerable opposition to the salvage sales. Manti-LaSal Forest Supervisor Janette Kaiser said that an accelerated salvage program was necessary to stop the expansion of a beetle infestation on the Wasatch Plateau. Citing the problems that the salvage sales would cause for various nesting birds and the Mexican spotted owl, Mark Clemens of the Sierra Club's Utah Valley Group countered that dead trees were part of a healthy forest. Kaiser expected the salvage sales would lose money, and Jerry Hansen, co-owner of the Hansen Lumber Company of Fairview, said that he did not expect a great many bidders. Builders could use dead and decaying spruce, he said, principally for low-value products like cabin logs and mining roof supports.[86]

In spite of the opposition, the Manti-LaSal began to offer trees for salvage sales in 1996 and continued offering them into 1997. In an attempt to facilitate the transition to a healthy ecosystem, forest managers undertook a number of measures. Instead of cutting the timber as rapidly as possible, the foresters tried to preserve wildlife and plant habitat and recreational values in the rust-and-gray-colored trees by spreading the sales over a ten- to twenty-year period, even though the beetle infestations would likely continue to spread over that time. In order to protect the watershed from other potentially more destructive types of yarding such as skidding, the Manti-LaSal required the companies to use helicopters to get the trees out. The helicopters transported the logs to Twelve Mile Flat, and from there trucks took them to at least three sawmills. These included a mill that lay between Mayfield and Gunnison operated by Satterwhite Log Homes of Longview, Texas, one operated at Fairview by Hansen Lumber Company, and one in Huntington operated by Cascade Lumber Company of Price.[87]

On the Manti-LaSal as on the Dixie, these sales encountered considerable opposition because of wildlife habitat and scenic destruction. In late 1996 and early 1997 the Sierra Club, the SUWA, and the Boulder Mountain Wilderness Committee tried to halt a salvage sale in Jacobs Swale on the Dixie on these grounds. At the expiration of the salvage sales act in January 1997, Idaho Senator Larry Craig proposed to renew the authorization, but Vice President Al Gore and others in the Clinton administration expressed serious reservations, and as of late 1999 it had failed to pass.[88]

As the restoration/salvage sales proceeded, new timber sale patterns emerged in southern Utah. Increasingly, larger lumber companies with headquarters outside the state began to bid on the timber. Although Kaibab Forest Products, with mills in Panguitch, Utah, and Fredonia, Arizona, tended to think of the Dixie National Forest as serving its interests, the law required the national forests to sell timber to the highest responsible bidder. As a result, a 4.5-million-board-feet sale on the Midway Face of the Cedar City Ranger District went to a Louisiana Pacific mill in Rexburg, Idaho, and a small helicopter sale near Panguitch Lake went to Boise Cascade of Boise. Boise Cascade intended to ship the logs by rail from Cedar City to its mill in Emmett, Idaho.[89]

In December 1994 Kaibab Forest Products announced that it would close its Fredonia mill and enlarge its Panguitch mill. Company president Don Olson blamed the closure on environmental lawsuits "coupled with conflicting laws and regulations" that "made it impossible to obtain the trees to supply . . . the mill." In spite of the blast at environmentalists, it seems likely that the company simply could not compete with larger, more efficient operations like Louisiana Pacific and Boise Cascade, particularly when the health of the ecosystem mandated helicopter logging. As Ken Rait pointed out, the Dixie foresters had tried several times to sell 22 million board feet of timber in the Sidney Valley, and Kaibab did not bid on the sale.[90]

More seriously, even in the absence of harvests, forests changed. Appellants against the Rainbow Meadows salvage sale on the Dixie National Forest argued that the harvest of diseased timber on the Dixie would cause irreparable damage. Moreover, the appeal said, "Once these trees are cut, the forest as it now exists cannot be replaced."[91] This statement was, of course, true, but the unstated assumption that if loggers did not take out the trees, the forest would remain in its current condition was probably false. The assumption underpinning the comment was a radical assertion of Frederic E. Clements's climax theory, which held that in the absence of climatic change the trees would flourish in a stable growth pattern.[92] The statement did not factor in the consequences of climatic change and fire suppression. Moreover, it ignored the views of competing theorists such as historian James Malin and ecologist Charles Kay, who argued in favor of dynamic and largely unpredictable change rather than for predictable change toward a climax condition.[93] In this context, even had Clements's theories been correct, the climate had changed from a cycle of relatively wet years during the early 1980s to a cycle of dry years during the late 1980s and early 1990s. Most significantly, since the late nineteenth century the public had demanded—and professional opinion had prescribed—fire suppression. Most observers believed, erroneously as it proved, that fire suppression would protect scenic and economic values and generate a healthy ecosystem.

More recently, ecologists, foresters, and others have recognized that this view is wrong. Thus, since the 1980s the Intermountain Region has allowed some fires to burn, and it has conducted prescribed burning in certain areas. Nevertheless, the outcry during the 1988 Yellowstone fires indicated public intolerance for widespread use of burning as a tool in areas of high economic and scenic value. Many forested districts on the Dixie, Fishlake, and Manti-LaSal lie in and near recreation areas, national parks and monuments, and home sites. No forester with even a modicum of political insight would allow fires to burn uncontrolled in such areas.

With continued fire suppression, old growth, old second growth, and overly dense forests like those in southern Utah will most likely experience stress and undoubtedly die as long as they are subjected to recurrent drought. Beetles, wildfires, and other agents of destruction will continue to threaten them. Nevertheless, the critics were right: the Forest Service need not have used below-cost salvage sales to control the infestations and revive the forests. In the absence of thinning in some other way (the means that come readily to mind are prescribed burning and wind throw), recurrent stress would have undoubtedly offered ideal conditions for beetle infestations or wildfires, which would thin the forests as well.

What does all of this mean for the goal of holistic ecosystem management in southern Utah and elsewhere in the Intermountain Region? Clearly, the Forest Service had mismanaged the national forests. That mismanagement, however, consisted neither in failing to introduce natural predators nor, since the drafting of plans during the late 1980s and early 1990s, in trying to support the business of lumber companies by getting out the cut. It consisted, rather, in following the public demand to preserve both amenity and economic values through fire suppression. This resulted in dense and temporarily beautiful but frequently unhealthy stands of trees.

After forest managers in southern Utah detected the beetle infestation, they responded with a plan they hoped would induce the development of a healthy ecosystem (I use the word *induce* rather than *restore,* the term that foresters use). Since human beings, both Native Americans and Euro-Americans, have influenced the ecosystems of southern Utah for at least 12,000 years, it seems unlikely that anyone could restore the system to a pristine condition or even determine what that was. Moreover, contrary to the climax theorists, it is unlikely that the ecosystem would have reached a static climax condition, even in the absence of climatic change. As others have argued, ecosystems are dynamic rather than static. They undergo constant change not only through human intervention but through unpredictable forces that operate according to natural laws. Change—often unpredictable change—is inevitable.[94]

That said, we must understand that the generation of foresters who served before Thompson, Taylor, and Kaiser, together with their allies in the commodity community, bear most of the blame for the fear with which critics of the Forest Service responded to proposed timber sales.

Throughout the Intermountain Region (indeed, throughout the national forest system), especially during the 1950s, 1960s, and 1970s, logging companies had cut far too much timber on the national forests. Most significantly, cutting prescriptions had frequently caused excessive damage to watersheds, wildlife habitat, and amenity values.[95] Under the circumstances, it is difficult to blame the critics for fearing that the same things might happen again, and compromise has been far too scarce. Moreover, the public seems to have a generally erroneous view of the relative damage done in national forests by logging and recreation. If studies such as those by Knight, Wilcove, and others are accurate, nonvehicle recreation causes an even greater threat to endangered species than logging or off-highway vehicles.

The managers of the national forests of southern Utah tried to find some way to reduce the density of the trees in the hope of generating a healthy ecosystem. Contrary to the charges of their critics, they understood the importance of amenity values, especially of recreation. On the eve of the new millennium a resolution of these conflicts seemed quite remote, and total management of healthy ecosystems seems to lie in a distant and perhaps unachievable future.

NOTES

1. Brian Stout, interview by Andrea Radke, October 27, 1993, 9–10, Charles Redd Center for Western Studies, Brigham Young University, U.S. Forest Service Oral History Project (hereafter referred to as CRC). For a general history of the Intermountain Region of the U.S. Forest Service, see Thomas G. Alexander, *The Rise of Multiple-Use Management in the Intermountain West: A History of Region 4 of the Forest Service,* U.S. Department of Agriculture, Forest Service Publication FS 399 (Washington, D.C.: Government Printing Office, 1987). For a general history of the Forest Service, see Harold K. Steen, *The U.S. Forest Service: A History* (Seattle: University of Washington Press, 1976). For a general history of the relationship between the Forest Service and timber management, see David A. Clary, *Timber and the Forest Service* (Lawrence: University Press of Kansas, 1986). For a study of timber management in the 1980s and early 1990s, see Paul W. Hirt, *A Conspiracy of Optimism: Management of the National Forests since World War Two* (Lincoln: University of Nebraska Press, 1994).

2. Hugh C. Thompson, interview by Sondra Jones, November 23, 1993, 13, CRC.

3. R. M. "Jim" Nelson, interview by Sondra Jones, Sparks, Nev., December 7, 1993, 13, 23, CRC.

4. *High Country News,* January 20, 1997.

5. Nelson, interview, 19–20; Gary Schaffran, interview by Andrea Radke, December 14, 1993, 5, CRC.

6. Nelson, interview, 21–22.

7. The Northern Region includes the forests of Montana, northern Idaho, and North Dakota; the Rocky Mountain Region consists of the forests of Wyoming east of the

Continental Divide, Colorado, South Dakota, Nebraska, and Kansas; the Southwestern Region covers New Mexico and Arizona; and the Intermountain Region includes the forests of Utah, Nevada, southern Idaho, and western Wyoming.

8. Kent Taylor, interview by David Wilson, March 26, 1993, 12–13, CRC. On the Sunbird Conference, see Hirt, *A Conspiracy of Optimism,* 283–84.

9. Thompson, interview, 18.

10. See Hirt, *A Conspiracy of Optimism,* 286–87, and a review of the manuscript by Stan Tixier. Mumma now serves as wildlife director for the state of Colorado.

11. Nelson, interview, 24.

12. George Olsen, interview by David Wilson, Ogden, Utah, November 12, 1992, 8, CRC.

13. Frank Carroll, interview by Andrea Radke, December 16, 1993, 21, CRC.

14. Annual Collection Statements FY 1970 and 1985 furnished by Philip Johnson of the regional office. Foyer Olsen, interview by Thomas G. Alexander, March 1984, 9, Historical Files, regional office.

15. Paul Nordwall, interview by Sondra Jones, October 7, 1993, 12–14, CRC.

16. Fred A. Kingwill, interview by Sondra Jones, October 23, 1993, 4–5, CRC.

17. For a summary, see Thompson, interview, 13.

18. At a meeting between environmentalists and commodity interests that I helped organize in southeastern Utah in 1990, several of the environmentalists said that they did not believe that public land management agencies ought to charge for recreation since the activities were essentially benign.

19. Richard L. Knight, letter to the editor, *High Country News,* January 18, 1998.

20. David S. Wilcove, David Rothstein, Jason Dubow, Ali Phillips, and Elizabeth Losos, "Quantifying Threats to Imperiled Species in the United States," *BioScience* 48 (August 1998): 610, 612.

21. Olsen, interview, 23–24; Tixier review of the manuscript.

22. *Provo Daily Herald,* April 13, 1996.

23. Taylor, interview, 7.

24. Ibid., 8–9.

25. Ibid., 9.

26. *Provo Daily Herald,* May 24, 1996.

27. Taylor, interview, 8.

28. *Salt Lake Tribune,* March 27, 1994.

29. Carroll, interview, 21. By 1997 the Forest Service had undertaken a program of reappraising the value of cabin lots and charging an increased fee. See *Provo Daily Herald,* September 16, 1997.

30. Carroll, interview, 22.

31. *High Country News,* December 9, 1996.

32. *Salt Lake Tribune,* July 17, 1993.

33. Olsen, interview, 14–15.

34. *Provo Daily Herald,* January 2, 1997.

35. *Salt Lake Tribune,* July 17, 1993.

36. Ibid., February 26, 1993. For O'Toole's view of the Forest Service, see Randal O'Toole, *Reforming the Forest Service* (Washington, D.C.: Island Press, 1988).

37. *Salt Lake Tribune,* April 12, 1996.

38. Ibid.

39. This discussion is based on the comments of Paul Hirt on a draft of the essay. Hirt himself believes that fees are appropriate.

40. *High Country News,* October 13, 1997.

41. *Salt Lake Tribune,* April 26, September 6, November 27, 1996; *Provo Daily Herald,* September 6, 1996.

42. Gene S. Jensen, interview by Sondra Jones, October 28, 1993, 15, CRC.

43. *Provo Daily Herald,* May 24, 1996.

44. Schaffran, interview, 6–7.

45. Often expressed as mmm. The *m*'s in this designation stand for thousands, thus m = one thousand, mm = one million, and mmm = one billion. A board foot is a piece of lumber one foot long, one foot wide, and one inch thick. Approximately two thousand board feet of lumber are needed for an average-size house.

46. *Provo Daily Herald,* May 2, 1993.

47. *Salt Lake Tribune,* September 29, 1994; *Provo Daily Herald,* September 19, 1994.

48. *Salt Lake Tribune,* September 29, 1994.

49. Ibid.

50. Taylor, interview, 4–6. As indicated before, it experienced only one relatively easy appeal from the plan.

51. In commenting on an earlier version of the essay, Paul Hirt asked me to be "a little skeptical about rationales for salvage logging" in the belief that "many times it is a new justification for the same old pressure to get logs to mills." Though this may well be true in some cases, I do not believe that it was in the case of the Fishlake or the Dixie. The Fishlake is not a major timber forest, since it has an allowable sale quantity of only about 3 million board feet per year. The Dixie's allowable sale quantity is only about 25 million board feet. In the case of the Dixie, the allowable sale quantity was a reduction from a previously higher level. Even with the salvage logging sales, the allowable sale quantity did not exceed that established in the forest plan, although it was higher than the 17 million board feet that Thompson thought advisable.

52. William H. Klein, "Foresters Have Right Plan to Battle Beetles," *Salt Lake Tribune,* December 5, 1993. In one case, between 1988 and 1993 the Forest Service, in collaboration with Utah's Department of Agriculture, and the USDA Animal and Plant Health Inspection Service sprayed stands in Wasatch Front Canyons with *Bacillus thuringiensis* to control an outbreak of gypsy moths. The bacteria, which occur naturally in the soil, kill the moths by attacking their digestive systems. By 1993 it appeared that the *thuringiensis* bacteria had eradicated the moths, and a follow-up study in 1997 by Ron F. Rockwell of the Carnegie Museum in Pittsburgh, under contract with the regional office, indicated that the moths had disappeared from the treated areas. See *Salt Lake Tribune,* December 3, 1993; *Provo Daily Herald,* August 9, 1997.

53. A panel of foresters, forestry professors, ecologists, and entomologists assessed this problem at a conference held at Sun Valley, Idaho, November 14–20, 1993. They concluded that the forests of the inland West would need preventive treatment over the next twenty years. See the draft in possession of the author. For a discussion of conditions in western Montana, see James R. Habeck, "Old-Growth Ponderosa Pine–Western Larch Forests in Western Montana: Ecology and Management," *Northwest Environmental Journal* 6 (1990): 271–92.

54. The term "maturity" is, of course, a silvicultural term having to do with the age at which a tree ceases to grow significantly and thus fails to add to the economic value of the forest.

55. The quotations are from Klein, "Foresters Have Right Plan." The statistics come from Thompson, interview, 8. On fire management by Indians, see Stephen W. Barrett, "Indians and Fire," *Western Wildlands* 6 (Spring 1980): 17–21, and George E. Gruell,

"Fire on the Early Western Landscape: An Annotated Record of Wildland Fires, 1776–1900," *Northwest Science* 59 (1985): 97–107.

56. Klein, "Foresters Have Right Plan"; Thompson, interview, 8.

57. Thompson, interview, 19.

58. Ibid., 5–6.

59. Ibid., 2–3. The view of Thompson's critics comes from a private conversation with Michael and Valerie Cohen, residents of Cedar City, following the delivery of an earlier version of this essay at the meeting of the American Association for Environmental History in Tucson, May 14–18, 1999.

60. Thompson, interview, 3.

61. Thompson, "Dixie Forest Manager Balances Demands," *Salt Lake Tribune*, September 5, 1993.

62. Klein, "Foresters Have Right Plan"; Thompson, interview, 8.

63. This is, as Paul Hirt pointed out in a critique of an earlier version of this essay, a silvicultural analysis. He argues that an "ecosystem view of the problem would be quite different." I would argue, however, that a wide range of different mixes of components can generate a healthy ecosystem. Under the circumstances, a mix of plants and animals viewed from a silvicultural perspective might or might not constitute a healthy ecosystem. Nevertheless, it is not unreasonable for foresters who value both the economic and amenity values of a forest to favor a stand of young and vigorous trees and to try to manage the ecosystem to create a healthy system that includes such a mix. This is particularly true if we assume that human beings are part of the ecosystem and that their intentions for that system are as valid as those of the other animals and plants in the system. See Stewart L. Pimm, *The Balance of Nature?: Ecological Issues in the Conservation of Species and Communities* (Chicago: University of Chicago Press, 1991), 83–134. Clearly, in such a system certain plants and animals would be disadvantaged. Some might disappear entirely from the system.

64. Thompson, interview, 2–4.

65. *Salt Lake Tribune*, March 26, 1993; Hugh C. Thompson, "Dixie Forest Manager Balances Demands in American Public's Land of Many Uses," *Salt Lake Tribune*, September 5, 1993.

66. Thompson, interview, 19.

67. *Salt Lake Tribune*, July 2, 1993. "Yarding" is the term used by foresters to describe the process of conveying the cut trees from the forest.

68. Thompson, interview, 7.

69. Ibid.

70. Jim Woolf, "Insect-Infested Trees at Cutting Edge of S. Utah Logging Issue," *Salt Lake Tribune*, July 2, 1993; *Provo Daily Herald*, July 7, 1993; Mike Gorrell, "Logging to Salvage Forest from Beetles Poses Puzzle," *Salt Lake Tribune*, October 18, 1993.

71. Thompson, interview, 19.

72. Ibid., 20. See also Ronald S. Wilson, Cedar City District Ranger, "A Status Report on Insect Activity on Dixie National Forest," n.d. (ca. fall 1993), copy in author's possession.

73. *Salt Lake Tribune*, December 5, 1993.

74. In a critique of a previous draft of the essay, Paul Hirt raised the question, "What *ecological* principle says that insect-damaged trees should be cut down and hauled out of the forest?" He considers this an economic objective. I would agree that it is an economic objective, but I would argue that it may also be an ecological objective, since forest officers expected to generate a healthy ecosystem, something they could not do

while the infested trees were still in place.

75. *Salt Lake Tribune*, December 5, 1993. For a treatment of fire policy, see Stephen J. Pyne, *Fire in America: A Cultural History of Wildland and Rural Fire* (Princeton, N.J.: Princeton University Press, 1982), and idem, *America's Fires: Management of Wildlands and Forests*, Forest History Society Issues Series (Durham, N.C.: Forest History Society, 1997); Habeck, "Old-Growth Ponderosa Pine."

76. Klein, "Foresters Have Right Plan."

77. *Salt Lake Tribune*, October 1, 1992.

78. Ibid., March 13, April 28, 1994.

79. Ibid., October 18, 1993. Mills on the forest were operated by Kaibab Forest Products in Panguitch, Utah, and Fredonia, Arizona.

80. Ibid., September 25, 1994.

81. Ibid., May 10, 1994.

82. *High Country News*, April 7, 1995.

83. *Salt Lake Tribune*, August 7, 1995.

84. Ibid., July 28, December 28, 1993, January 25, June 16, November 26, 1994, November 14, 19, 1995; *Provo Daily Herald*, July 31, 1993.

85. *Salt Lake Tribune*, December 16, 1993. The phrase "saw timber" is, of course, a silvicultural term, but it seemed justified to national forest officers perhaps because any forest ecosystem can thrive in a number of conditions, including one designed to produce at least some saw timber.

86. Ibid., April 2, 1996.

87. *Provo Daily Herald*, April 3, 1997.

88. Ibid., January 3, 1997, November 8, 1996; *Salt Lake Tribune*, January 3, 1997.

89. Thompson, interview, 10.

90. *Salt Lake Tribune*, December 3, 1994.

91. Ibid., October 1, 1993.

92. "Climax—The final stage of a succession which continues to occupy an area as long as the climate remains unchanged. The succession leading to a climax represents adjustment to changing conditions, and the climax itself the ultimate adjustment to a condition of relative stability." See B. W. Allred and Edith S. Clements, eds., *Dynamics of Vegetation: Selections from the Writings of Frederic E. Clements* (New York: H. W. Wilson, 1949), 280.

93. See James C. Malin, Introduction, in Robert P. Swierenga, ed., *History and Ecology: Studies of the Grassland* (Lincoln: University of Nebraska Press, 1984), and idem, *The Grassland of North America: Prolegomena to Its History* (Lawrence, Kans.: Privately printed, 1956); Charles Kay, "Politics Tangles with Science," *High Country News*, September 15, 1997. For a recent survey of research and theory on the questions such as resilience, variability, and other problems raised by the views of Clements and others, see Pimm, *The Balance of Nature?* esp. 371–89, which summarize the conclusions of the various studies.

94. For a popular discussion of this point of view, see James Gleick, *Chaos: Making a New Science* (New York: Penguin, 1987). For discussions of Clements's point of view, see Donald Worster, *Nature's Economy: A History of Ecological Ideas* (New York: Cambridge University Press, 1977), 210–20; Pimm, *The Balance of Nature?*; and as it applies to colonial New England, William Cronon, *Changes in the Land: Indians, Colonists, and the Ecology of New England* (New York: Hill and Wang, 1983), 10–11.

95. For discussions of these matters, see Alexander, *The Rise of Multiple-Use Management*; Hirt, *A Conspiracy of Optimism*; and Clary, *Timber and the Forest Service*.

8

Biopolitics

A Case Study of Political Influence on Forest Management Decisions, Coronado National Forest, Arizona, 1980s–1990s

PAUL W. HIRT

An endangered species controversy has raged on a national forest in Arizona for more than ten years, pitting biologists against astronomers, environmentalists against a university, government agency against government agency, and even staff members within agencies against each other. The issue has received national and international media attention, especially after evidence of politically manipulated biological studies led to congressional oversight hearings and a U.S. General Accounting Office (GAO) investigation. Like the better known "owls versus jobs" controversy in the Northwest, this environmental battle has been superficially viewed as "squirrels versus 'scopes" because an endangered subspecies of red squirrel helped delay and downscale construction of a massive observatory complex on a mountaintop in southern Arizona. During the Bush administration, the controversy garnered special attention when Secretary of the Interior Manuel Lujan remarked, "Do we have to save every subspecies? . . . Nobody's told me the difference between a red squirrel, a black one or a brown one."[1] Lujan's comment is especially revealing since he was ultimately responsible for administering the Endangered Species Act. As the secretary of the interior should have known, the Endangered Species Act does quite purposefully apply to subspecies.

Forest lands all over the Southwest face a multitude of wildlife habitat problems, especially as logging and other forms of resource extraction continue to shrink the already tiny portion of old growth forest that remains in the region. Beginning in 1995, protection for the Mexican spotted owl, which is largely dependent on old growth habitat, led to a sudden and significant decline in logging on national forests in the Southwest, similar to what happened with the northern spotted owl in the Pacific Northwest earlier in the decade. Species dependent on these rare, old, biologically diverse, and remote forestlands often make it onto the federal endangered species list because their habitats are so restricted after a century of intense logging in the region. It is the *habitat,* then, that is endangered; the condition of the individual species that live there serves as an indicator of the health and integrity of the forest habitat.

When the U.S. Fish and Wildlife Service designates a plant or animal species as threatened or endangered, the nation's strongest environmental protection law—the Endangered Species Act—comes into play. The act forbids federal agencies to take any action that increases the risk of extinction of a listed species, it forbids private citizens or businesses from causing harm or death to the species, and it requires the Fish and Wildlife Service to develop conservation strategies for recovering that species to viable population levels. There is very little wiggle room in this act. It is one of the only conservation laws that effectively draws a line in the sand that says, Thus far and no farther. Because the Endangered Species Act is so strong and so clear in its intent, it often finds its way to the center of many nature preservation battles. While resource industries and real estate developers can often exploit loopholes in weaker laws designed to preserve wetlands or reduce soil erosion or protect water quality, they have a harder time sidestepping the Endangered Species Act. This act, then, becomes the focus of many battles. Although the act is designed to protect the larger habitat that supports endangered species, the individual species themselves become the issue in the eyes of the media and the general public, as

RED SQUIRRELS, BLACK SQUIRRELS, BROWN SQUIRRELS... I CAN'T TELL THE DIFFERENCE..DO WE HAVE TO SAVE EVERY SUBSPECIES?

Cartoon by Fitzsimmons. Courtesy *Arizona Daily Star*

in owls versus loggers, fish versus dams, or squirrels versus 'scopes. Opponents of the Endangered Species Act benefit from such simplistic characterizations.

Because the stakes are high, the process of listing a species as threatened or endangered can be highly contentious. Even more critical, after a species is listed, every proposed activity or development on federal lands that might affect the population or habitat of that species must be evaluated by the Fish and Wildlife Service and either approved, denied, or modified to ensure that it poses no additional risk to the survival of that species. While the Endangered Species Act requires federal agencies to consider the social and economic repercussions of protecting each endangered species, it also requires the biological documents assessing the species' population, habitat, and level of risk to be scrupulously science-based, and it requires the Fish and Wildlife Service to choose a course of action that does not increase the species' risk of extinction. Basing final decisions on biological rather than political or economic considerations is an admirable goal when truly rare and endangered ecosystems are at stake. Unfortunately, this goal is often hard to achieve when there are economic interests at stake. When species preservation comes into conflict with economic development, drawing that line in the sand becomes increasingly difficult. Politics can exert considerable influence over scientific documents and over the work of scientists themselves. As former Forest Service timber sale planner Jeff DeBonis has said, "There is tremendous pressure on timber planners, wildlife biologists, hydrologists and others to suppress their professional expertise and modify their documentation to ensure that short-term political expediencies . . . are met."[2] At that point the rational model of decision making unravels, and the decision arena becomes simply a power struggle. This is what happened on Mount Graham.

In 1984 a consortium of astronomers, led by the University of Arizona's Steward Observatory, asked the U.S. Forest Service for permission to develop an extensive astrophysical complex on half a dozen summits and ridges atop remote Mount Graham (also known as the Pinaleño Mountains), a large block fault mountain range within the Coronado National Forest in southeastern Arizona. Conservationists opposed the development from the beginning because it would crisscross a unique forest habitat of very limited size (more on this below) with roads, buildings, giant observatories, parking lots, dormitories, and other facilities. The Forest Service generally favored the proposal. Arizona politicians at the federal level unanimously supported it, as did local politicians in the small towns at the base of the mountain. Opponents included a large and diverse collection of environmental groups, sport hunting clubs, conservation biologists, Native American activists, and student groups.

Until the Mount Graham red squirrel was listed as endangered in 1987, the Forest Service had sole decision-making authority over the project since it would all be built on national forest land. Once the red squirrel and the Endangered Species Act entered the picture, decision-making authority became shared in a messy way between the Forest Service and the Fish and Wildlife Service. The interaction of all these players—university astronomers, the Forest Service, conservationists, politicians, the Fish and Wildlife Service, the Arizona Game and Fish Department, and other groups—makes a fascinating, complicated case study of "biopolitics" in federal forest management. The controversy eventually became so heated and events so scandalous that lawyers, lobbyists, the U.S. Justice Department, Congress, and a host of federal and state authorities all got dragged into it through lawsuits and investigations. In a collection of essays on the controversy published in 1995 by the University of Arizona Press, appropriately titled *Storm over a Mountain Island,*[3] one of the essayists, a Washington, D.C., journalist, wrote:

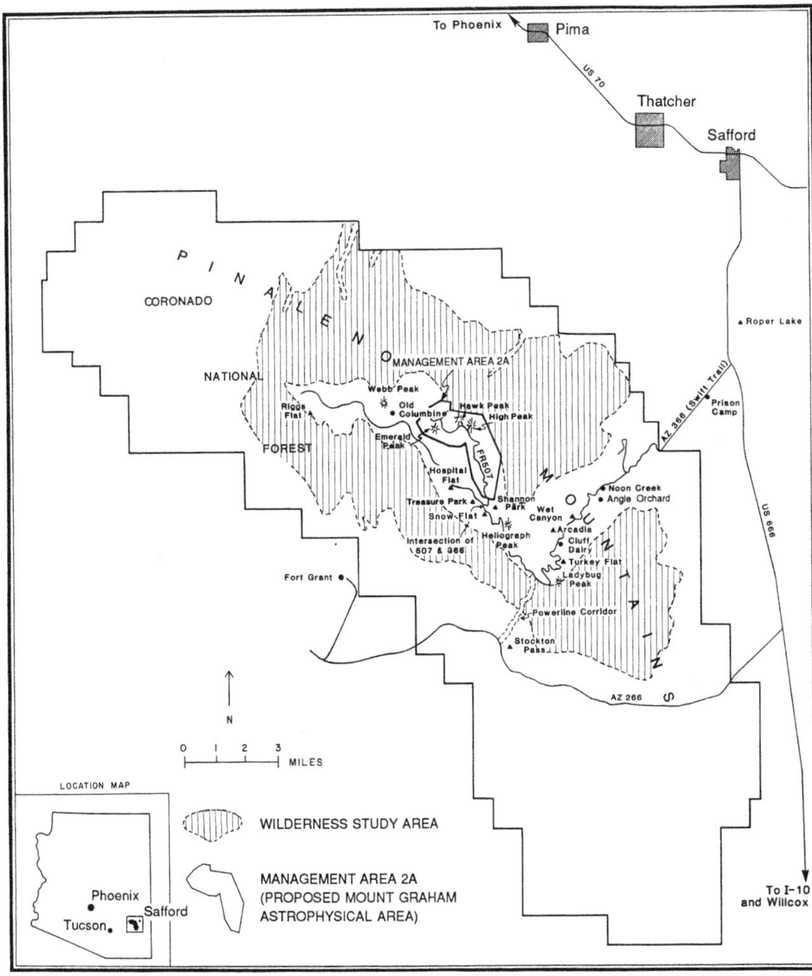

Coronado National Forest boundary encompassing Mount Graham (the Pinaleño Mountains) in southeastern Arizona. The proposed astrophysical development area is shown in the center. The Wilderness Study Area and High Peak, Hawk Peak, and Emerald Peaks are also shown.

"In the annals of environmental warfare, the battle over Arizona's Mt. Graham has been as nasty as any: an exercise in recrimination and venom equal to the strife over the snail darter and the spotted owl."[4] The Mount Graham battle was an epic environmental confrontation that provides a textbook example of the complex nature of federal land management and of the ways politics and science interact. It exemplifies the problems encountered by public land managers who try to enforce environmental protection mandates that conflict with economic development proposals, and it offers important lessons for anyone in the private sector interested in how environmental protection mandates get implemented or circumvented.[5]

Overview

The Mount Graham controversy centers on a small pocket of spruce-fir forest at the highest elevations of the Pinaleño Mountains—a sky island of coniferous forest in a sea of Sonoran and Chihuahuan desert.[6] Popularly referred to as Mount Graham, this rugged massif towers a vertical mile and a half above the Gila River Valley of southeastern Arizona and attains an elevation of 10,700 feet—a thousand feet higher than the next highest mountain in the area. Mount Graham is one of over a dozen sky island mountain ranges managed as a group by the Coronado National Forest. As a whole, the Coronado hosts probably the greatest diversity of plant and animal species of any unit in the national forest system.[7]

The tiny, fragile, six-hundred-acre pocket of pure spruce-fir forest found above 10,000 feet elevation on Mount Graham is the southernmost spruce-fir forest in North America and is a living biological relict of the Pleistocene. Many of the plants and animals associated with this Canadian-type forest ecosystem, which has been segregated from similar forest habitats since the retreat of the great continental glaciers 12,000 years ago, have evolved in isolation from other related gene pools for many thousands of years. Three mammals, including the now famous Mount Graham red squirrel, are endemic to the mountain (they exist there and nowhere else) as well as more than a dozen plants and invertebrates. Besides endemic species, there are unique or rare associations of plants and animals, some of which reach the limits of their northern or southern continental distributions on Mount Graham. The mountain hosts at least seventeen species with formal protected status under state and federal endangered species programs, with many additional candidate species under consideration.[8]

In addition to these unusual ecological characteristics, the mountain has qualities attractive to astronomers: dark skies, a relatively arid climate, and physical accessibility. Since the early 1980s, the Steward Observatory at the University of Arizona (UA) in Tucson has coveted Mount Graham as a potential site for an astrophysical observatory. Although the university has telescopes on four other mountain peaks surrounding Tucson, three on Coronado National Forest land already,[9] it wanted

The Mount Graham red squirrel. Courtesy Arizona Game and Fish Department; photograph by Bob Miles

to develop Mount Graham as an additional site primarily because Mount Graham offers a darker sky for optical viewing than existing sites in southern Arizona and because another large observatory site would bring additional research dollars and prestige to the UA.[10] The Bureau of Business and Economic Research at Arizona State University in Phoenix in 1985 claimed that astronomy added $84 million per year to the Arizona economy: "As host to the largest collection of research-quality telescopes in the world, Arizona has benefited from astronomy's transition into big business."[11] Astronomy, like many research sciences, can be viewed as an academic industry. Astronomers at major research institutions are in the business of producing information, the quality and quantity of which largely determine their ability to attract additional public and private funding to continue or expand their research programs. Importantly, the more "viewing time" available to an astronomy department at a university, the more that department can attract prestigious researchers and effectively compete for research grants. In the business of astronomy, additional special privileges accrue to institutions that develop and host a telescope site. The demand for good telescope sites far exceeds the supply, so a university that successfully controls a desirable mountaintop can attract a consortium of collaborators that will help share the costs of development and maintenance. These collaborators also pay the host institution various forms of rent for the privilege of using that site, which often includes reserving for the host institution a block of viewing time on the nonhost telescopes, thus multiplying research opportunities for the host.[12] Consequently, a new, large site for astronomy in southern Arizona, controlled by the UA, offered multiple benefits to the university and to the state as a whole. Mount Graham, although a relatively wet site for the arid Southwest with only 54 percent of nights suitable for astronomy due to cloud cover and humidity,[13] nevertheless offered an attractive and profitable opportunity for the UA.

In the early 1980s the UA gathered together a consortium of prestigious partners backing its proposal to develop Mount Graham, each with a stake in southern Arizona or engaged in a collaborative project with the UA. By 1986 these partners included the National Optical Astronomy Observatories (NOAO), the Smithsonian Astrophysical Observatory, and the Vatican Observatory (all three with branch offices in Tucson, where the UA is headquartered), as well as the Max Planck Institut für Radioastronomie in Germany, the University of Chicago, and Ohio State University, each of which was a partner with the UA on telescopes then in the development stage. Other potential partners considered joining the consortium over the following decade but eventually pulled out, in part due to cost and in part due to the controversy, including the Universities of Texas, Illinois, Pittsburgh, and Toronto.[14] NOAO pulled out of the project after 1987, the University of Chicago pulled out a little later, and the Smithsonian faded into the background after federal decision makers deemed its pet project far too large and environmentally damaging to be allowed on the mountain. The UA, however, persisted wholeheartedly with its plans.

Early Days of the Controversy: Forest Service Land Management Plans

In 1982, during the early stages of the Coronado National Forest's preparation of a land and resource management plan, the Safford District ranger, Cecil Sims, received a letter from astronomers at the UA expressing interest in Mount Graham as a site for an astronomical observatory. That same year the Arizona Game and Fish Department, under contract to the Fish and Wildlife Service, initiated a study of three endemic mammal subspecies on Mount Graham, preparatory to possible recommendation for federal listing as endangered species. One of those animals was the Mount Graham red squirrel (*Tamiasciurus hudsonicus grahamensis*), a morphologically, behaviorally, and genetically distinct cousin of the familiar red squirrel of northern conifer forests.[15] These independent actions in 1982 were destined for collision. The area desired by astronomers exactly coincided with the most important habitat for the squirrel and the ecologically unique portion of the mountain—the small pocket of old growth, spruce-fir forest above 10,000 feet elevation. In 1984 the Arizona Game and Fish Department completed its preliminary research on the red squirrel, the UA submitted a proposal to develop an astrophysical complex on Mount Graham, and the Coronado National Forest began preparing management alternatives and an Environmental Impact Statement (EIS) for its comprehensive forest plan. Despite the fact that the astrophysical proposal represented a tentative "wish list" and virtually no studies had been done regarding its environmental impact, the Forest Service recommended full observatory development in its draft forest plan of 1985.[16] This included up to eighteen telescopes, a dormitory, water and sewer systems, a garage, a machine shop, a powerhouse, access roads, parking areas, and other facilities, along with an aerial tramway and a power line up the side of the mountain through a Wilderness Study Area.[17] Yet, under a section titled "Environmental Consequences," the draft forest plan presented only three short paragraphs assessing in vague terms the potential impacts of this massive development, to wit: recreation in the 3,500-acre area slated for astrophysical development would be partially restricted, scenic resource objectives for the mountain would be harder to meet, and forest wildlife would experience some unspecified additional impacts.[18]

Intense criticism from conservationists and the expectation that the red squirrel would shortly be listed as an endangered species motivated the forest supervisor, Robert Tippeconnic, to suspend these draft recommendations and announce that he would prepare a separate EIS for the observatory proposal. Although the supervisor's initial approval in the draft forest plan was now moot, his recommendation nevertheless set a significant precedent. The agency appeared willing to accept the

UA's proposal, so the astronomers continued full-scale planning, with the expectation of receiving some form of approval shortly. This proved to be significant in subsequent months and years as the UA assumed approval was inevitable and continually referred to the environmental analysis as a mere "delay," characterizing the decision-making process as nothing more than annoying red tape. To the UA and its supporters the issue became not *whether* the observatory would be built but *when*. This failure to adequately assess environmental impacts from the start shaped expectations in a way that made it difficult for the agency to properly perform its duties.[19]

The process of developing a separate EIS for the Mount Graham observatory grew extremely complicated because of the complexity of the astrophysical proposal, the unique features of the mountain, and intensifying public interest. Nevertheless, in just over a year (October 1986) the Coronado produced its draft Mount Graham EIS for public comment. The new Proposed Action called for a much more conservative development of five telescopes to be built on only one peak on Mount Graham, called High Peak. No one liked the Forest Service Proposed Action. The UA wanted approval for more than five telescopes on more than one peak. Backing the university, the entire Arizona congressional delegation jointly sent a letter to Washington, D.C., Forest Service Chief Dale Robertson supporting the observatory and urging him to approve a larger allocation than just the five telescopes on High Peak. Conversely, conservationists, the Arizona Game and Fish Department, and the Fish and Wildlife Service wanted no telescopes at all. The Arizona Game and Fish Department accused the Forest Service of "bias," while the Interior Department's Office of Environmental Review warned of a "resource conflict of national significance" and a possible appeal of the Proposed Action.[20]

The Endangered Species Act and the Fish and Wildlife Service Enter the Fray

The designation of the Mount Graham red squirrel as endangered on June 3, 1987, interrupted the Forest Service's preparation of a final EIS.[21] The Endangered Species Act requires consultation with and approval from the Fish and Wildlife Service before any proposed federal action affecting a protected species or its critical habitat may be approved. Approval of the observatory by the Forest Service constituted such a federal action; thus the Fish and Wildlife Service had to sign off on it. The proposal to list the squirrel as endangered was published in the *Federal Register* on May 21, 1986, with formal designation following about a year later.[22] University lobbyists at first opposed the listing of the squirrel, claiming that they could adequately protect the animal by implementing their own habitat conservation plan in conjunction with building the telescopes.[23] After the squirrel was listed, however, they changed tactics and approved the listing but continued to tout themselves as the endangered squirrel's ticket to recovery. Criticizing the Forest Service for having logged portions of the squirrel's habitat in the past and criticizing the Arizona Game and Fish Department for having introduced in the 1940s the larger, huntable Abert's squirrel, which may compete with the red squirrel for food, the UA claimed that a collection of telescopes in critical habitat would provide the perfect institutional base for monitoring the animal and managing its recovery.[24]

Preservation of the squirrel quickly became the main focus. A simplistic, symbolic battle between squirrels and 'scopes suddenly overshadowed the larger question of whether the observatory was an appropriate land use for that mountaintop. The media especially loved the dichotomy. To conservationists the red squirrel served as an indicator species of a unique, isolated, and endangered spruce-fir forest habitat. Protecting the red squirrel from extinction meant protecting a fragile, nationally significant natural area from continued human modification. When working properly, the Endangered Species Act really serves as an endangered *habitat* act. In 1987 well-known ornithologist and retired Kofa National Wildlife Refuge Manager Gale Monson founded a group of over six dozen prominent scientists opposed to the observatory—a veritable Who's Who of southwestern ecologists. The following is an excerpt from their statement:

> It has become increasingly clear that [Mount Graham] is indeed a priceless biological museum. . . . The development of an astrophysical observatory is, in our opinion, incompatible with the goal of preserving natural conditions and evolutionary processes. . . . Mt. Graham should be recognized for its value to the natural sciences and accorded its due respect by the public and managing agencies. Economic benefits and considerations of academic prestige should weigh low on the scale of values with which a management strategy is developed.[25]

By July 1990 the number of Ph.D. cosigners of this statement had climbed to well over two hundred. But advocates of the observatory saw things differently. They viewed the squirrel and the Endangered Species Act as a potential threat to their telescope project and therefore sought to trivialize the whole conflict by characterizing it simply as squirrels versus telescopes. Eventually this dichotomy—like owls versus jobs—dominated public perceptions, forcing everyone involved to respond to the issue in these terms.

In 1986 news of this looming clash between biologists and astronomers had reached Washington, D.C., where political support for the observatory was building. Perhaps reflecting the UA's fear that the Endangered Species Act might block the project, officials in the Washington office of the Fish and Wildlife Service improperly delayed the designation of "critical habitat" for the red squirrel well beyond the legally mandated deadline. Concurrent with the proposal to list the squirrel as endangered on May 21, 1986, the Fish and Wildlife Service had recommended designation of critical habitat. The Endangered Species Act requires the Fish and Wildlife Service to designate critical habitat at the same time that it lists a species (or within one year of the date that critical habitat was proposed) but allows for a one-year extension of the deadline if necessary to gather additional data.[26] In this case, however, designation was delayed until January 5, 1990, three and a half years after the Fish and Wildlife Service proposed it and a year and a half beyond the legally mandated deadline.[27] The Sierra Club Legal Defense Fund (SCLDF) cited this failure to designate critical habitat as one of several complaints in its 1989 lawsuit. Just prior to a federal district court ruling on this complaint, however, the Fish and Wildlife Service finally met its obligation. The federal government never offered an explanation for this delay, but SCLDF attorney Mark Hughes charged that the action was deliberately calculated to facilitate approval of the observatory. Timely critical habitat designation would have made it impossible to approve the observatory, he argued, since the Endangered Species Act implementing regulations require the agency to stringently avoid any action resulting in "the destruction or adverse modification of critical habitat."[28] The project as conceived at that time would have affected dozens of acres of that rare, remnant spruce-fir forest, including as much as 10 percent of the highest quality habitat.

Political Manipulation of Biological Reports

The procedures for approving projects under Endangered Species Act guidelines are somewhat different from those guiding Forest Service decision making. In this case, the Forest Service, cooperating with the Arizona Game and Fish Department, the Fish and Wildlife Service, and the UA, prepared a Biological Assessment of the red squirrel and the potential effects of the Forest Service's Proposed Action (five telescopes on High Peak) on its survival.[29] After completion of the Assessment in July 1987, the Forest Service initiated formal "consultation" with the Fish and Wildlife Service under Endangered Species Act guidelines in which the Fish and Wildlife Service would prepare a Biological Opinion determining whether or not the Proposed Action significantly increased the Mount Graham red squirrel's risk of extinction. On August 31, 1987, the Fish and Wildlife Service completed its Biological Opinion and transmitted it to the Forest Service, which in turn transmitted it to the UA for review before releasing it to the public.[30] The Fish and Wildlife Service had concluded: "the implementation of the preferred alternative, which includes establishment of an astrophysical facility on the proposed High Peak site (Mount Graham), is likely to jeopardize the continued existence of the endangered Mt. Graham red squirrel."[31]

By law the Fish and Wildlife Service must offer "reasonable and prudent alternatives" to a Proposed Action when such a "jeopardy" decision is rendered. The Biological Opinion indicated four alternatives in order of preference. The agency's most preferred alternative was to place the telescopes on other existing developed observatory sites in Arizona. Its least preferred alternative (number 4) allowed the UA to build four telescopes on the High Peak site—similar to the Forest Service's 1986 Proposed Action minus one telescope.[32]

The Fish and Wildlife Service justified alternative number 4 by claiming that the jeopardy could be mitigated by tight land-use restrictions and closing and reforesting the jeep trail to Emerald Peak, a nearby peak the UA dearly wanted to develop.[33] An excerpt from the 1987 Opinion (page 38) says this about the Emerald Peak site: "The destruction of habitat on Emerald Peak for siting an observatory would have greater detrimental impacts than the proposed siting on High Peak. *Furthermore, those impacts on Emerald Peak could not be reduced below jeopardy with reasonable and prudent alternatives.*" (Emphasis added. This statement is crucial, as the reader will see in a moment.)

Although the four telescopes on High Peak appeared to be a reasonable compromise, in fact, the political manipulation of biological documents had already begun. In sworn testimony in 1990 the head of the Fish and Wildlife Service field office in Phoenix, Sam Spiller, claimed that his boss in the Fish and Wildlife Service regional office in Albuquerque "directed" him to provide an allocation on High Peak *even before any studies were completed.*[34] In subsequent testimony before Congress, the primary author of the Biological Opinion, Lesley Fitzpatrick, also of the Phoenix office, stated: "Prior to the Forest Service request for formal consultation, I was informed by my supervisor, Mr. Sam Spiller, that our Regional Office in Albuquerque wanted to ensure that there was an alternative allowing development of an observatory in the Pinaleño Mountains—also known as Mt. Graham—in the Biological Opinion that would be developed. *If we could not so provide, the Biological Opinion would be prepared by another office*" (emphasis added).[35] The Endangered Species Act requires decisions in Biological Opinions to be based *only* on the best available biological data. Predetermined conclusions are a clear violation of this mandate.

The Fish and Wildlife Service approval of a High Peak observatory turned out to be inadequate for the UA's needs, anyway. The astronomers insisted that Emerald Peak be included in any allocation. Prior to the circulation of the 1987 Biological Opinion, the UA had repeatedly informed the Forest Service "of the necessity for optical telescope locations in the Emerald Peak area." Seeing that a High Peak allocation was the only alternative the federal agencies were willing to consider, the university angrily charged in a letter to the regional forester that the agency had failed to address its proposal, leaving the university "with no alternative but to urgently request that you direct the Region and Forest to address the viable proposal." In essence, the university was demanding approval for a land allocation that the two federal agencies had already indicated they were unwilling to accept.[36]

Arizona Senator Dennis DeConcini (D-Ariz.), Representative Jim Kolbe (R-Ariz.), and other members of the Arizona delegation went to bat for the UA, pressuring both federal agencies to cooperate with the university and to "expedite" an acceptable allocation. In fact, on October 12, 1987, just six weeks after the Fish and Wildlife Service circulated the Biological Opinion for internal review, DeConcini boasted on a radio talk show that he had "convinced the Forest Service just recently to expand their environmental study to include Emerald Peak. . . . They were not going to do that and we asked them to do that under an expedited procedure; and I met with Sotero Muniz, the director of the Forest Service for this region, just last week." DeConcini expressed optimism that the project would go forward, indicating he was "prepared to fight" the opposition: "I'm committed to it, and I'll do anything I can, including trying to change the law . . . to let it happen."[37] Later, in October 1988, he would follow through on his promise to intervene legislatively on behalf of the UA.

In response to these pressures from the UA and the Arizona congressional delegation, the Forest Service suspended biological consultation, withheld the 1987 Biological Opinion from public release, and began the analysis process anew. Hoping to cover his bases this time, Coronado Forest Supervisor Tippeconnic directed the UA to submit a "minimum viable observatory" proposal for analysis in an "expanded" Biological Assessment. The UA complied on October 2, 1987, with a proposal for a seven-telescope observatory to be built on both High and Emerald Peaks, including space for a very large instrument array called an interferometer that had twice been disallowed by the federal agencies because of its extensive environmental impact.[38] The interferometer comprised a collection of six huge radio dishes to be moved around on rail tracks to nine separate cleared locations spread out across a mile and a quarter of forest.

Against the wishes of the consulting biologists, the forest supervisor accelerated this second Environmental Assessment in response to UA complaints about delays.[39] Just six months later in February 1988 the agency published its expanded Biological Assessment.[40] Two of the four authors, the two not directly employed by the Forest Service, distanced themselves from the document as soon as it was released. One coauthor, Arizona Game and Fish Department biologist Barry Spicer, author of the original study and proposal to list the squirrel in 1985, claimed the document "was so rushed, we have not been able to do a thoroughly reliable job." Another coauthor, Dr. Peter Warshall of the UA's Office of Arid Lands Studies and the principal researcher for the original 1986 draft EIS, charged that the Forest Service had unilaterally edited the findings, softening the assessment of the impact of the observatory. He submitted a lengthy critique of the document.[41]

Despite these concerns about the integrity of the 1988 expanded Assessment, the Forest Service reinitiated formal consultation with the Fish and Wildlife Service on February 17, 1988. On April 18, 1988, the

Phoenix field office of the Fish and Wildlife Service completed a second Biological Opinion and circulated it for review to the Forest Service, which then once again shared the document's contents with the UA while withholding them from the public. To the UA's great consternation, the document reiterated the conclusions of the 1987 Opinion; that is, the UA's minimum viable observatory (now seven telescopes on High and Emerald Peaks) would negatively impact too many acres of critical forest habitat, foreclose too much opportunity to regenerate past losses of habitat, and pose an unacceptable additional risk of extinction for the red squirrel. The Opinion again offered four telescopes on High Peak as the least preferable alternative and required the closure and reforestation of an old jeep trail to Emerald Peak to allow habitat recovery there.[42]

This second Opinion was never released to the public, however. Instead, in response to a UA request for another extension of the consultation period, the Forest Service and the Fish and Wildlife Service shelved the document. Flexing every ounce of political muscle available, the UA pressured the Fish and Wildlife Service to negate its findings and come up with something more acceptable. The crucial lobbying event came six weeks later on June 3, 1988, when representatives of the UA met privately with the regional director and assistant director of the Fish and Wildlife Service, Michael Spear and Jim Young, and with the regional director of the Forest Service, Sotero Muniz, at the Tucson airport to drive home their unalterable commitment to the Emerald Peak site. None of the actual authors of the Biological Assessments or Opinions attended the meeting, nor were they invited. Nevertheless, immediately thereafter Spear ordered his biologists in Phoenix to provide an Emerald Peak allocation, regardless of their earlier conclusions—a second violation of the Endangered Species Act. The field biologists dutifully but unhappily rewrote their document so it would include an allocation on Emerald Peak.[43] On July 14, 1988, the agency formally released to the public this flawed Biological Opinion, which contained a new "reasonable and prudent alternative #3" that approved *three telescopes on Emerald Peak,* along with a host of mitigation measures designed to lessen their impact. One of the authors of this rigged Opinion, Lesley Fitzpatrick, hoped the Draconian mitigation measures would serve as a "poison pill" to make the Emerald Peak alternative untenable to the UA. Since this July 1988 Opinion was not supported by previous biological studies, environmentalists immediately suspected foul play and announced their intention to appeal the decision.[44]

But the Emerald Peak alternative offered the UA an acceptable foot in the door since the Emerald Peak site by itself was large enough to eventually support all seven telescopes, including the interferometer. With this preliminary, biologically unjustified approval the UA hired the Mercedes of Washington, D.C., lobbying firms—Patton, Boggs and Blow—and went to Congress for help. Accusing the biologists of bias and the environmentalists of stonewalling, lobbyists for the UA asked the Arizona delegation to pass a law that would endorse the seven-telescope proposal and deem the requirements of applicable environmental laws "satisfied" so that opponents could not delay their project any longer. John Moag, the Patton, Boggs and Blow lobbyist, frankly stated to a Tucson reporter that "the university did what all of us do, *all of us*, appeal to a decision-making body to make a decision favorable to it. Environmentalists do it every day, business does it every day, labor does it every day."[45] In this case, however, the university requested exemption from environmental laws because it did not like the findings arrived at by government scientists or the procedures designed to ensure the integrity of their research. Chairman of the House Interior Committee Morris Udall explained to the same Tucson reporter that the whole reason Congress had passed the Endangered Species Act in 1973 was to get important biological decisions out of the hands of politicians and into the hands of scientists where they belong.[46] But the economic stakes were perceived as too high to leave this land-use allocation to resource professionals. As a UA economic development coordinator gloomily predicted, "Arizona's astronomy industry is at a critical turning point. If the Mount Graham International Observatory is not completed, the in-state industry will decline significantly over the next ten years and the state will experience a brain drain that will have a significant economic impact for Arizona."[47]

Congress Intervenes

In early August 1988 Democratic Senator DeConcini, in consultation with lobbyist John Moag and university attorneys, drafted legislation that would have approved construction of a seven-telescope observatory on Emerald Peak, "notwithstanding any other act, law, rule, or regulation," and would have eliminated many of the mitigation measures in the Biological Opinion and exempted the project from administrative and judicial appeals by declaring all environmental laws "satisfied."[48] Arizona-based environmentalists, with some help from their national counterparts, tried to block its passage, arguing that the 1988 Biological Opinion that approved development on Emerald Peak was unsupportable and illegal. But when queried by legislators, the regional and Washington offices of the Fish and Wildlife Service stood by the Opinion's conclusions. Meanwhile, university lobbyists argued that further delays might cause the UA's astronomy consortium partners to take their telescopes elsewhere, with a resultant loss to Arizona's astronomy industry. Defending legislative intervention, UA Vice President Allan Beigel stated: "The fact is that unless either the Congress or the U.S. Forest Service calls a halt to the continuing delays on the Mount Graham International Observatory proposal, some of the telescope projects that are now part of the plan will be lost. The Arizona congressional delegation is aware of this. We hope that the Forest Service is too."[49]

Republican Congressman Kolbe, whose district includes Mount Graham, reiterated that the UA "will start losing collaborators if some decision is not reached by September. . . . By awaiting the final tortuous and protracted twists of the process, the project will be crushed under the weight of the bureaucratic delay."[50] The implication in both of these statements is that bureaucracy was to blame for the so-called delays rather than the UA's refusal to accept the judgment of the professionals, that unnecessary red tape and not legitimate concerns held up the project. Project proponents subsequently used this spin on the decision process to great advantage in the legislative arena and in the court of public opinion.

A communication between Forest Service Chief Robertson and the Arizona delegation apparently fueled the UA's fear of further delays. In a subsequent investigation, the General Accounting Office discovered that the chief had privately informed the delegation in 1988 that the project would probably end up as a "no-go" decision if it had to go through the normal environmental review and approval channels, and thus special legislation might be necessary to get it off the ground.[51] This, of course, led to a redoubled effort by the Arizona delegation to pass special legislation approving the observatory before the October congressional recess. To avoid hearings and to facilitate quick passage of a Mount Graham bill, the Arizona delegation decided to append it as a last-minute amendment—a rider—to the Arizona-Idaho Conservation Act (AICA), an unrelated bill then making its way through the final stages of approval. The delegation saw the AICA as having enough momentum to overcome any opposition the attached Mount Graham rider might encounter. (Legislative riders, especially attached to appropriations bills, became a favorite vehicle for anti-environmental legislation in the 1980s and 1990s.)

Senators DeConcini and John McCain shepherded the AICA through the Senate, while Congressmen Udall and Kolbe worked the House. The only sticking point was the fact that all endangered species legislation had to be routed through the House Merchant Marine and Fisheries Committee. There, lobbyists for the National Wildlife Federation, the Sierra Club, and other environmental groups tried but failed to get the rider removed. At best they could only convince Walter Jones (D-N.C.), the committee chairman, and Gerry Studds (D-Mass.), a subcommittee chairman, to demand revisions to the rider to make it slightly more consistent with the (faulty) 1988 Biological Opinion. In the floor debate on the bill, Jones stated: "Quite frankly, earlier drafts of this measure could have been misinterpreted as waiving parts of the Endangered Species Act . . . additional discussions with the staff of the Merchant Marine and Fisheries Committee resulted in considerable alteration of the Mount Graham language. These changes were essential for receiving my support of this legislation."[52]

The AICA, with the Mount Graham rider attached as Title VI, passed in October 1988, becoming Public Law 100-696. It established a Mount Graham International Observatory on 120 acres surrounding Emerald Peak and required the Forest Service to immediately issue a special-use permit to the UA for construction of the first three of seven telescopes *subject to the terms and conditions of the 1988 Biological Opinion*. This last clause came at the insistence of Jones and Studds. The National Environmental Policy Act and Section 7 of the Endangered Species Act regarding biological consultation were deemed "satisfied" by the Act, effectively insulating the Forest Service EIS and the Fish and Wildlife Service Biological Opinion from appeals or litigation. An additional four telescopes were "tentatively" approved by Congress but remained subject to applicable environmental laws—for the time being.

In the end, Congress was misled into supporting a local pet project based on the conclusions of a federal document that turned out to have been illegally manipulated and biologically unsound. Many members of Congress thought that by supporting the Mount Graham rider they were simply endorsing a sound Biological Opinion and cutting red tape.

Politicians usurped the Forest Service's decision-making authority and eliminated appeal procedures that could have corrected the flaws forced into the decision process by special interest pressure. (Despite the tremendous outside pressures on the Forest Service, the agency actually opposed the legislation because it circumvented the agency's decision-making authority.)[53] The UA had successfully characterized the formal environmental review process as an unnecessary delay and the Mount Graham affair as a trivial contest between squirrels and a two-hundred-million-dollar observatory. The larger environmental policy issue of preserving biological diversity and the larger procedural policy issue of protecting the integrity of scientific studies took a back seat in this case to a political agenda driven primarily by economic considerations.

Environmentalists File an Appeal, and Politicians Intervene Again

The biopolitics did not end in 1988. By the summer of 1989 the university was poised to begin construction of a new access road to Emerald Peak. On May 4, 1989, the Arizona Wildlife Federation and other appellants filed a lengthy appeal of the Forest Service road construction permit, citing a failure to implement all the mitigation measures required in the Biological Opinion, as well as a violation of Section 4 of the Endangered Species Act for failing to designate critical habitat within two years after proposing it. Moreover, a sudden precipitous decline in the red squirrel population constituted "new information," which the plaintiffs argued should require reinitiation of biological consultation as stipulated in the Biological Opinion that Congress had enacted into law. If environmentalists could not block the project, they at least hoped to ensure that every environmental protection clause in the Biological Opinion would be enforced. Seeing substantial merit in the appeal, the regional office of the Forest Service recommended a stay of construction from May until July pending a decision on the appeal. The regional forester then sent copies of his memo to the UA and the Arizona congressional delegation, among others, and forwarded his decision to the Washington office for approval.[54] Richard Flannelly, appeals officer in the Washington office of the Forest Service, told a Tucson newspaper reporter on May 16, 1989, that the stay was likely to be granted, and ground-disturbing activities would be halted until at least July.[55]

But proponents of the observatory had other plans. Upon hearing of the stay recommendation Senators DeConcini and McCain and Congressman Kolbe angrily demanded a meeting with the Forest Service for May 18, at which McCain's aide Mike Jimenez advised there would be some "ass-chewing." The new Coronado forest supervisor, James Abbott (who had taken over from Tippeconnic that year), was also ordered to Washington for the briefing. Abbott, Flannelly, and the Forest Service's threatened and endangered species coordinator, Glen Contrerez, attended along with DeConcini, McCain, Kolbe, and three members of the legislators' staffs. According to Flannelly, McCain "read the F.S. staff the riot act." The GAO quotes Flannelly reporting that "during McCain's oration he severely chastised the F.S. for dragging its feet. He said he had an understanding with the Chief and he was very upset that the understanding was not being honored as evidenced by the Regional Forester's recommendation to grant the stay." The senator went on to say that he perceived the stay as a "delay," and he wanted the staff to tell the chief that he, McCain, felt it should not be approved: "The message better get back to the Chief," McCain concluded. At one point during the meeting McCain allegedly told James Abbott that if he did not cooperate on this project "he would be the shortest tenured Forest Supervisor in the history of the Forest Service." That, at least, was Flannelly's recollection of events as told to GAO investigators.[56] When this story later reached the press, McCain denied threatening Abbott's job, and Abbott denied feeling threatened.[57] Nevertheless, an aide felt obliged to apologize to Abbott the next day for the senator's harsh words and offered to arrange for some additional funding for desired recreation projects on the Coronado and Mount Graham in particular.[58] This kind of back-door funding of a special project for an individual forest circumvents the normal federal agency budgeting process and in this case served as a combination warning and reward. Later that year, DeConcini and Kolbe attached another rider to the House and Senate appropriations bills ordering the Forest Service to spend several hundred thousand dollars for recreation development and biological studies on Mount Graham. The bill provided no *additional* money; it simply required the agency to shift funds from other projects to Mount Graham. This is one small example of how easily Congress can manipulate federal agencies through their budgets.[59]

On May 19, the day after the meeting described above, Abbott, Flannelly, Contrerez, and another Forest Service officer, Denny Schweitzer, met with the Forest Service chief, who confirmed his previous "understanding" with McCain and directed Flannelly to deny the stay. Flannelly issued the denial that day.[60] Four days later, after returning to Tucson, Supervisor Abbott wrote to DeConcini, McCain, and Kolbe thanking them for the opportunity to "visit" with them, apologizing for not informing them about the stay recommendation, and promising to do his "level best to carry out the intent of the legislation through the permit to the university." Abbott ended his letter with a reference to the delegation's offer to get the Coronado some special funding for recreation projects: "Per David Steele's [an aide to DeConcini] request we will soon be sending a specific list of the recreation and habitat improvement projects and costs that we can accomplish in FY 90. I believe that progress in these areas can help replace some of the recreational opportunity lost due to closures implemented to protect the squirrel."[61] Abbott also enclosed a copy of a letter he sent to his staff the same day that provided "guidance" to them on how to approach the Mount Graham issue: "We must remember that congress passed the law (PL 100-696) providing for construction of three telescopes . . . the congressmen involved clearly expect this to be carried out and it is understandable that the university wishes to proceed with dispatch."[62]

With this level of cooperation, the UA was able to complete the new road to Emerald Peak before winter snows arrived, despite the continuing violations of the mitigation requirements in the Biological Opinion. In December the Coronado National Forest unveiled a recreation plan for Mount Graham that included new trails and a visitors center. The Safford District ranger predicted to a local newspaper reporter that the $530,000 needed to implement the recreation plan would be provided in the next fiscal year: "I think the federal government sees a need to make those changes, especially in light of the restrictions placed on the use of the mountain with the Arizona-Idaho Conservation Act."[63]

Litigation, Revelations, and Congressional Reconsideration

Between 1986 and 1989 the red squirrel population had suffered a devastating decline due to drought conditions that affected its food sources. Its estimated numbers dropped from an already critically low level of approximately 326 in 1986, to a little over 200 individuals in 1988, to about half that in the spring of 1990. Anything less than 350–500 reproductive individuals was considered below the "minimum viable"

A new road to Emerald Peak, hastily cut into the forest in red squirrel habitat in the fall of 1989.

population size and critically endangered.[64] Environmentalists argued in federal district court in late 1989 and early 1990 that the population decline, the invalid Biological Opinion, and the fact that critical habitat had just recently been designated (with incorrect boundaries) provided just cause for halting construction, reopening biological consultation, and preparing a new Biological Opinion. The UA argued that congressional law (the AICA) mandated construction of the first three telescopes regardless of any changed conditions. On March 26, just before the 1990 construction season was about to open, the federal district court sided with the Sierra Club and issued a 120-day injunction against construction, citing "serious issues" that had been raised by the plaintiffs. The judge further ruled that the AICA did *not* preclude reconsultation, but he sidestepped the responsibility for ordering reconsultation and instead asked Congress to take a second look at its approval of the project. In issuing his injunction, Judge Alfredo Marquez said, "What's the harm in saying there are some serious issues here. . . . Congress, why don't you take another look at it?"[65] Environmentalists were cautiously elated. The university immediately appealed to the Ninth Circuit Court.[66]

Portions of the court depositions from Sam Spiller and Lesley Fitzpatrick of the Fish and Wildlife Service describing their boss's predetermined demands regarding the Biological Opinion had also

Cartoon by Fitzsimmons. Courtesy *Arizona Daily Star*

been released to the media earlier that spring, precipitating a flood of angry responses from editorial boards about the importance of honest science. An editorial in the *Phoenix Gazette* titled "Mount Graham Scandal" proclaimed, "We haven't liked the Mount Graham telescope project all along. After hearing that federal biologists were told to skew their reports, we like it less." Other editorials bolstered Marquez's call for congressional reconsideration. The *Arizona Republic* approved of Marquez's decision, saying, "It is critical to the credibility of the [biological] study that it be seen as unbiased and scientifically sound."[67] This issue of scientific integrity is a position the UA, as an institution of higher education, should have backed wholeheartedly. But the fear of construction delays weighed far more heavily in importance. The university continued to vigorously oppose any reevaluation that would delay construction.

On April 4, just nine days after the district court injunction, Udall formally announced that his Interior Committee would sponsor an oversight hearing, saying, "The Mount Graham law was premised on the Biological Opinion. Some parties have raised serious questions about its validity. This hearing will focus on that issue. We hope to get some answers."[68] In response, Senators DeConcini and McCain called for a GAO investigation. McCain told a reporter, "I think most people [in

Arizona's delegation] realized we need to basically comply with the judge's opinion, which was that he wanted Congress to have another look at it. It's something we need to do."[69]

In the meantime, the national environmental organizations began paying increasing attention to this Arizona controversy, especially after Interior Secretary Manuel Lujan's incautious remarks in May 1990 about the Endangered Species Act being "too tough."[70] Fudged data, political intervention, and an interior secretary's suggestion that the Endangered Species Act should be weakened represented a national challenge to the implementation of environmental protection laws. Approaching congressional allies, Arizona environmentalists sought legislation that would reinstate the full application of Endangered Species Act regulations to the project and, if possible, repeal the 1988 Mount Graham rider. But Udall's position as Interior Committee chair and his reputation as an environmental leader inhibited the success of this tactic, since Udall did not support efforts to tinker with the rider he had supported. Instead, members of Congress, in a game of pass the hot potato, put pressure on the federal agencies to rectify the situation. In May 1990 Congressmen Walter Jones and Gerry Studds of the Merchant Marine and Fisheries Committee wrote to the chief of the Forest Service and the director of the Fish and Wildlife Service indicating that reinitiation of consultation certainly seemed appropriate under the current circumstances and that the AICA did not waive either agency's responsibilities for reconsultation. In part, their letter stated, "One of the reasons we supported the AICA was the fact that this language providing for the reinitiation of consultation was included in the Biological Opinion. Nothing in the AICA waives, weakens or in any way undermines the legal effect of this language."[71]

A month later, on June 8, 1990, Representative Claudine Schneider (R-R.I.) and nine other members of the House Merchant Marine and Fisheries Committee, Interior Committee, and Science, Space, and Technology Committee sent a similar letter to Forest Service Chief Dale Robertson expressing deep concern over the agency's continued failure to reinitiate consultation and urging a halt to construction pending the results of a new Biological Opinion.[72] The request for a halt to construction had become all the more urgent because a week after the Jones/Studds letter, on May 15, the Ninth Circuit Court of Appeals overturned the lower court's injunction. The appeals court, in a two-paragraph order based only on a perusal of the main arguments of the case, decided that significant new information did not exist and lifted the injunction, although it still directed the lower court to proceed with final hearings on the merits of the case.[73] Judge Marquez's provision of four months of breathing space for Congress to reconsider its approval of the observatory had vanished.

In the meantime, Forest Service lawyers had joined forces with the university, arguing that the AICA required telescope construction to proceed.[74] Opposing reconsultation, the Forest Service indicated it would approve the UA's special-use permit for telescope construction even if reconsultation were authorized. The letter to Chief Robertson from Congresswoman Schneider pointedly took issue with the agency's position, concluding that "it would be a violation of the Endangered Species Act and its implementing regulations were the Forest Service to permit the University of Arizona to proceed with construction before any reinitiated Section 7 consultation has been completed."[75] In hopes of avoiding further confrontation, the Arizona congressional delegation asked the UA to voluntarily hold off construction until completion of the oversight hearings scheduled for late June. The UA agreed but indicated it would proceed with construction after the hearings, regardless of the outcome.

The June 26, 1990, oversight hearings on the 1988 Biological Opinion climaxed this debate over reconsultation. The GAO kicked off the hearing with a report of its three-month investigation into the preparation of the Opinion. It had concluded that reasonable and prudent alternative number 3 of the Opinion was "not supported by prior biological studies," that Congress had approved the Mount Graham rider with a less than full understanding of the facts, and that because of new information and changed conditions "an updated biological opinion is warranted."[76] Previously unanimous support for the observatory among the Arizona congressional delegation wavered. The following day a headline in a Tucson newspaper read: "Telescope support splintering: Udall among four lawmakers calling for building delay."[77] Udall wanted construction halted until a new Biological Opinion could be written (which might take up to ninety days), but the UA insisted it could not afford to lose another construction season and would only voluntarily hold off for thirty days.[78] In response, Senators McCain and DeConcini asked the Fish and Wildlife Service to "update" the Biological Opinion within thirty days so as not to burden the university with additional delays. The Fish and Wildlife Service agreed instead to set up a blue ribbon panel to review all the facts and allegations and then determine within thirty days whether or not it agreed with the GAO that a new Biological Opinion was warranted.[79]

On August 6, 1990, the special Fish and Wildlife Service panel of national experts released its findings, *which concluded in even stronger terms than those of the GAO,* that the Emerald Peak alternative was biologically insupportable and biological consultation should be reinitiated, a new Opinion written, and critical habitat designation reevaluated for possible expansion of its boundaries.[80] In the wake of this unequivocal report, members of the Arizona delegation, with the notable exception of Kolbe, finally agreed that construction should be halted and the irregularities corrected. Mo Udall, in a press statement issued August 6, stated: "I fully accept today's report of the Fish and Wildlife Service special review team . . . and I think it outlines the only avenue available under the law to resolve the conflicts that must be resolved concerning the Mount Graham telescope project." Reflecting on the AICA, Udall said, "Supporters of the Endangered Species Act never would have cleared the Arizona-Idaho Conservation Act for passage if we believed that it undermined the integrity of the Endangered Species Act or precluded reinitiation of consultation under the set of circumstances contained in today's Fish and Wildlife Service report."[81] The same day, all five Republicans in the Arizona delegation issued their own statement expressing similar sentiments: "We have always believed that the Arizona-Idaho Conservation Act contemplated the possibility of reinitiation of consultation where new information has been found." Admitting that significant new information existed, the congressmen agreed to "defer" to the Fish and Wildlife Service. At the same time, they expressed the hope that additional studies would be expedited in order to keep the project moving forward without delay. Finally, they promised to "do whatever we can to help keep the telescope partners intact and committed to an Arizona project."[82] In fact, in a Tucson press conference that day Kolbe actually advocated *continued construction of the observatory during reconsultation.* Of course, this would obviate the need for reconsultation.

It is important to note that none of the congressmen proposed amending the AICA in spite of the fact that the injunction had been lifted and the university had the legal right to proceed with construction; indeed, the UA announced that it *would* proceed despite the findings of the Fish and Wildlife Service report. UA spokesman Steve Emerine

Cartoon by Fitzsimmons. Courtesy *Arizona Daily Star*

acknowledged that the university planned to award a construction contract the next day, and "construction could begin whenever the contractor can physically get his employees and equipment up the mountain." Asked whether that might look bad from a public relations standpoint, Emerine responded, "We're simply going to begin to do what we've had the right to do."[83] Udall's press statement the day before had said, "Of course, I am sure the university understands that sending construction equipment up Mount Graham at this time may well be viewed by many as inflammatory, and I would urge the university to exercise restraint in this regard."

A rational person at this point would expect reconsultation to occur. But despite this overwhelming evidence of the need to reconsult, the Washington office of the Forest Service and President Bush's Justice Department clung to the view that the AICA exempted the first three telescopes from any further review under Endangered Species Act regulations. UA lobbyists had found new allies in Washington. Accompanying the August 6 Fish and Wildlife Service report was a short memorandum from the agency's assistant solicitor noting that the Justice Department had instructed the agency not to reinitiate consultation. The Justice Department, suddenly playing a key role, asserted that because of the SCLDF lawsuit it was now responsible for interpreting the AICA. The

Fish and Wildlife Service disagreed with the Justice Department's interpretation of the law and wanted to reinitiate biological consultation, but the Justice Department insisted on representing the official position of the administration.[84] An internecine battle ensued, with intensive negotiations primarily between Richard B. Stewart of the Justice Department and Dave Hardy, a solicitor for the Department of the Interior, representing the Fish and Wildlife Service. Stewart finally prevailed. On August 23, 1990, a press release announced: "The U.S. Departments of Agriculture and Justice have concluded that the [AICA] does not provide for additional consultation with the Fish and Wildlife Service." The Forest Service issued a construction permit to the UA the next day.[85]

About a week later, the SCLDF lost a second attempt to block construction in the federal district court under Judge Marquez. Marquez apparently felt chastened by the Ninth Circuit Court of Appeals decision in May to lift his injunction and consequently denied the environmentalists' request for another restraining order: "I think that based on what the Ninth Circuit said and did after I granted the original injunction, that the district courts [*sic*] should limit itself to what it considers to be noncompliance with the Biological Opinion requirements."[86] Marquez further refused to order reconsultation, arguing defensively that "Congress has that updated study, and it is my present feeling that if Congress wants to stop the project . . . it is up to Congress since they have all the information that we have at this point to take some action . . . it certainly isn't up to me or the district courts to tell Congress whether they should authorize the project or not."[87] The SCLDF immediately entered an emergency appeal before the Ninth Circuit.

In frustration, Congressman Studds introduced a bill on September 12 to halt construction, order a new Biological Opinion, and require the university and the Forest Service to conform to whatever it determined.[88] No cosponsors were found in the Senate, however, and even in the House it became immediately clear that Mo Udall, powerful and respected chairman of the Interior Committee, would not support it. The day Studds introduced his bill, Udall issued a press release saying, "I very much respect Congressman Gerry Studds' intentions as reflected in the bill he has introduced. . . . In all candor, however, I must say that I doubt that his legislation will be enacted." Throwing the hot potato right back to the courts, Udall said the pending SCLDF appeal was a more appropriate vehicle for solving the problem. "It is my hope that the court will rule against the Department of Justice so that the obstacle to getting a sound Biological Opinion can be removed and the Endangered Species Act can perform as intended."[89] It is unclear why Udall thought the court would suddenly change its mind, especially considering Judge Marquez's comment that the ball was in Congress's court now. In effect, his opposition to the Studds bill and the decision of all the Arizona delegation members to let the Justice Department ruling stand put observatory construction on a fast track.

The SCLDF's appeal to the Ninth Circuit Court proved to be another case of a reversal of fortunes. On September 5, a hundred or so protesters camped out near the summit of Mount Graham, threatening civil disobedience against construction activities. A tree-cutting crew had arrived in Safford, Arizona, at the base of Mount Graham and was scheduled to be escorted under police guard to the summit at 5:00 A.M. the following morning. SCLDF attorney Mark Hughes called the appeals court in San Francisco to inform the court that time had run out. The court in turn called the Forest Service in Safford, Arizona, to verify that timber cutting was scheduled for the next day, and when it received an affirmative answer two of the three judges presiding over the case voted to issue a temporary ten-day injunction against construction. Two weeks later, on September 18, the court extended the injunction to December, when a full hearing was scheduled.[90] Once again, environmentalists were elated, while the university petitioned for a reconsideration, claiming that millions of dollars would be lost and that its partners might withdraw from the project if it were delayed.[91] UA lobbyists contacted the appellate judges, and ten days later, on a virtually unchanged record and without any explanation, one of the two judges who had voted for the injunction changed his mind, thus lifting the temporary restraining order. Immediately, the UA signed contracts, hired construction crews, and rushed up the mountain to clear the sites for the first two telescopes and a support facility before winter snows and the December court hearing arrived.[92]

By November the UA had accomplished its site-clearing objectives despite a month of civil disobedience by Earth First!ers and others in which authorities arrested or detained dozens of people blocking the path of bulldozers, chaining themselves to gates, and climbing to the tops of trees slated for felling.[93] The university hoped this action would render the appeal moot. As its new Washington, D.C., attorney, David Todd, put it to a Tucson reporter: "What was really going on here was an attempt to prevent habitat destruction, and the truth of the matter is that those trees are already gone, and you can't undo it. Our position is that fundamentally the case is mooted." SCLDF lawyer Mark Hughes responded, "I think going ahead and cutting old growth forest that the squirrels rely on for survival, in the absence of any study that accurately describes the impact of the project on the squirrels, is completely unconscionable. I think they did it purely out of hubris, and I can't conceive that the cutting in any way strengthens their [legal] position."[94] The Ninth Circuit Court heard oral arguments from both sides on December 13 and early the next year ruled in favor of the university.

Cartoon by Len Boro. Courtesy *Phoenix Gazette*

Telescope Construction, Eco-sabotage, and Site Selection Blunders

At over 10,000 feet elevation, Mount Graham gets a considerable amount of snow in the winter, making it difficult and expensive, if not impossible, to do construction work during this period. After a tense wait, the UA began construction of its first two telescopes in the late spring of 1991. Two years later, in the summer of 1993, the UA/Vatican optical telescope and the UA/Max Planck submillimeter telescope were completed. Protests never ended, however. That same summer of 1993 the radical environmental group Earth First! held its annual national gathering (the Round River Rendezvous) on Mount Graham. Three hundred people showed up for a week of camping, music, strategy sessions, and training in nonviolent civil disobedience. Earth First! had been involved in Mount Graham protests since the mid-1980s but became especially prominent after the passage of the AICA in 1988, when mainstream groups turned their attention mainly to administrative appeals and the courts. Many mainstream environmentalists thought the issue was by then a lost cause, leaving Earth First!, the Student Environmental Action Coalition (SEAC), and the Maricopa Audubon Society almost alone in the trenches. Also at this time, simmering opposition from traditional Apaches, who consider Mount Graham a sacred mountain, boiled into a major public protest led by a group called the Apache Survival Coalition, resulting in lawsuits of their own. From this point on, the avant-garde of protesters comprised a coalition of environmentalists, Indian activists, and social justice advocates.[95]

Although only minor acts of vandalism occurred during and after the Round River Rendezvous, Earth First!'s reputation for sporadic acts of sabotage, or "ecotage," led the Arizona Department of Public Safety to spend $208,304 on law enforcement during the Earth First! gathering. Virtually all the protest was confined to staged rallies and an illegal march up the UA's closed and gated observatory access road. In Tucson, twenty-five SEAC protesters temporarily occupied the UA administration building, leading to several arrests. Much to the university's consternation, these kinds of protests, which began in 1987, continued unabated through at least 1997. Protesters hoped, perhaps futilely, that the UA might still abandon the project and that the area could be reforested. Barring that, the environmentalists hoped to ensure that the environmental protection measures outlined in the 1988 Biological Opinion would be faithfully implemented. As past experience suggested and as events to come would show, neither the UA nor the federal agencies could be counted on to implement the Biological Opinion faithfully.

When the UA insisted that the Emerald Peak area had to be included in its astrophysical permit zone, one of its main reasons was that it offered better viewing conditions than High Peak, especially in regard to wind turbulence. Years of testing on High Peak, Emerald Peak, and other promontories supposedly indicated to the astronomers the clear superiority of the Emerald Peak site for astronomical viewing. In 1992, four years after securing the Emerald Peak site through a legislative rider, a study by UA astronomer Neville Woolf indicated that image-distorting wind turbulence on Emerald Peak had been "severely underestimated." As a consequence, UA astronomers asked the Forest Service in March 1993 to allow them to build their third and largest telescope on another peak, also in critical red squirrel habitat, more than half a mile away from the site where the Biological Opinion required the three telescopes to be clustered. Protests immediately erupted. This time environmentalists were not alone, however. After four months of wrangling, the Fish and Wildlife Service formally notified the Forest Service that a move to the newly proposed peak was outside the legislative authorization (meaning the AICA did not exempt the site from environmental laws) and thus would require "reconsultation." To the astronomers, that meant months or years of studies and no assurance that the site would be approved, so after some unsuccessful additional lobbying they withdrew their request in November 1993.[96]

Rather than go back to the original site specified in the Biological Opinion, however, the astronomers continued to push for two alternative sitings, one of them on peak 10,477 about a thousand feet to the east of the required cluster area. Instead of insisting on compliance with the Biological Opinion and the law, the Fish and Wildlife Service finally capitulated and approved construction of the third telescope on peak 10,477, stating that this new site was arguably within the area authorized by the AICA and that it fulfilled the requirement to "cluster" the three telescopes. Environmentalists disagreed and once again filed an administrative appeal in an attempt to make the federal agencies comply with their own regulations.[97]

While the appeal wound its way through bureaucratic channels, events on the mountain turned ugly. On December 6, 1993, just after the Fish and Wildlife Service approved development on peak 10,477, the Forest Service issued a permit to the UA to begin clearing trees from the site. Knowing there would be protests, the UA quickly and quietly sent tree cutters up the mountain the next day, December 7 (ironically, Pearl Harbor Day), for what environmentalists called a "sneak attack" on Emerald Peak. Before public word of the approval to develop peak 10,477 reached the media, over two hundred old growth spruce and fir trees in "critical habitat" were cut from the telescope site and from a new six-hundred-foot-long access road. They were eventually sold for cordwood. Later that month vandals put grinding compound in the gear case of a front-end loader and a snowblower near the construction site, purportedly causing $20,000 worth of damage to the front-end loader. Local law enforcement officials suspected Earth First!ers because of the nature of the vandalism, but no suspects were ever arrested.[98] In April SEAC occupied the offices of Tucson-based Research Corporation, an investor in the third telescope and a relatively new company founded by former UA President Jack Schaeffer. Some protesters were arrested. The company sought a restraining order against members of SEAC and other groups opposed to the telescopes and sued several protesters to recover "damages" for lost work time and added security expenses.[99]

Meanwhile, the Forest Service had denied the environmental appeal, thus clearing the way for opponents of the project to go to court. In May 1994 eighteen environmental groups signed on as plaintiffs in a suit to challenge the legality of the move to peak 10,477. Of course, many people thought the issue was moot now that the forest had been cleared, but opponents felt the issue was a matter of principle if nothing else, and they persevered. The old lawsuits filed by environmentalists during and after the congressional hearing in 1990 had by this time all been settled in favor of the UA. Congressional commitment to redress the wrongs perpetrated during the writing of the Biological Opinion had faded away. A lawsuit filed by the Apaches for violation of the American Indian Religious Freedom Act had also been lost. With two telescopes built, a third site cleared, and all lawsuits settled in its favor, the UA was feeling confident. Unfortunately, the tables were about to turn.

While a national gathering of SEAC members camped out in protest on Mount Graham in mid-June 1994,[100] Judge Marquez mulled over the environmentalists' latest lawsuit. Based on what he called "persuasive arguments" made by the plaintiffs and "serious questions" about the recent move to peak 10,477, Marquez finally issued a ten-day restraining order on July 12, 1994. Once again, astronomers decried this "delay" during the summer construction season, claiming that millions of dollars could be lost during only a few weeks of down time. After further review, Marquez found the UA's argument unconvincing and issued a permanent injunction against further work on peak 10,477. He agreed with the plaintiffs that the third telescope site was clearly outside the cluster area required by the AICA and ruled that any further construction required formal reconsultation with the Fish and Wildlife Service and the satisfaction of all environmental laws.[101] Environmentalists were elated; the UA was stunned.

On August 2 UA attorneys appealed Marquez's decision to the Ninth Circuit Court of Appeals in San Francisco, asking for an emergency order lifting Marquez's injunction so work could continue on the site that summer. The appeals court denied the emergency request and scheduled a full hearing for November of that year. Not until April of the following year, however, did the Ninth Circuit Court judges finally issue their decision. They agreed with Marquez and the environmentalists that peak 10,477 was not authorized by the AICA and therefore not exempted from environmental laws.[102]

One More Round with a Rider in Congress

Once again unable to achieve its goals through normal channels in compliance with existing laws and unwilling to accept the decision of the courts, the UA turned one more time to the Arizona congressional delegation for legislative relief. It took quite a few months, causing the university the loss of yet another construction season, but finally in December 1996 Republican Congressman Jim Kolbe, coauthor of the 1988 AICA rider, succeeded in attaching another rider to an appropriations bill giving the UA permission to continue construction on peak 10,477. Unfortunately for the UA, this appropriations bill was vetoed by President Bill Clinton during the big budget standoff that year. Among his reasons for the veto, Clinton mentioned several "objectionable legislative riders" dealing with the environment. Mount Graham was only one of a whole collection of riders, including several concerned with logging in Alaska's Tongass National Forest, revisions to the Endangered Species Act, and other public land issues.[103] Kolbe did not give up, though. With the help of Arizona Senator John McCain, the Mount Graham rider survived the budget compromises and finally sailed through attached to a spending bill of $160 billion passed by Congress on April 25, 1996. The one-sentence rider authorized work to continue on the third telescope on peak 10,477, thus nullifying the district court's and the Ninth Circuit Court's decisions.[104] The project proponents had triumphed once again via legislative intervention.

For a while during the summer of 1996 litigation to protect the Mexican spotted owl, which lives in the forests on Mount Graham, held up the project, but only briefly. The spotted owl nests in mixed conifer forest at lower elevations than the observatory, so Judge Carl Muecke in Phoenix, who had placed an injunction on all logging in old growth in Arizona and New Mexico, allowed the UA to complete tree clearing for the third telescope.[105] By the end of the 1996 construction season, 800 cubic yards of concrete had been poured for the foundation of the 122-foot-tall large binocular telescope on peak 10,477.[106] By 1999, the first three telescopes were either completed or well on the way toward completion. The UA had its foothold on the mountain.

Conclusions

Despite the intricate facade of rational decision-making procedures for natural resource management, political considerations often exercise a controlling influence over biological decisions. While the biological data clearly indicated that the observatory should not be approved if environmental laws were to be satisfied, neither the Forest Service nor the Fish and Wildlife Service had the backbone to make the appropriate decisions, considering the project's tremendous political support. Instead, each agency approved small allocations—a foot in the door for the astronomers—against the better judgment of its field biologists and then tossed the hot potato back to the other agency for final disposition. Eventually, political interventions overrode both agencies' discretion. When it later appeared that the legislative rider had been hasty and inappropriate, the previously proud architects of the AICA tried to pass the buck back to the federal agencies and to the courts to rectify the improprieties rather than take responsibility for resolving the problems that they themselves created.

Legislative intervention in this case was facilitated by the university's success in characterizing the issue as a few squirrels versus advanced technology, academic status, and jobs for a rural county. The real issue—whether to preserve a unique, endangered ecosystem or preserve the UA's prominence in the field of astronomy—became obscured by the trivial if colorful dualism of squirrels versus 'scopes. Characterizing environmental review procedures as delays and shifting the fundamental decision question from *whether* the observatory should be built to *when* it would be built helped the university. Controlling the definition of the issue thus played a significant role in determining outcomes.

Legislative intervention was further facilitated by the aura of respectability lent to the project by virtue of its being proposed by a university. One congressional staffer quoted anonymously by a Tucson reporter observed that "the University of Arizona acted no differently than a mining company or an oil company would. They wanted access to public lands for a development project"; but because the UA saw itself as an institute of higher learning with irreproachable motivations, the hostile reception to its proposal shocked project proponents. "Of course we'll do the right thing," the congressional staffer parroted university officials when asked about environmental impacts. "We don't screw up the environment, we're the university."[107] Federal agencies often share this deference toward university officials, as when Fish and Wildlife Service Regional Director Michael Spear told Chairman Studds at the congressional oversight hearing in 1990 that if the proposal was a housing development or some other ordinary use he would not have approved it. Studds angrily asked, "What is the legal basis for making that kind of distinction?" Spear correctly answered, "I don't think there is a legal basis for that kind of distinction."[108]

Members of Congress and the public expect honest and accurate environmental studies from federal agencies, especially in controversial forest management decisions. Conservation leaders in Congress bristle at the thought of federal agencies masking, suppressing, predetermining, or otherwise distorting scientific information in order to make controversial forest development appear more palatable. When Congress approved the AICA, they believed a biologically sound opinion backed their action. Unfortunately, as Udall noted in his September 12 press release, "the project is now lacking such an opinion." Statements of key congressmen indicate that an exemption from Endangered Species Act regulations would not otherwise have been granted. Scientific findings had to be massaged to get the necessary support for legislation. Subsequently, that legislation blocked remedial responses from the Fish and Wildlife Service and Forest Service when the fudged documents were called into question. This suggests a continuing need for caution when legislative intervention is contemplated, especially for situations in which short-term benefits (academic prestige) are weighed against possible long-term irretrievable losses (species extinction).

Finally, there is much debate these days about the function and value of public participation in federal environmental decision making. Some argue that the whole exercise is superfluous, a rational facade that ties agencies in knots or simply masks predetermined decisions. Certainly, public participation takes time and often produces results of questionable value. But it can be argued that information disclosure requirements are a positive aspect of public participation, since they help ensure honesty and accountability in decision making. In this case, even though the AICA rendered the eventual disclosures ineffective, public access to documents and judicial review provided a remarkable level of accountability. At the very least, they make critical analyses such as this possible.

NOTES

1. Lujan's comments were first quoted in the *Denver Post* (May 10, 1990), followed by the *San Francisco Chronicle* (May 12, 1990, A1) and the *New York Times* (May 12, 1990). Editorials in Arizona criticizing the interior secretary's comments appeared in the *Scottsdale Progress* ("Endangered Secretary," May 12, 1990, A12), the *Arizona Daily Star* ("Lujan's Ignorance," May 15, 1990, A10), the *Arizona Republic* ("Watt's That Again?" May 15, 1990, A8), and the *Phoenix Gazette* ("Lujan's Odd Candor," May 15, 1990, A10).

2. Jeff DeBonis is also the founder of the Forest Service newsletter *Inner Voice*. DeBonis's quote is from Peter Montgomery, "Science Friction: Playing Politics with Scientific Research," *Common Cause* (November–December 1990): 26.

3. Conrad A. Istock and Robert S. Hoffman, eds., *Storm over a Mountain Island: Conservation Biology and the Mt. Graham Affair* (Tucson: University of Arizona Press, 1995). This book was the product of a workshop, organized by the University of Arizona and the Smithsonian Institution, both sponsors of the Mount Graham observatory, held in Tucson in 1989 at the height of the controversy. The papers were collected for publication in 1990, so most contributions are now out of date. Moreover, the editors were outspoken supporters of the telescopes. There are no chapters by anyone active with the opposition and no sympathetic portrayals of the environmental position on the controversy. In fact, a professor from the University of Washington who spoke at the symposium, Donald Grayson, wrote a severely critical letter to Conrad Istock afterward, characterizing the symposium as a "whitewash" and calling the university's role in the affair "as ugly a display of institutional selfishness in the face of real environmental needs as I have ever witnessed" (Donald K. Grayson to Conrad A. Istock, n.d. [1989], in the author's possession). Grayson is also quoted in an article by Norma Coile, "Prof Blasts, Others Defend Mt. Graham Talks," *Tucson Citizen*, December 4, 1989.

4. M. Mitchell Waldrop, "The Long, Sad Saga of Mt. Graham," in Istock and Hoffman, eds., *Storm over a Mountain Island*, 9. This chapter was originally published as an article under the same title in *Science*, June 22, 1990, 1479–81.

5. For a discussion of Mount Graham in the context of this uneasy relationship between science and politics, see Montgomery, "Science Friction," 24–29.

6. The best overall discussion of these southwestern sky islands is Frederick R. Gehlbach, *Mountain Islands and Desert Seas: A Natural History of the U.S.-Mexican Borderlands* (College Station: Texas A&M University Press, 1981, 1993). See also *Biodiversity and Management of the Madrean Archipelago: The Sky Islands of the Southwestern United States and Northwestern Mexico,* U.S. Department of Agricultural General Technical Report RM-GTR-264 (Fort Collins, Colo.: Rocky Mountain Forest and Range Experiment Station, September 1995); and Russell Davis, "The Pinaleños as an Island in a Montane Archipelago," in Istock and Hoffman, eds., *Storm over a Mountain Island,* chap. 8.

7. *Final EIS, Coronado National Forest Plan* (Albuquerque, N.M.: USDA, Forest Service, 1986). This EIS stated: "The diversity of plants and animals found on the Coronado is unique in the National Forest system," and it recognized sixty-four species as needing special protection (5, 54, 68). A 1978 U.S. Fish and Wildlife Service report identified southeastern Arizona as having the greatest density of mammal species (*species* density, not population density) in the United States. The report also identified eight "unique and nationally significant wildlife ecosystems" in Arizona, all but one occurring in southeastern Arizona. See *Unique Wildlife Ecosystems of Arizona* (Albuquerque, N.M.: U.S. Fish and Wildlife Service, 1978).

8. For a detailed discussion of the unique biological and geophysical qualities of Mount Graham, see *Final Environmental Impact Statement, Proposed Mt. Graham Astrophysical Area,* EIS 03-05-88-1 (Albuquerque, N.M.: USDA, Forest Service, November 1988), 76–110.

9. The University of Arizona has observatories on Mount Lemmon and Mount Bigelow in the Santa Catalina Mountains of the Coronado National Forest and shares an observatory site with the Smithsonian Astrophysical Observatory on Mount Hopkins in the Santa Rita Mountains of the Coronado National Forest. It also uses Kitt Peak sixty miles southwest of Tucson as well as observatories in Chile and Hawaii. The UA also is a lead sponsor of the Hubble Space Telescope.

10. For the perspective of a UA astronomer, see Peter A. Strittmatter, "Astronomy and the Pinaleño Mountains," in Istock and Hoffman, eds., *Storm over a Mountain Island,* chap. 5. Strittmatter was the director of the UA's Steward Observatory and a staunch advocate of developing Mount Graham as an astrophysical site.

11. Arizona State University, College of Business, Bureau of Business and Economic Research, *Arizona Business* 32, no. 4 (June 1985): 1.

12. An example of this rent/barter relationship between astronomy research institutions is as follows: In the very earliest stages of the proposal to develop Mount Graham (1984–86), the Smithsonian Astrophysical Observatory agreed to act temporarily as the lead proponent (on behalf of the UA) because of its prestige and "excellent relations" with the Forest Service. A memo from Irwin Shapiro, director of the observatory, dated February 20, 1985, stated: "Because of our long and fruitful relation with the University of Arizona . . . [we] agreed to continue to help the University to develop Mt. Graham in every way we could . . . (such help could include, if the University wishes, a continuation of SAO's present role as nominal landlord for the purpose of dealing with the Forest Service . . .)." In exchange, the Smithsonian suggested the UA provide it "access to a certain amount of observing time on the University of Arizona–Max Planck Institut für Radioastronomie submillimeter telescope." Later, as the host of the Mount Graham site, the UA would be on the receiving end of such barter arrangements with other collaborators.

13. Between 1984 and 1987 the National Optical Astronomy Observatories (NOAO) consortium, funded by the National Science Foundation and also based in Tucson, considered placing a very large telescope (then in its planning stages) on Mount Graham rather than on the superior but more expensive site of Mauna Kea in Hawaii. For planning purposes, the NOAO funded a two-year study comparing viewing quality on the two sites: see NOAO Advanced Development Program, "Comparison Study of Astronomical Site Quality of Mount Graham and Mauna Kea," Technology Development Program Report No. 10 (March 1987). A summary of data on page 30 shows that 73 percent of nights on Mauna Kea were suitable for astronomy, compared to only 54 percent on Mount Graham. For the important requirement of low atmospheric humidity, Mount Graham registered an average of 73 percent humidity as against Mauna Kea's 25 percent. The NOAO, not surprisingly, chose Mauna Kea as the preferred site.

14. Istock and Hoffman, eds., *Storm over a Mountain Island,* vii.

15. Robert B. Spicer, James C. deVos Jr., and Richard Glinski, "Status of the Mount Graham Red Squirrel, *Tamiasciurus hudsonicus grahamensis* (Allen), of Southeastern Arizona," Arizona Game and Fish Department Contract No. 14-16-0002-82-216 (August 1985).

16. *Draft Environmental Impact Statement, Proposed Coronado National Forest Plan,* EIS 03-05-85-1 (Albuquerque, N.M.: USDA, Forest Service, May 1985), 19, 23, table 2.

17. Steward Observatory, University of Arizona, communication to Cecil Sims, Safford District Ranger, "Site Development Plan for Mount Graham," SO-PLAN-84-1, June 11, 1984, 1. The aerial tramway proposal actually originated with the forest supervisor and not the astronomers (*Draft Environmental Impact Statement,* 113).

18. *Draft Environmental Impact Statement,* 113.

19. A GAO Record of Interview with Safford District Ranger Cecil Sims, April 19, 1990, records Sims as saying "the fix was in," guaranteeing project approval from the start. The interview notes come from the GAO's spring 1990 investigation into the Mount Graham controversy and were obtained through a Freedom of Information Act request by Robin Silver, M.D. (in the author's possession).

20. Referenced letters and comments reprinted in *Public Comments and Forest Service Response to the DEIS, Proposed Mt. Graham Astrophysical Area, Pinaleno Mountains, Coronado National Forest* (Albuquerque, N.M.: USDA, Forest Service, 1988), 619–20, 670–71, 710–12, 245–421.

21. *Federal Register* 52 (June 3, 1987): 20,994.

22. *Federal Register* 51 (May 21, 1986): 18,630–34.

23. At an FWS-sponsored public hearing on the proposal to list the Mount Graham red squirrel as endangered, UA Vice President for Research Laurel Wilkening asked the agency to halt the listing process and allow the UA to substitute its own conservation plan. The UA's plan would serve "as an alternative to listing the squirrel," she said, adding that the plan "would do everything and more" that federal listing would accomplish. See Jim Erickson, "UA Asks U.S. to Drop Rare Squirrel from Endangered List," *Arizona Daily Star,* August 27, 1986; see also Waldrop, "The Long, Sad Saga," 13–14.

24. The University of Arizona's Office of Public Information produces a periodic newspaper titled *Lo Que Pasa.* Volume 11, number 12 (November 23, 1987) featured a four-page insert boosting the Mount Graham project. An article by Sharon Kha, then director of the Office of Public Information, and an article by Conrad Istock, then head of the Ecology and Evolutionary Biology Department (an advocate of the observatory), advanced the arguments referred to in the text above. See also Istock's chapter 2 in Istock and Hoffman, eds., *Storm over a Mountain Island* for similar arguments.

25. Gale Monson, "Scientists for the Preservation of Mt. Graham," 8831 N. Riviera Drive, Tucson, AZ 85737 (statement and list of cosigners in the author's possession). See also Gene Varn, "Scientists Join to Fight UA Bid to Build on Peak," *Arizona Republic*, August 25, 1987, B1.

26. Endangered Species Act, 16 U.S.C. §1533(a)(3) and §1533(b)(6). See also the Fish and Wildlife Service's own summary of these obligations in the *Federal Register* 55 (January 5, 1990): 427.

27. *Federal Register* 55 (January 5, 1990): 425–29.

28. 50 C.F.R. §402.02. For the Sierra Club Legal Defense Fund argument, see "Appellants' Opening Brief," in *Mt. Graham Red Squirrel, et al., v. Clayton Yeutter, et al.*, on appeal from the U.S. District Court for the District of Arizona, Ninth Circuit, No. 90-16125, 43–45.

29. *Mount Graham Red Squirrel: A Biological Assessment of Impacts, Proposed Mt. Graham Astrophysical Project, Coronado National Forest* (Albuquerque, N.M.: USDA, Forest Service, July 1987).

30. Communication from Fish and Wildlife Service Southwest Regional Director Michael Spear to Forest Service Southwest Regional Director Sotero Muniz, Draft Biological Opinion, 2-21-86-F-75 (August 31, 1987).

31. Ibid., 5; see also p. 29: "The establishment of an astrophysical observatory on High Peak does result in an increase in the existing level of jeopardy to the red squirrel through the further loss of habitat that is essential to its survival."

32. Ibid., 32–36.

33. References to closing and reforesting Forest Road 669 to Emerald Peak are found in several places in both the Forest Service's 1987 Biological Assessment and the Fish and Wildlife Service's 1987 Biological Opinion. Page 5 of the Assessment says, "Forest Road 669 would be closed to public access and revegetated." Page 10 of the Opinion repeats the same phrase, and page 31 states that under all of the various reasonable and prudent alternatives, the Forest Service must "immediately close FR 669."

34. Deposition of Sam Spiller, in *Mt. Graham Red Squirrel, et al., v. Clayton Yeutter, et al.*, January 11, 1990, 93–94. Also see Sam Spiller's testimony before Congress in "Joint Oversight Hearing on Preparation of the 1988 Biological Opinion Regarding Mt. Graham Red Squirrels," National Parks and Public Lands Subcommittee of the House Interior Committee and Subcommittee on Fisheries and Wildlife Conservation of the House Merchant Marine and Fisheries Committee, 101st Cong., 2d sess., June 26, 1990, 55–59. (These page numbers correspond to the original transcript of the hearing and may be different in the final published version.)

35. Lesley Fitzpatrick, testimony before U.S. Congress in "Joint Oversight Hearing," 60–66; the quote is from page 61.

36. Quotes come from a letter written by UA Vice President Laurel Wilkening to Sotero Muniz, Regional Forester, August 17, 1987.

37. Safford, Arizona, radio station KFM, October 12, 1987. A transcript of the interview is in the author's possession.

38. A copy of the UA letter to the Forest Service outlining this seven telescope minimum viable observatory is found in the Forest Service's Final EIS, *Public Comments and Forest Service Response to the DEIS, Proposed Mt. Graham Astrophysical Area*, 544–51. The Forest Service's 1986 draft Mount Graham EIS and the 1987 Biological Assessment and Opinion explicitly rejected development alternatives that included the interferometer.

39. In a letter to the Forest Service requesting the new study, the UA said, "The University specifically requests that this be done within the current decision schedule.

Further delays in the decision process not only unnecessarily increase the cost to all concerned, but gravely impact the viability of the entire proposal" (Laurel Wilkening to Sotero Muniz, August 17, 1989, 3). Coronado Forest Supervisor Tippeconnic told the GAO that "there was great pressure for speed of decision-making and approvals" during this period (GAO Record of Interview with Bob Tippeconnic, May 7, 1990). Fish and Wildlife Service Regional Director Michael Spear told Congress: "In this particular case, one thing we knew for sure was that time was of the essence. We heard that everywhere" (testimony of Michael Spear, "Joint Oversight Hearing," 87).

40. *Biological Assessment (Expanded) on the Mt. Graham Red Squirrel* (Albuquerque, N.M.: USDA, Forest Service, February 1988).

41. Reporter Gene Varn quotes the dissenting biologists in an article titled "Mount Graham Data Termed Incomplete: Coauthor Faults Report on Squirrels," *Arizona Republic*, March 2, 1988, B1.

42. The GAO uncovered this Biological Opinion in its investigation. It never became a public document and is simply referred to as the April 18, 1988, FWS draft Biological Opinion.

43. Deposition of Sam Spiller, 65–69. See also the GAO's testimony presented at the June 26, 1990, congressional oversight hearing: pages 6–7 of the GAO report states, "According to the FWS Regional Director and FWS biologists, as well as Forest Service and university officials, the [Emerald Peak] alternative resulted from a meeting between the Regional Director, Regional Forester, and university officials wherein the university officials rejected the alternatives of locating the facility either outside the Mt. Graham area or on High Peak. The FWS biologists informed us that they disagreed with the Regional Director's decision to add the third alternative and did so only at his specific direction."

44. The GAO written testimony presented at the June 1990 oversight hearing repeatedly underscored that the Emerald Peak alternative was not supported by previous biological studies (see 2, 4–8).

45. John Moag, quoted by Charles Bowden, "How the UA Knocked Off Mt. Graham," *City Magazine* 4, no. 1 (January 1989): 34. Two years later, John Moag wrote an article for the Washington Legal Foundation's newsletter, *Legal Backgrounder*, titled, "It Is Time to Amend the Endangered Species Act" (October 19, 1990, 1–6), in which he discussed the Mount Graham issue as a case in point for why the Endangered Species Act should be amended. The Washington Legal Foundation bills itself as "America's largest pro–free enterprise public interest law and policy center."

46. Morris Udall, paraphrased by Bowden, "How the UA Knocked Off Mt. Graham," 31.

47. Brian C. Catts and Lynn Witten, Office of Community and Public Service, University of Arizona, "Mt. Graham International Observatory: An Analysis of the Economic Impact," revised March 30, 1990, 4.

48. This draft bill, dated August 8, 1988, was leaked to reporters in early September and appeared in its entirety under the title "Proposed Draft Legislation" in the *Eastern Arizona Courier*, September 7, 1988.

49. Allan Beigel, letter to the editor, *Arizona Daily Star*, August 14, 1988. Beigel was responding to an editorial critical of the legislative intervention that ran in the *Star* on August 7 titled "End Run: Or, Circumvent and Conquer on Mount Graham."

50. Representative Jim Kolbe, letter to the editor, *Arizona Daily Star*, August 14, 1988.

51. GAO interview notes with Richard Flannelly, Forest Service appeals officer in Washington, indicate that "Sen. McCain said in his [Flannelly's] presence that the For-

est Service told him to get legislation if he wanted this project. . . . Mr. Flannelly understood, in this context, that the Chief told McCain that if the project went through the normal process it would probably result in a 'no-go' decision: hence the need for legislation." Copies of GAO interview notes obtained through a Freedom of Information Act request by Robin Silver, M.D., a Maricopa Audubon Society activist (in the author's possession).

52. *Congressional Record,* 134, daily ed. (October 20, 1988): H10553.

53. "UA Tactic Is Decried by Agencies," *Arizona Republic,* August 5, 1988, B6.

54. The Arizona Wildlife Federation appeal and request for a stay of construction were received by the regional forester on May 4, 1989. On May 12 Deputy Regional Forester R. Forrest Carpenter approved the stay ("Reply to 1570-1 [LMP]") and sent copies of his decision to the appellants, the Coronado National Forest, the University of Arizona, the Arizona congressional delegation, and others.

55. Jim Erickson, "Forest Service Likely to Delay Mt. Graham Scope Project until End of July," *Arizona Daily Star,* May 16, 1989.

56. GAO Memo of Interview with Richard Flannelly.

57. "Job Threat Allegation Denied by McCain, Forest Service Official," *Arizona Daily Star,* July 29, 1990, 8A.

58. GAO Memo of Interview with Richard Flannelly.

59. For a fuller historical analysis of how Congress directs Forest Service actions through the budgeting process, see Paul W. Hirt, *A Conspiracy of Optimism: Management of the National Forests since World War Two* (Lincoln: University of Nebraska Press, 1994).

60. The Washington office's denial of the stay request, "Reply to 1570-1 (LMP)," is dated May 19, 1989.

61. Coronado Forest Supervisor James Abbott to Senator Dennis DeConcini, "Reply to 2670," May 23, 1989, in the author's possession.

62. Memorandum, Forest Supervisor James Abbott to Coronado National Forest staff, "Reply to 2710," May 23, 1989.

63. "Forest Service Outlines Plan for Recreational Opportunities," *Eastern Arizona Courier,* December 23, 1989, A1.

64. Paul Ehrlich stated in a deposition for the Sierra Club lawsuit: "The current population level of the Mt. Graham Squirrel is substantially below what I consider necessary to ensure against the extinction of the species." Peter Warshall's 1989 deposition added, "The most recent census conservatively estimated that the squirrel population has a 95% probability of being somewhere between 87 and 145 adults. This is below every reasonable estimate of the minimum population size needed to sustain the Squirrel."

65. *Arizona Daily Star,* March 27, 1990, 1A.

66. Ibid., March 28, 1990, 1A.

67. *Phoenix Gazette,* February 8, 1990, A10, March 28, 1990; *Arizona Republic,* April 2, 1990, February 8, 1990, B5; *Arizona Daily Star,* February 7, 1990, 1B.

68. *Arizona Daily Star,* April 5, 1990, 1A.

69. Ibid.

70. "Mount Graham Red Squirrel vs. Observatory Construction," *Sierra Club National News Report,* April 24, 1990, 7. A *New York Times* article on May 12, 1990, quotes Interior Secretary Manual Lujan and representatives of the National Wildlife Federation, the Wilderness Society, the Natural Resources Defense Council, and Greenpeace responding to Lujan's comments.

71. Chairmen Walter Jones (D-N.C.) and Gerry Studds (D-Mass.) to F. Dale Robertson, Chief of the Forest Service, and John Turner, Director of the Fish and Wildlife Service, May 9, 1990. The letter also says, "Subsequent to the passage of the AICA, the Fish and Wildlife Service designated critical habitat for the Mt. Graham red squirrel. . . . Moreover, additional population monitoring studies on the squirrel have suggested that its population has continued to decline significantly since the issuance of the original Biological Opinion for the observatory. Either one of these above events could serve as the basis for the reinitiation of consultation under Section 7 regulations. Again, nothing in Title VI of the AICA waives this. Our committee would never have agreed to the language if it had."

72. Representatives Claudine Schneider, Wayne Owens, Constance Morella, Tom Lantos, Frank Pallone, James Scheuer, Jim Saxton, Thomas Carper, Roy Dyson, and George Hochbrueckner to Dale Robertson, Chief of the Forest Service, June 8, 1990.

73. *Mt. Graham Red Squirrel, et al., v. Clayton Yeutter, et al.,* No. 90-15463, May 15, 1990.

74. Larry Bradfish, attorney for the Forest Service in U.S. District Court, consistently argued that Congress mandated alternative number 3 of the 1988 Biological Opinion and that no reconsideration of the project either by agencies of the federal government or by the courts was allowable under the law. At a March 26, 1990, hearing on a preliminary injunction before Judge Marquez, Bradfish argued, "In adopting reasonable and prudent alternative three, Congress essentially insulated that reasonable and prudent alternative from judicial scrutiny, from legal challenges by the plaintiffs" (p. 34 of the court transcript).

75. Claudine Schneider to F. Dale Robertson, June 8, 1990, 2. This letter quoted the implementing regulations of the Endangered Species Act as follows: "After initiation or reinitiation of consultation required under Section 7(a)(2) of the Act, the Federal agency and any applicant shall make no irreversible or irretrievable commitment of resources with respect to the agency action which has the effect of foreclosing the formulation or implementation of any reasonable and prudent alternatives which would avoid violating section 7(a)(2)."

76. GAO written testimony submitted at the joint oversight hearing, June 26, 1990, 2.

77. *Tucson Citizen,* June 27, 1990, 1C.

78. UA spokesman Steve Emerine told a reporter that it was critical for the UA to begin construction that summer: "We feel strongly that if we lose this entire season, we're going to have some real problems with holding the project together" (*Arizona Republic,* August 7, 1990, A4).

79. *Tucson Citizen,* July 4, 1990, 1C. DeConcini, McCain, and Representative John Kyl (R-Ariz.) reported that they had convinced the Fish and Wildlife Service to comply with their request to update the Opinion within thirty days. Udall is reported as having declined to join the request for a thirty-day update, saying that such a short-term study would not accomplish the task needed. Tom Smylie of the Fish and Wildlife Service regional office in Albuquerque said the agency would do a thirty-day review but that "thirty days is not a reconsultation"; it would not fall under the aegis of the Endangered Species Act but "would be a separate thing that they requested."

80. Memorandum, U.S. Fish and Wildlife Service, Washington Office, to Regional Director, Fish and Wildlife Service, Albuquerque, "Refer to: FWS/DHC/BFA," August 1, 1990 (released August 6). This document is also referred to as the "Biological Update Team Findings and Recommendations." See pp. 1, 15.

81. Press statement of Representative Morris K. Udall, issued from his Washington

office, August 6, 1990.

82. Press statement of Representatives Kolbe, Jay Rhodes, Jon Kyl, and Bob Stump and Senator John McCain, August 6, 1990.

83. *Tucson Citizen*, August 8, 1990.

84. Memorandum, Charles Raynor, Assistant Solicitor, Fish and Wildlife, to Deputy Director, Fish and Wildlife Service, re: "Biological Update/Red Squirrel," August 6, 1990. The memo stated in part: "I have been instructed to advise you that Richard Stewart, who is the Assistant Attorney General for Environment and Natural Resources, Department of Justice, and the Solicitor have agreed that . . . the Department of Justice has the authority to manage the conduct of litigation concerning the interests of the United States . . . [and] the Fish and Wildlife Service should not take any action with regard to the recommendations of the Study Team concerning reinitiation of consultation on this project."

85. USDA Office of Public Affairs, News Release, "Forest Service Will Authorize Mt. Graham Telescope Construction," August 23, 1990. See also *Arizona Daily Star*, August 25, 1990, 1B.

86. U.S. District Court for the District of Arizona, "Motion for a Temporary Restraining Order," No. CIV 89-410 CLO ACM, August 27, 1990, 51.

87. Ibid., 4, of the court transcripts. See also *Arizona Daily Star*, "Judge Won't Block Mt. Graham Construction," August 29, 1990, 1B.

88. U.S. Congress, House, "A bill to provide for a revised biological opinion . . . " 101st Cong., 2d sess., H.R. 5603 (September 12, 1990).

89. Representative Morris Udall, Committee on Interior and Insular Affairs news release, "Udall Statement on Congressman Studds' Mt. Graham Bill," September 12, 1990.

90. *Arizona Republic*, September 6, 1990, B3, September 8, 1990, C138, September 19, 1990, A1; *Arizona Daily Star*, September 19, 1990, 1A; *Tucson Citizen*, September 19, 1990, 1C.

91. *Arizona Daily Star*, September 22, 1990, 1B; "UA Pays for Standstill," *Arizona Daily Wildcat*, September 11, 1990.

92. *Arizona Daily Star*, October 3, 1990, 1A, October 4, 1990, 1B; *Arizona Daily Wildcat*, October 1, 1990, 1.

93. *Arizona Daily Star*, October 3, 1990, 1A; *Arizona Daily Wildcat*, October 3, 1990, 3, October 5, 1990.

94. "Tree-clearing Settles Scope Fight, UA Says," *Arizona Daily Star*, December 9, 1990, 1B.

95. The best discussion of the rise of the Apache protest is the article by Carol Ann Bassett, "Fighting for the Heavens," *Tucson Weekly*, April 22–28, 1992, 6–7, 10–11, 34. A similar story by Bassett appeared in *High Country News* at about the same time. Elizabeth Barker covered a visit by Apache elders to UA telescope partners in Italy and Germany in 1991 titled "Lobbyists Visit Europe to Protest Continuation of Mt. Graham Project," *Arizona Summer Wildcat*, July 9, 1991, 1, 3. The Apache protest has been covered widely in the local, national, and international media. For more information and reprints of many newspaper articles, contact the Apache Survival Coalition's web site at www.planet-peace.org/mt_graham/

96. Jim Erickson, "UA to Drop Controversial Site Plan for Graham Scope," *Arizona Daily Star*, November 5, 1993, 1A, 13A; idem, "Third UA Scope on Mt. Graham to Rival the Best," *Arizona Daily Star*, February 6, 1994, 1B, 12A; Joseph Barrios, "UA Researching Alternative Site for Telescope," *Arizona Daily Wildcat*, November 8, 1993, 1, 4.

97. Erickson, "Third UA Scope." See also Dr. Robin Silver (Maricopa Audubon Society) to Jack Ward Thomas, Chief of the U.S. Forest Service, February 23, 1994 (in author's possession).

98. "$20,000 Damage Caused to Mt. Graham Machinery," *Arizona Daily Star,* January 11, 1994, 1B, 2B.

99. Kimberly Peterson, "SEAC Sued for Protest," *Arizona Daily Wildcat,* April 13, 1994, 1.

100. "Group Opposed to Observatory Starts Camp-in on Mt. Graham," *Arizona Daily Star,* June 12, 1994.

101. Jim Erickson, "18 Groups Sue to Halt Work on Mt. Graham," *Arizona Daily Star,* May 26, 1994, 1B–2B; idem, "UA Told to Stop Cutting Trees on Mt. Graham," *Arizona Daily Star,* July 13, 1994, 1B, 3B; idem, "Graham Scope Work Barred Pending Further Study," *Arizona Daily Star,* July 29, 1994, 1B, 3B.

102. "UA Files Motion to Resume Mt. Graham Work," *Arizona Daily Star,* August 3, 1994; Jim Erickson, "U.S. Court Halts UA Telescope Work at Mount Graham," *Arizona Daily Star,* August 25, 1994, 1B–2B; Bob Egelko, "Court Requires New Review for Telescope Relocation," *San Jose Mercury News,* April 26, 1995.

103. Eric Eyre, "Federal Budget Standoff Delaying Telescope Project," *Arizona Daily Wildcat,* January 25, 1996, 1.

104. Jim Erickson, "Kolbe Gets UA Mount Graham Scope Financing," *Arizona Daily Star,* April 26, 1996, 1B–2B and "UA Telescope Construction to Resume," *Arizona Daily Star,* June 18, 1996, 1B–2B. See also www.planet-peace.org/mt_graham/fitz3–14–96.html for the Apache Survival Coalition's protest against Representative Kolbe's 1996 Mount Graham rider.

105. Jim Erickson, "Spotted Owls Hold Up Mt. Graham Scope," *Arizona Daily Star,* June 25, 1996.

106. Jim Erickson, "Two Groups Added to UA Scope Team," *Arizona Daily Star,* February 22, 1997, 1B–2B.

107. Quoted by Bowden, "How the UA Knocked Off Mt. Graham," 31.

108. Testimony of Michael Spear, "Joint Oversight Hearing," 104.

9

Epilogue

Seeing the Forest Not for the Trees: The Future of Southwestern Forests

HAL K. ROTHMAN

Forests in the southwestern United States have become more valuable as scenery than as timber. . . . Wait! *Forests in the southwestern United States have become more valuable as scenery than as timber.* In the most mythic part of the nation, a transformation as radical as the Industrial Revolution itself has changed the value of one of the commodities essential to the growth of the American nation. Traditional extractive natural resource use—timber cutting—has been replaced by recreational use of forests. The evidence is everywhere. In the Las Vegas Valley, where 19,775 new homes were begun in 1998, a check of the lumber piles next to construction sites all over the valley reveals that most of the timber that goes into these houses comes from Canada. In 1994 no county in Arizona or New Mexico derived even 35 percent of its employment from the combination of mining, logging, farming, and ranching. The vast expanses of national forests in the Southwest, the ones with mythic names like Gila, Pecos, and Coconino, are being used for something other than timber.

The shift in use and meaning of forests is part of a transformation as significant as any in the last two hundred years. We've crossed a fault line as great as the Industrial Revolution itself, a shift as big as the one that drove Ned Ludd and his friends in the late eighteenth century to their hopeless destruction of the machines that had displaced them. We've crossed from industrial to postindustrial almost without noticing, except for the bleating of the Rust Belt and the rise of the right-to-work Sun Belt, which if you count George W. Bush's claim to be a Texan, has elected every president since Lyndon B. Johnson. We've come through the portal and out on the other side, in the heart of the Microchip Revolution. Land and water will look the same, but they'll be forever valued differently.

The catalyst for this transformation came from a minuscule piece of silicon called the microchip. This little piece of information-storing material caused the radical, inexorable, fundamental, and overnight transformation of the world economy. The Microchip Revolution—the constellation of changes associated with the rise of this funny little piece of equipment—changed life as we know it. Microchips created the so-called Information Age, in which knowledge and the ability to manipulate it genuinely became power, and it dramatically increased the range and significance of the service economy. Without the microchip, the revolutions of the past twenty years in everything from manufacturing and banking to health technologies simply could not have taken place.

This transformation created much wealth and many consequences. Cyber transactions now move capital around the globe, creating what the noted cultural critic Christopher Lasch recognized as a transnational class of monied individuals who are essentially stateless in their sense of national obligations. This boundary-free wealth has few obligations to people and places, and its incredible mobility can be a threat to the stability of individual places and national economies. The tremendous exporting of manufacturing to developing countries, from the *maquiladora* corridor in Mexico to the sweatshops in the Far East that got Kathie Lee Gifford in so much trouble, and the 1999 protests against the World Trade Organization in Seattle are only poignant examples of a trend that has profound implications for anyone dependent on a paycheck. The result has been that some skills—especially white-collar, information-based skills—have gained in value, while others such as industrial skills have been deprived of value. "Foreman says these jobs are going," sang rock-and-roller Bruce Springsteen in the early 1980s, "and they ain't never coming back." And right he was.

For corporations, the transition to a smaller, less permanent, more transient, less well paid workforce seemed like an easy strategy. The adage "cut the fat" easily translated into slashing the payroll, especially when the work could be outsourced or shipped overseas at much lower cost. Larger companies bought out smaller ones, creating conglomerates that dwarfed even the huge consolidations of the 1960s. Stockholders rubbed their hands together with glee as profits fattened and the DOW rose, but there were unintended consequences of corporate actions that CEOs rarely considered.

Like all its predecessors, this new revolution selects winners and losers. Some sectors of the economy fare marvelously. The educated, the highly educated upper end of the service sector—everyone from attorneys to consultants of all stripes—find opportunities galore at fees that reward their choice of career. The technically proficient—and here skilled health care professionals stand out, along with the denizens of Silicon Valley—also benefit greatly. They are their skill; they sell their knowledge, and it is transferable from the equipment of one employer to another. These two groups lead the pack. Wage workers of all kinds, agricultural workers, and semiskilled and unskilled bear the brunt of this change. Their skills have simply lost value. Stocking bottles in a 7-Eleven never had the upward mobility of industrial work, not to mention the job security, wages, and benefits.

One of the hardest hit industries has been the timber industry. A combination of factors, from the protests over the old growth forests in California in response to the buyout of timber companies by multinationals and corporate raiders to the spotted owl controversy throughout the West, has highlighted the difficulties traditional timber extraction has faced. Simply put, American timber, once the source of billions of board feet of lumber and thousands of well-paying, dependable jobs, is no longer viable in the market. Independent timber cutters, the "mom-and-pop" sawmills that for three generations in the Southwest sustained the industry, disappeared as technology and circumstances drove large corporations to include lands they had previously eschewed in their plans. The Forest Service, itself squeezed in a changing culture, has been forced to address challenges to its traditional behavior from both outside and inside the agency. Timber sales that lose money have inspired ire, and the appointment of Jack Ward Thomas as chief of the Forest Service placed a critic of the agency's business-as-usual policy at its helm. In a changing economy, the Forest Service, arguably the least flexible of federal agencies as a result of what former Forest Service historian David Clary has called its religious devotion to the sustained-yield principles of its founder, Gifford Pinchot, has been forced either to become introspective or to hold to an older value system, shouting about its relevance and using its small but historically effective constituency to defend it.

The forests of the Southwest presaged many of these issues. While not exactly given over to recreation by the end of World War II, they were far less valuable for timber cutting than forests in other parts of the country. A combination of aridity, steep grades, large cutover areas, and second- and sometimes third-generation growth meant that southwestern timber, while important in the region, never received the direct focus of the Forest Service. The Southwest was an afterthought in the timber-producing model of national forests. The Forest Service's tropes come from Region 1, the Inland Northwest, the home of the great fire of 1910, the place where many members of the first two generations of agency leadership earned their spurs. The Southwest has neither the timber nor the mythic claim to grand majestic forests of the northern tier.

In another way, the Southwest was the Forest Service's test case, the place where it learned about recreation and even wilderness. Marginal to begin with, the forests of the Southwest were linked to the myths of the region, the promulgation of an original American identity that separated the nation from its past and its long-held self-perception of inferiority to Europe. The result was a revolution in recreation that baffled the Forest Service and its Progressive Era value system and tied its political hands.

In one respect, this battle was a management question between rivals in the federal government. The Forest Service shared mission and constituency with other federal agencies, and the choice to use forests for timber was determined by the conditions of Progressive America and its acutely felt sense of scarcity. In another construction, forests could be used for other purposes. As Americans acquired the wealth, freedom, technology, and infrastructure that permitted travel and the cultural predisposition that made that activity an affirmation of American culture, forests could be construed in different ways.

Before World War II, the Forest Service's rival for management of recreation was the National Park Service. Founded in 1916, a decade after the Forest Service, the Park Service marketed a sedate and unthreatening uplifting leisure, an affirmation of American culture through the vistas of glorious nature. Its holdings included the most spectacular—and, some said, the least commercially viable—public lands in the nation. The Park Service developed a well-earned reputation as intrepid; led by a marketing professional, the agency intuited the culture of leisure that followed World War I and built its constituency by catering to the public. With able leadership and aggressive politics, the Park Service quickly became a savvy competitor, adaptable to the changing winds of the 1920s. Its frustrated detractors called the agency "inchers" for its oft-used tool of acquiring some land and then coming back for more. The Park Service and the Forest Service quickly became rivals that grappled to implement their programs on the same land. Usually the Park Service sought to acquire sections of national forests.

With the exception of the Olympic National Park controversy in the 1930s, most of the great battles between the Forest Service and this upstart cousin took place in the Southwest. These were largely scrapes over mission and constituency, turf battles, really, but they forced the Forest Service to come to grips with recreation in the American dreamscape long before such concerns were important in the heavy timber-producing regions. Inspired by Aldo Leopold, whose work culminated in the first administratively designated wilderness areas such as the Gila in southwestern New Mexico, the Maroon Bells near Aspen, Colorado, and parts of the Pajarito Plateau in northern New Mexico, to later efforts to accommodate recreational users, forest managers in the Southwest took the lead in Forest Service recreation.

In an agency with a clearly defined mission of timber management, the lead in recreation proved an internal disadvantage. The marginal position of Southwest forests in the agency's larger market-oriented scheme made decisions and ideas that emanated from it problematic in the broader Forest Service cosmology. As a result, the Southwest became a dead end through the 1950s, a place where forest managers got stuck in career-ending positions. Only after 1960, with the articulation of the doctrine of multiple use and sustained yield, the idea that the forest had many purposes, of which timber cutting theoretically was only one, and the passage of the Wilderness Act, which designated areas of roadless acreage to be managed for their natural rather than commercial values, did southwestern forests become central to the agency.

A two-generation-long revolution in American behavior that grew up hand in hand with the automobile, roads, and other technologies that freed Americans from other historic strictures inspired this change. Beginning with the automobile revolution of the 1920s and roads such as the mythic Route 66 that followed, a broader array of Americans traveled in greater numbers to the Southwest than ever before. The elite train traveler, once the target of national park promotions, was replaced by the automobile camper, who was not confined to the corridors created by rails but instead could move widely around the Southwest. Many even traveled off the main roads, and the large expanses of forest often sheltered them in designated and impromptu campsites. By the end of World War II, Americans of all backgrounds were accustomed to traveling their country by automobile. This decision forced the Forest Service to address its fundamental ambivalence about recreational use in national forests.

The ambivalence stemmed from the clear-cut mission Gifford Pinchot carved for his agency. From its founding, the Forest Service was first and foremost about timber, about making sure the nation had an adequate supply over time. In contrast, recreation was insignificant. Agency decisions made clear the zeal with which the Forest Service pursued Pinchot's goals. In 1919 the agency willingly gave up the Grand Canyon to the Park Service because it didn't want to be as deeply ensconced in the recreation business as managing the Grand Canyon demanded. The two Forest Service ranger stations along the South Rim were seven miles east and seven miles west of El Tovar and the tourist developments that surrounded it. The Park Service located its operations at the South Rim, near El Tovar, in the middle of the tourist facilities.

The pattern was established for the Forest Service. Recreation was a second-tier activity. In the 1920s the Forest Service eschewed development, opposing roads, hotels, and other amenities for the forests while claiming that its visitors were more serious than the frivolous National Park Service crowd. An elitism and even smugness permeated this perspective, and it made change to a recreational perspective harder than it needed to be for the Forest Service. Even after World War II and as late as the 1960s, the Forest Service really didn't regard recreation as a significant part of its mission. Despite their marginal timber, southwestern forests were deemed timber producers. Other uses lagged, even under multiple use.

Then came the deluge. Following 1950, the use of recreational land grew exponentially. National forest lands bore more than their share of the brunt of this growth; campgrounds filled, garbage cans overflowed, forest fires blazed as people tossed cigarette butts out car windows. Recreation altered the natural landscape just as logging and ranching had, but in a more localized manner. The impact was widely distributed, but national policy lacked the mechanisms to effectively respond to it.

The Southwest experienced a disproportionate share of this rapid growth, but it did not lead to a comprehensive or responsive recreation policy. Between 1946 and 1961 the number of visits to national forests increased from about 18 million to almost 102 million; in Arizona alone, visitation jumped from 1,716,900 visitors in 1946 to 13,495,400 in 1964. New Mexico forests experienced similar growth in recreational use. Although better positioned than other regions as a result of the marginality of the timber industry, southwestern foresters came up in the same culture as the rest of the agency. They too were taught to emphasize timber production and grazing; they too spent their days solving the problems of local constituencies and paying little attention to visitors from afar. They didn't really understand the needs of recreationists, and they did not attempt the kind of sophisticated analyses of visitors and their values that later became a hallmark of federal agencies.

The growth of the 1950s presaged the larger recreation revolution that occurred after 1965. A combination of factors, including innovations such as Gore-Tex, freeze-dried foods, and lightweight camping gear that made the outdoors easier to use, a new set of values that promoted a relationship with nature as one of the pinnacles of authentic experience, and the growth of the regional population all contributed to the transformation of southwestern forests. Tourists coveted the majestic beauty of the Southwest above all other regions of the nation. In 1965 Arizona forests received 8,797,100 visitor-days, equivalent to a twelve-hour stay by each visitor; by 1984 the average visitor-days had tripled. Again, parallel growth occurred in New Mexico national forests. Despite these new pressures from tourists and recreationists, the Forest Service lacked a strategy to respond to visitor demands and failed to address the long-term implications of the change in public behavior.

The passage of the Wilderness Act in 1964 became another catalyst for the transformation of the forests. The Wilderness Act fulfilled one of Aldo Leopold's and Robert Marshall's most fervently held goals: the establishment of untrammeled reserves designated by law as wilderness to be managed as wildland. With this congressional mandate, the Forest Service was forced to respond. It initially selected less accessible, high elevation land with little or no timber or from which timber was difficult to extract, such as the Pecos in New Mexico, for designated wilderness.

From the start, wilderness area designations posed two problems. Wilderness attracted many visitors because it symbolized American uniqueness or exceptionalism, and young Americans of the 1960s and 1970s sought authentic experience in nature to prove themselves. The Forest Service, still lacking the framework to manage wilderness, balked at the new restrictions imposed by the Wilderness Act. The law tied agency hands, its detractors complained, with powerful legal restrictions that limited the options from which grassroots managers could choose to make viable decisions.

Despite these management difficulties, wilderness designations and the 1970s Roadless Area Review and Evaluation (RARE I and II) guaranteed that the southwestern forests would be devoted principally to recreation rather than to timber cutting and grazing during the remainder of the twentieth century. The further into the high desert, the less likely that timber cutting was an integral part of forest management. The forests proximate to urban centers—from Flagstaff to Santa Fe—experienced considerable recreational day use; the more remote forests and their wilderness areas served a smaller but devoted constituency that tended to stay longer and that especially valued the pristine nature of the wildland the forests harbored. The Southwest became the premier wilderness region in the United States. By 1984 New Mexico contained more than 1.4 million acres of wilderness in national forests, and Arizona topped 1.3 million acres. The Forest Service was in the recreation business in the American Southwest even if its officials perpetuated the practices, standards, and ideology of the federal agency's original Progressive Era mission.

Clearly, recreation enjoyed a prominent if underacknowledged position in the southwestern forests; their rise to prominence within the agency began in the 1970s and followed a changing economic climate. The combination of Vietnam War–era inflation and the OPEC oil embargo brought postwar prosperity to a screeching halt and unmasked the aging of the American industrial infrastructure. Industrial America was forced to compete against newer facilities and cheaper labor overseas, and the result was disastrous. Between 1974 and 1997 the real value of American wages declined every year. The nation even lost entire industries. Once the United States made televisions. The last place in the nation to assemble them was the Zenith plant in Springfield, Missouri. By 1998 the same building housed the mail order inventory for Bass Pro Shops, a recreational tourist entity that catered to the tastes of blue-collar America.

The overlay of Bass Pro Shops atop a Zenith facility clearly illustrates one of the largely overlooked transitions of the Microchip Revolution. As Americans increasingly became creatures of popular culture and as the world became linked through a range of technologies that to date has culminated in the Internet, a new world formed. Its tenets were clear if a little baffling.

In this new world of individual expression, experiences became currency, and entertainment replaced conventional culture. This equation has had enormous ramifications for national forests throughout the country, but especially in the Southwest. In the national dreamscape, the rise of outdoor experience as commodity sent thousands streaming to the Southwest, the most mythic region of the nation and the destination of the greatest number of tourists. Not only did visitors seek to find themselves and to prove their hardiness against real nature, they also sought to possess the amorphous culture of the self. Nature promised self-realization in a world of inauthentic and fabricated experience. Against the backdrop of the American Indian and Hispano past and with the imprimatur of wilderness, forest recreation attained great cachet.

Ski resorts became prominent manifestations of this new impulse. Skiing had been a specialty sport until the 1950s, a province of the hardy and self-affirming who challenged nature on its own terms. After the 1950s the technological revolution allowed skiers in high-tech gear to keep warm on the slopes, ride high-speed lifts to maximize their cold-weather pleasures, and diversify their exhilaration via snowboards and shaped-edge skis. The array of ski technologies first made skiing popular and sexy and then more affordable to a wider segment of the population. All that remained was skiing easy enough for inexperienced outdoor recreationists, those who admired and sought to emulate the Buddy Werners, Jean-Claude Killys, and Spider Sabiches of the world but who knew that they lacked the skill and conditioning for head-long charges down a mountain. The national forests helped provide that venue.

Ski areas in national forests have been common since the late 1940s but have only been seen as problematic in the 1980s and 1990s. As skiing became a mass activity, in fact, a daily perquisite for people in towns like Santa Fe and Flagstaff, it gobbled up larger quantities of forest. Snowmaking became a threat to water sources. Since most southwestern skiing took place on Forest Service land by permit, it demanded a greater amount of agency attention. By 1980 the Forest Service had introduced regional recreation coordinators into the Southwest and Rocky Mountains, but these new-style forest managers were anomalies in an agency still primarily committed to timber and grazing interests. Nevertheless, their presence foreshadowed a transition of epic proportions in the national forests.

As recreational users became a more important constituency, traditional users became progressively more unhappy. Statutory restrictions on timber cutting and grazing, especially in the ecologically fragile Southwest, played an enormous role in creating the sagebrush rebellion in 1979–80 and its ancillary, the county movement. Angry traditional forest and range users felt abandoned and neglected in a changing world. Their short-sighted solution was to claim priority because of "culture and custom," kick out the federal managers, assert local control, and, like the Luddites, try to hold back the tides of change. The tension became thick, the animosity fierce. In 1993 an armed Nye County, Nevada, commissioner named Dick Carver drove a bulldozer at a Forest Service employee who tried to prevent Carver from opening a closed road. In 1995 Guy Pence, a forest supervisor, was standing in his home when a pipe bomb blew up in his van, which was parked in his driveway. After more than a decade of hostility, in 1999 Humboldt-Toyaibe Forest Supervisor Gloria Flora resigned in protest because of the threats made against her and other federal managers. The anachronistic voice of the few who had long made their living from the forests of the Southwest sounded in fury.

For the Forest Service, the increasing violence of the sagebrush rebellion should have served as a signal that times had changed and the agency needed new supporters. Committed to its admirable mission of decentralization and local autonomy, it lacked the ability to react. Recreationists who flooded the national forests in all seasons made likely supporters, especially when faced with the prospect of the privatization called for by the county movement and a sure step toward higher user fees. Even with the evident hostility forest managers confronted, they often failed to grasp the importance of the recreational constituency. The result as the century ended was management chaos in the southwestern forests. Traditional users were fewer in number, more militant, and less central to the agency. Recreationists, enormous in number and creating a series of ecological problems of their own, remained diffuse. The result too often became forests managed for the benefit of the few, the long-term users, at the expense of the broader constituency.

The future is clear. The Southwest remains the home of one of the most prominent American myths of reinvention in a romantic natural landscape. The impact of technology, wealth, and leisure has meant a windfall for operators and a massive transformation in the patterns of national forest use. Little about this picture will change in the immediate future. The baby boomers, the largest cohort in American history, are beginning to retire in great numbers. They have more discretionary spending power, ample leisure time, and better health than any previous generation. Even more, many stand to be beneficiaries of the greatest cross-generational transfer of wealth in human history, as their parents, the generation who came to maturity after World War II, pass on their assets. They will travel, by car and especially by recreational vehicle, to see the natural beauty and fascinating antiquities of the Southwest's forestlands.

Likewise, the upcoming generations of Americans and even Europeans will covet the nation's forests. These "echo-boomers," possessed of the birthright of leisure in a popular culture–based society, represent the future. From mountain bikers to Extreme Games participants, they will continue to flock to southwestern forests, winter and summer, to entertain themselves and others while experiencing a sense of freedom and exhilaration in nature. They too will grow in numbers and in wealth; they too will make their needs felt across the forests of the region. Like their predecessors, extractive users, these newer users will alter the forest environment.

Added to the hikers, campers, hunters, and other recreational users, these baby boomers and echo-boomers represent a deluge. They are already the dominant users of southwestern forests, and both land managers and traditional extractive users fear their presence. Their prominence is likely to increase; timber cutting and grazing are rapidly fading in significance. Like the Bureau of Reclamation (now called the "Bureau of Recreation" by satirists), the Forest Service faces a critical problem. It needs to recognize and respond to its new mission, at the same time keeping track of and maybe even mending relationships with its anachronistic traditional users. This is a difficult mission, especially for an agency with the powerfully established traditions of the Forest Service. Yet it is not really a choice. The southwestern forests have become playgrounds. The more than 90 percent of westerners who live in the urban centers of the region and visitors from the rest of the country are their primary users. Forest managers and users, new and old, will have to learn to share the forests and minimize their impact on them in the twenty-first century. It's a new and very different world.

Contributors

THOMAS G. ALEXANDER is Lemuel Hardison Redd Jr. Professor of History at Brigham Young University. Professor Alexander is the author of more than a dozen books, including *Utah: The Right Place* (Layton, Utah, 1995) and *The Rise of Multiple-Use Management in the Intermountain West: A History of Region 4 of the Forest Service* (Washington, D.C., 1987), and numerous articles in journals such as the *Western Historical Quarterly*, the *Utah Historical Quarterly*, and the *Pacific Historical Review*.

SUZANNE S. FORREST is an independent historian with a Ph.D. in history from the University of Wyoming. She is also the author of *The Preservation of the Village: New Mexico's Hispanics and the New Deal* (University of New Mexico Press, 1989, reprinted in 1998). Her chapter addresses many of the natural resource issues facing New Mexicans since World War II.

ARTHUR R. GÓMEZ has been the History Program manager for the National Park Service, Intermountain Support Office–Santa Fe (formerly Southwestern Region) since 1992. He is the author of the acclaimed *Quest for the Golden Circle: The Four Corners and the Metropolitan West, 1945–1970* (University of New Mexico Press, 1994) and other books as well as articles in numerous historical journals such as the *New Mexico Historical Review* and the *Mining History Journal*.

DIANA HADLEY is an environmental historian in Tucson. She works as an editor and translator of Spanish historical documents for Documentary Relations of the Southwest at the Arizona State Museum of the University of Arizona. She has degrees in archaeology and history from Washington University and the University of Arizona. She specializes in the history of land use and the history of ecological change in the greater Southwest. She has co-authored two book-length ethnoecological studies of Aravaipa Canyon and Bonita Creek for the Bureau of Land Management and a book-length report for the U.S. Forest Service entitled *Land Use History of the San Rafael Valley, Arizona (1540–1960)*, and she is currently working on an environmental history of the border area of southeastern Arizona and southwestern New Mexico for the Forest Service. She is co-editor with Thomas H. Naylor and Mardith K. Schuetz-Miller of *The Presidio and Militia on the Northern Frontier of New Spain: A Documentary History*, Volume 2, Part 2: *The Central Corridor and the Texas Corridor, 1700–1765*, for the University of Arizona Press (1997).

JOHN HERRON recently received a Ph.D. in history at the University of New Mexico. Formerly a Forest Service firefighter, or "Hot Shot," Herron has been studying environmental history for more than a decade and is the coeditor of *Human/Nature: Biology, Culture, and Environmental History* (University of New Mexico Press, 1999) and several articles on environmental history.

PAUL W. HIRT is an associate professor of history at Washington State University. He is the author of *A Conspiracy of Optimism: Management of the National Forests since World War Two* (University of Nebraska Press, 1994) and numerous journal articles on environmental history. Dr. Hirt is also an environmental activist and served as the executive director of the Sky-Island Alliance in the early 1990s.

CHRISTOPHER J. HUGGARD is an associate professor of history at North-West Arkansas Community College, serves as a historian for the National Park Service, and is the editor of the *Mining History Journal*. He has published several essays on the environment of the Southwest in journals such as the *Journal of the West* and the *New Mexico Historical Review*. He has a forthcoming book, *Kennecott: Labor, Environment, and Open-pit Mining in the Southwest* (University of New Mexico Press).

HAL K. ROTHMAN is a professor of history at the University of Nevada, Las Vegas. He is the author of *Devil's Bargain: Tourism in the Twentieth Century American West* (University Press of Kansas, 1998), winner of the 1999 Spur Award for Best Contemporary Nonfiction from the Western Writers of America, *On Rims and Ridges: The Los Alamos Area since 1880* (University of Nebraska Press, 1992), *Preserving Different Pasts: The American National Monuments* (University of Illinois Press, 1989), and other books and articles. He is also the editor of *Environmental History,* the journal of the American Society of Environmental History and the Forest History Society.

DUANE A. SMITH is a professor of history at Fort Lewis College in Durango, Colorado, and codirector of the Center for Southwest Studies there. Author of more than twenty-five books, Dr. Smith is best known for his *Rocky Mountain Mining Camps: The Urban Frontier* (Indiana University Press, 1967) and *Mining America: The Industry and the Environment, 1800–1980* (University Press of Kansas, 1987).

Index

Note: Page numbers in italics indicate figures.